An Introduction to Dynamic Meteorology

SECOND EDITION

This is Volume 23 in
INTERNATIONAL GEOPHYSICS SERIES
A series of monographs and textbooks
Edited by ANTON L. HALES

A complete list of the books in this series appears at the end of this volume.

AN INTRODUCTION TO DYNAMIC METEOROLOGY

Second Edition

JAMES R. HOLTON
Department of Atmospheric Sciences
University of Washington
Seattle, Washington

ACADEMIC PRESS New York San Francisco London
A Subsidiary of Harcourt Brace Jovanovich, Publishers

ACADEMIC PRESS, INC.
111 Fifth Avenue, New York, New York 10003

United Kingdom Edition published by
ACADEMIC PRESS, INC. (LONDON) LTD.
24/28 Oval Road, London NW1 7DX

Library of Congress Cataloging in Publication Data

Holton, James R
An introduction to dynamic meteorology. Second edition.

Bibliography: p.
Includes index.
1. Dynamic meteorology. I. Title.
QC880.H65 1979 551.5'153 79-12918
ISBN 0-12-354360-6

PRINTED IN THE UNITED STATES OF AMERICA

80 81 82 9 8 7 6 5 4 3 2

Contents

PREFACE ix
PREFACE TO FIRST EDITION xi

Chapter 1 **Introduction**

1.1 The Atmospheric Continuum 1
1.2 Physical Dimensions and Units 2
1.3 Scale Analysis 4
1.4 The Fundamental Forces 5
1.5 Noninertial Reference Frames and "Apparent" Forces 10
1.6 Structure of the Static Atmosphere 17
Problems 22
Suggested References 25

Chapter 2 **The Basic Conservation Laws**

2.1 Total Differentiation 27
2.2 The Vectorial Form of the Momentum Equation in Rotating Coordinates 29
2.3 The Component Equations in Spherical Coordinates 31
2.4 Scale Analysis of the Equations of Motion 35
2.5 The Continuity Equation 40
2.6 The Thermodynamic Energy Equation 44

2.7 Thermodynamics of the Dry Atmosphere 47
Problems 52
Suggested References 53

Chapter 3 **Elementary Applications of the Basic Equations**

3.1 The Basic Equations in Isobaric Coordinates 54
3.2 Balanced Flow 56
3.3 Trajectories and Streamlines 64
3.4 Vertical Shear of the Geostrophic Wind: The Thermal Wind 68
3.5 Vertical Motion 71
Problems 74

Chapter 4 **Circulation and Vorticity**

4.1 The Circulation Theorem 78
4.2 Vorticity 83
4.3 Potential Vorticity 87
4.4 The Vorticity Equation 92
4.5 Scale Analysis of the Vorticity Equation 95
Problems 98
Suggested References 100

Chapter 5 **The Planetary Boundary Layer**

5.1 The Mixing Length Theory 102
5.2 Planetary Boundary Layer Equations 105
5.3 Secondary Circulations and Spin-Down 112
Problems 116
Suggested References 118

Chapter 6 **The Dynamics of Synoptic Scale Motions in Middle Latitudes**

6.1 The Observed Structure of Midlatitude Synoptic Systems 120
6.2 Development of the Quasi-Geostrophic System 126
6.3 Idealized Model of a Developing Baroclinic System 140
Problems 144
Suggested Reference 145

Chapter 7 **Atmospheric Oscillations: Linear Perturbation Theory**

7.1 The Perturbation Method 147
7.2 Properties of Waves 147
7.3 Simple Wave Types 152
7.4 Internal Gravity (Buoyancy) Waves 159

7.5 Rossby Waves 165
Problems 168
Suggested References 172

Chapter 8 **Numerical Prediction**

8.1 Historical Background 173
8.2 Filtering of Sound and Gravity Waves 175
8.3 Filtered Forecast Equations 178
8.4 One-Parameter Models 181
8.5 A Two-Parameter Model 183
8.6 Numerical Solution of the Barotropic Vorticity Equation 186
8.7 Primitive Equation Models 198
Problems 211
Suggested References 213

Chapter 9 **The Development and Motion of Midlatitude Synoptic Systems**

9.1 Hydrodynamic Instability 214
9.2 Baroclinic Instability: Cyclogenesis 216
9.3 The Energetics of Baroclinic Waves 227
9.4 Fronts and Frontogenesis 236
Problems 244
Suggested References 246

Chapter 10 **The General Circulation**

10.1 The Nature of the Problem 248
10.2 The Energy Cycle: A Quasi-Geostrophic Model 250
10.3 The Momentum Budget 260
10.4 The Dynamics of Zonally Symmetric Circulations 267
10.5 Laboratory Simulation of the General Circulation 274
10.6 Numerical Simulation of the General Circulation 281
10.7 Longitudinally Varying Features of the General Circulation 288
Problems 293
Suggested References 294

Chapter 11 **Stratospheric Dynamics**

11.1 The Observed Mean Structure and Circulation of the Stratosphere 296
11.2 The Energetics of the Lower Stratosphere 298
11.3 Vertically Propagating Planetary Waves 300
11.4 Sudden Stratospheric Warmings 303
11.5 Waves in the Equatorial Stratosphere 307
11.6 The Quasi-Biennial Oscillation 313
11.7 The Ozone Layer 317
Problems 321
Suggested References 322

Chapter 12 **Tropical Motion Systems**

12.1 Scale Analysis of Tropical Motions 324
12.2 Cumulus Convection 330
12.3 The Observed Structure of Large-Scale Motions in the Equatorial Zone 343
12.4 The Origin of Equatorial Disturbances 351
12.5 Tropical Cyclones 357
Problems 360
Suggested References 361

Appendix A **Useful Constants and Parameters** 363

Appendix B **List of Symbols** 364

Appendix C **Vector Analysis** 368

Appendix D **The Equivalent Potential Temperature** 371

Appendix E **Standard Atmosphere Data** 373

Answers to Selected Problems 375

Bibliography 379

INDEX 383

Preface

In preparing this revised version of "An Introduction to Dynamic Meteorology" I have been motivated by two primary objectives:

(1) The incorporation of numerous pedagogical improvements based on suggestions from many colleagues as well as my own experience.
(2) The inclusion of material based on advances in our knowledge which have occurred since the appearance of the first edition.

Nearly fifty percent of the text has been completely rewritten for this edition. Treatment of the fundamental fluid dynamics necessary for understanding large-scale atmospheric motions has been consolidated into the first five chapters. This group of chapters includes expanded treatments of atmospheric thermodynamics and of the planetary boundary layer. The basics of modern dynamical meteorology are presented in Chapters 7–12. This group of chapters includes an entirely new chapter on the dynamics of the stratosphere as well as extensive revision and updating of the remaining chapters.

In response to many requests, a number of additional problems have been included which span a wide range in difficulty. Answers to selected problems are included and a number of appendixes have been added for the aid of the student.

I am indebted to a large number of colleagues who have suggested improvements in the text. I wish especially to thank Professor Jacques Van Mieghem for his thoughtful critique of the first six chapters of the first edition, and Dr. Duane Stevens and Mr. John Knox for their helpful comments on the manuscript of the present edition. Finally, I wish to thank Ilze Schubert for her expert help in preparing the manuscript.

Preface to First Edition

During the past decade the rapid advances in the science of dynamic meteorology made during the 1940s and 1950s have been consolidated. There now exists a reasonably coherent theory for the development of midlatitude storms, as well as for the overall general circulation of the atmosphere. Therefore, the subject can now be organized and presented in textbook form.

In this book dynamic meteorology is presented as a cohesive subject with a central unifying body of theory—namely the quasi-geostrophic system. Quasi-geostrophic theory is used to develop the principles of diagnostic analysis, numerical forecasting, baroclinic wave theory, energy transformations, and the theory of the general circulation.

Throughout the book the emphasis is on physical principles rather than mathematical elegance. It is assumed that the reader has mastered the fundamentals of classical physics, and that he has a thorough knowledge of elementary differential and integral calculus. Some use is made of vector calculus. However, in most cases the vector operations are elementary in nature so that the reader with little background in vector operations should not experience undue difficulties.

Much of the material included in this text is based on a two-quarter course sequence for seniors majoring in atmospheric sciences at the University of

Washington. It would also be suitable for first-year graduate students with no previous background in meteorology.

The actual text may be divided into two main sections. The first section, consisting of Chapters 1–6, introduces those fundamentals of fluid dynamics most relevant for understanding atmospheric motions, primarily through consideration of a number of idealized types of atmospheric flow. I have found that nearly all the material in these six chapters can be presented successfully at the senior level in about 30 lectures. The second main section, consisting of Chapters 7–11, contains the core of modern dynamic meteorology with emphasis on the central unifying role of the quasi-geostrophic theory. These chapters contain more material than can easily be covered in a senior level course. However, I have attempted to arrange the material so that sections containing more advanced results can be omitted in elementary courses. On the other hand, for more intensive graduate level courses the material presented here might be supplemented by readings from original sources.

In addition to these two main sections, the book contains a concluding chapter which stands somewhat alone. In this final chapter I have attempted to review the current status of the dynamics of the tropical atmosphere. This chapter is by necessity somewhat speculative; however, the field of tropical meteorology is too important to omit entirely from a textbook on dynamic meteorology.

A short annotated list of suggested references for further reading is presented at the end of most of the chapters. I have limited the references in most cases to books and review articles which I have found to be particularly useful. No attempt has been made to provide extensive bibliographies of original sources. In the reference lists books are referred to by author and title, papers by author and date of publication. The complete references are listed in the bibliography at the end of the book.

An Introduction to Dynamic Meteorology

SECOND EDITION

Chapter 1 Introduction

1.1 The Atmospheric Continuum

Dynamic meteorology is the study of those motions of the atmosphere which are associated with weather and climate. For all such motions the discrete molecular nature of the atmosphere can be ignored, and the atmosphere can be regarded as a continuous fluid medium, or *continuum*. The various physical quantities which characterize the state of the atmosphere—pressure, density, temperature, and velocity—are assumed to have unique values at each point in the atmospheric continuum. Moreover, these *field variables* and their derivatives are assumed to be continuous functions of space and time. The fundamental laws of fluid mechanics and thermodynamics which govern the motions of the atmosphere may then be expressed in terms of partial differential equations involving the field variables.

The general set of partial differential equations governing the motions of the atmosphere is extremely complex, and no general solutions are known to exist. To acquire an understanding of the physical role of atmospheric motions in determining the observed weather and climate it is necessary to develop models based on systematic simplification of the fundamental governing equations. As we shall see in later chapters the development of models appropriate to particular atmospheric motion systems requires careful consideration of the scales of motion involved.

1.2 Physical Dimensions and Units

The fundamental laws which govern the motions of the atmosphere are expressed in terms of physical quantities (field variables and coordinates) which depend on four dimensionally independent *properties*: length, time, mass, and thermodynamic temperature. The dimensions of all atmospheric field variables may be expressed in terms of multiples and ratios of these four fundamental properties. To measure and compare the scales of atmospheric motions a set of units of measure must be defined for the four fundamental properties.

In this text the international system of units (SI) will be used almost exclusively. The four fundamental properties are measured in terms of the SI *base units* shown in Table 1.1. All other properties are measured in terms of SI *derived units* which are units formed from products and/or ratios of the base units. For example, velocity has the derived units of meter per second ($m\,s^{-1}$). A number of important derived units have special names and symbols. Those which are commonly used in dynamic meteorology are indicated in Table 1.2. In addition, the *supplementary unit* designating a plane angle—the radian (rad)—is required for expressing angular velocity ($rad\,s^{-1}$) in the SI system.[1]

Table 1.1 *SI Base Units*

Property	Name	Symbol
Length	Meter (metre)	m
Mass	Kilogram (kilogramme)	kg
Time	Second	s
Temperature	Kelvin	K

In order to keep numerical values within convenient limits it is conventional to use decimal multiples and submultiples of SI units. Prefixes used to indicate such multiples and submultiples are given in Table 1.3. The prefixes of Table 1.3 may be affixed to any of the basic or derived SI units except the kilogram. Since the kilogram already is a prefixed unit decimal multiples and submultiples of mass are formed by prefixing the gram (g), not the kilogram (kg).

[1] Note that the *hertz* measures frequency in *cycles* per second not in radians per second.

Table 1.2 *SI Derived Units with Special Names*

Property	Name	Symbol
Frequency	Hertz	Hz (s^{-1})
Force	Newton	N (kg m s^{-2})
Pressure	Pascal	Pa (N m^{-2})
Energy	Joule	J (N m)
Power	Watt	W (J s^{-1})

Although the use of non-SI units will generally be avoided in this text there are a few exceptions worth mentioning:

(1) In some contexts the time units minute (min), hour (h), and day (d) may be used in preference to the second in order to express quantities in convenient numerical values.

(2) For work in dynamic meteorology the kilopascal (kPa) is the preferred SI unit for pressure. However, most meteorologists are accustomed to using the millibar (mb), which is equal to 100 Pa or 0.1 kPa. Thus for the reader's convenience we will generally give pressures in kilopascals followed by the equivalent in millibars—e.g., standard sea level pressure equals 101.325 kPa (1013.25 mb). However, to conform with conventional meteorological practice upper level maps will be referred to using the millibar (e.g., the 500-mb surface).

(3) In discussing observed temperatures we will generally use the Celsius temperature scale, which is related to the thermodynamic temperature scale as follows:

$$T_c = T - T_0$$

where T_c is expressed in degrees Celsius (°C), T is the thermodynamic temperature in kelvins (K), and $T_0 = 273.15$ K is the freezing point of water

Table 1.3 *Prefixes for Decimal Multiples and Submultiples of SI Units*

Multiple	Prefix	Symbol
10^6	Mega	M
10^3	Kilo	k
10^2	Hecto	h
10^1	Deka	da
10^{-1}	Deci	d
10^{-2}	Centi	c
10^{-3}	Milli	m
10^{-6}	Micro	μ

on the kelvin scale. From this relationship it is clear that one kelvin unit equals one degree Celsius.

1.3 Scale Analysis

Scale analysis, or scaling, is a convenient technique for estimating the magnitudes of various terms in the governing equations for a particular type of motion. In scaling, typical expected values of the following quantities are specified: (1) the magnitudes of the field variables; (2) the amplitudes of fluctuations in the field variables; and (3) the characteristic length, depth, and time scales on which these fluctuations occur. These typical values are then used to compare the magnitudes of various terms in the governing equations. For example, in a typical midlatitude synoptic[2] cyclone the surface pressure might fluctuate by 2 kPa (20 mb) over a horizontal distance of 2000 km. Designating the amplitude of the horizontal pressure fluctuation by δp and the horizontal scale by L, the magnitude of the horizontal pressure gradient may be estimated by substituting $\delta p = 2$ kPa and $L = 2000$ km to get

$$\left(\frac{\partial p}{\partial x}, \frac{\partial p}{\partial y}\right) \sim \frac{\delta p}{L} = 1\ \text{kPa}/10^3\ \text{km} \quad (10\ \text{mb}/10^3\ \text{km})$$

Pressure fluctuations of similar magnitudes occur in other motion systems of vastly different scale such as tornadoes, squall lines, and hurricanes. Thus, the horizontal pressure gradient can range over several orders of magnitude for systems of meteorological interest. Similar considerations are also valid for derivative terms involving other field variables. Therefore, the nature of the dominant terms in the governing equations is crucially dependent on the horizontal scale of the motions. In particular, motions with horizontal scales of a few kilometers or less tend to have short time scales so that terms involving the rotation of the earth are negligible, while for larger scales they become very important.

Because the character of atmospheric motions depends so strongly on the horizontal scale, this scale provides a convenient method for classification of motion systems. In Table 1.4 examples of various types of motions are classified by horizontal scale for the spectral region from 10^{-7} to 10^7 m. In the following chapters scaling arguments will be used extensively in developing simplifications of the governing equations for use in modeling various types of motion systems.

[2] The term *synoptic* designates the branch of meteorology which deals with the analysis of observations taken over a wide area at or near the same time. This term is commonly used (as here) to designate the characteristic scale of the disturbances which are depicted on weather maps.

Table 1.4 *Scales of Atmospheric Motions*

Type of motion	Horizontal scale (m)
Molecular mean free path	10^{-7}
Minute turbulent eddies	10^{-2}–10^{-1}
Small eddies	10^{-1}–1
Dust devils	1–10
Gusts	10–10^{2}
Tornadoes	10^{2}
Cumulonimbus clouds	10^{3}
Fronts, squall lines	10^{4}–10^{5}
Hurricanes	10^{5}
Synoptic cyclones	10^{6}
Planetary waves	10^{7}

1.4 The Fundamental Forces

The motions of the atmosphere are governed by the fundamental physical laws of conservation of mass, momentum, and energy. In Chapter 2 we will show how these principles can be applied to a small volume element of the atmosphere in order to obtain the governing equations. However, before deriving the complete momentum equation it is useful to discuss the nature of the forces which influence atmospheric motions.

Newton's second law of motion states that the rate of change of momentum of an object referred to coordinates fixed in space equals the sum of all the forces acting. For atmospheric motions of meteorological interest, the forces which are of primary concern are the pressure gradient force, the gravitational force, and friction. These *fundamental* forces are the subject of the present section. If, as is the usual case, the motion is referred to a coordinate system rotating with the earth, Newton's second law may still be applied provided that certain *apparent* forces, the centrifugal force and the Coriolis force, are included among the forces acting. The nature of these apparent forces will be discussed in Section 1.5.

1.4.1 The Pressure Gradient Force

We consider a differential volume element of air, $\delta V = \delta x\, \delta y\, \delta z$, centered at the point x_0, y_0, z_0 as illustrated in Fig. 1.1. Due to random molecular motions, momentum is continually imparted to the walls of the volume element by the surrounding air. This momentum transfer per unit time per unit area is just the *pressure* exerted on the walls of the volume element by the surrounding air. If the pressure at the center of the volume element is

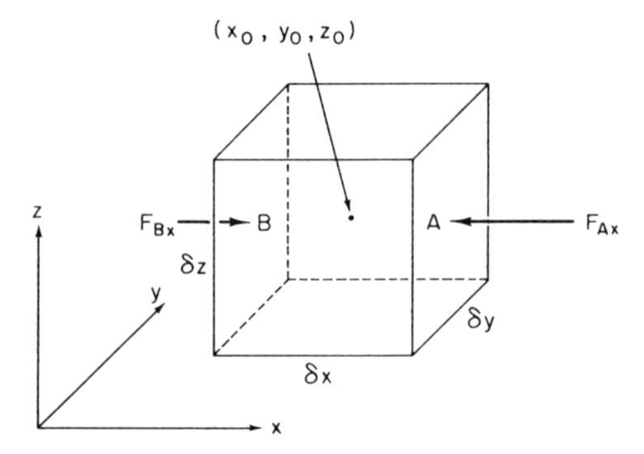

Fig. 1.1 The x component of the pressure gradient force acting on a fluid element.

designated by p_0, then the pressure on the wall labeled A in Fig. 1.1 can be expressed in a Taylor series expansion as

$$p_0 + \frac{\partial p}{\partial x}\frac{\delta x}{2} + \text{higher-order terms}$$

Neglecting the higher-order terms in this expansion, the pressure force acting on the volume element at wall A is

$$F_{Ax} = -\left(p_0 + \frac{\partial p}{\partial x}\frac{\delta x}{2}\right)\delta y\,\delta z$$

where $\delta y\,\delta z$ is the area of wall A. Similarly, the force acting on the volume element at wall B is just

$$F_{Bx} = +\left(p_0 - \frac{\partial p}{\partial x}\frac{\delta x}{2}\right)\delta y\,\delta z$$

Therefore, the net x component of the pressure force acting on the volume is

$$F_x = F_{Ax} + F_{Bx} = -\frac{\partial p}{\partial x}\,\delta x\,\delta y\,\delta z$$

The mass m of the differential volume element is simply the density ρ times the volume: $m = \rho\,\delta x\,\delta y\,\delta z$. Thus, the x component of the pressure gradient force per unit mass is

$$\frac{F_x}{m} = -\frac{1}{\rho}\frac{\partial p}{\partial x}$$

Similarly, it can easily be shown that the y and z components of the pressure gradient force per unit mass are

$$\frac{F_y}{m} = -\frac{1}{\rho}\frac{\partial p}{\partial y} \qquad \text{and} \qquad \frac{F_z}{m} = -\frac{1}{\rho}\frac{\partial p}{\partial z}$$

so that the total pressure gradient force per unit mass is

$$\frac{\mathbf{F}}{m} = -\frac{1}{\rho}\nabla p \tag{1.1}$$

It is important to note that this force is proportional to the *gradient* of the pressure field, not to the pressure itself.

1.4.2 The Gravitational Force

Newton's law of universal gravitation states that any two elements of mass in the universe attract each other with a force proportional to their masses and inversely proportional to the square of the distance separating them. Thus, if two mass elements M and m are separated by a distance $r \equiv |\mathbf{r}|$ (with the vector $\mathbf{r}$ directed toward m as shown in Fig. 1.2), then the force exerted by mass M on mass m due to gravitation is

$$\mathbf{F}_g = -\frac{GMm}{r^2}\left(\frac{\mathbf{r}}{r}\right) \tag{1.2}$$

where G is a universal constant called the gravitational constant. The law of gravitation as expressed in (1.2) actually applies only to hypothetical "point" masses since for objects of finite extent $\mathbf{r}$ will vary from one part of the object to another. However, for finite bodies (1.2) may still be applied if $|\mathbf{r}|$ is interpreted as the distance between the centers of mass of the bodies. Thus, if the earth is designated as mass M and m is a mass element of the atmosphere,

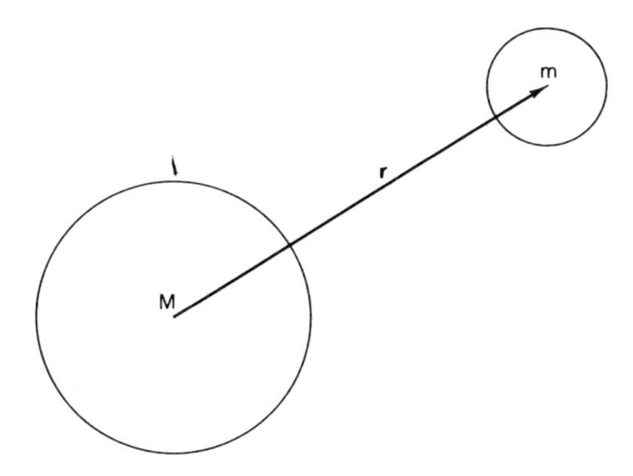

Fig. 1.2 Two spherical masses whose centers are separated by a distance r.

then the force per unit mass exerted on the atmosphere by the gravitational attraction of the earth is

$$\frac{\mathbf{F}_g}{m} \equiv \mathbf{g}^* = -\frac{GM}{r^2}\left(\frac{\mathbf{r}}{r}\right) \tag{1.3}$$

In dynamic meteorology it is customary to use as a vertical coordinate the height above mean sea level. If the mean radius of the earth is designated by a, and the distance above mean sea level is designated by z, then neglecting the small departure of the shape of the earth from sphericity, $r = a + z$. Therefore, (1.3) can be rewritten as

$$\mathbf{g}^* = \frac{\mathbf{g}_0^*}{(1 + z/a)^2} \tag{1.4}$$

where $\mathbf{g}_0^* = -(GM/a^2)(\mathbf{r}/r)$ is the value of the gravitational force at mean sea level. For meteorological applications, $z \ll a$ so that with negligible error we can let $\mathbf{g}^* = \mathbf{g}_0^*$ and simply treat the gravitational force as a constant.

1.4.3 The Friction or Viscosity Force

Although a complete discussion of the viscosity force would be rather complicated, the basic physical concept can be illustrated quite simply. We consider a layer of incompressible fluid confined between two horizontal plates separated by a distance l as shown in Fig. 1.3. The lower plate is fixed and the upper plate is moving in the x direction at a speed u_0. The fluid particles in the layer in contact with the plate must move at the velocity of the plate. Thus, at $z = l$ the fluid moves at speed $u(l) = u_0$, and at $z = 0$ the fluid is motionless. We find that the force tangential to the upper plate required to keep it in uniform motion is proportional to the area of the plate,

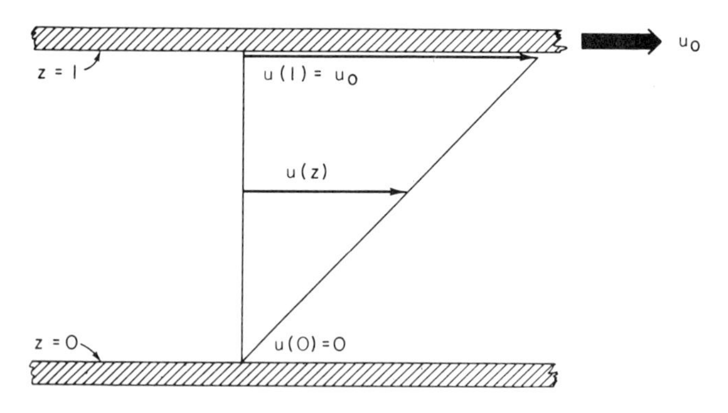

Fig. 1.3 One-dimensional steady-state viscous shear flow.

the velocity, and the inverse of the distance separating the plates. Thus, we may write $F = \mu A u_0/l$, where μ is a constant of proportionality, the dynamic viscosity coefficient.

This force must just equal the force exerted by the upper plate on the fluid immediately below it. For a state of uniform motion, every horizontal layer of fluid must exert the same force on the fluid below. Therefore, taking the limit as the fluid layer depth goes to zero, we may write the viscous force per unit area, or *shearing stress*, for this special case as

$$\tau_{zx} = \mu \frac{\partial u}{\partial z}$$

where the subscripts indicate that τ_{zx} is the component of the shearing stress in the x direction due to vertical shear of the x velocity component.

From the molecular viewpoint this shearing stress results from a net downward transport of momentum by the random motion of the molecules. Because the mean x momentum increases with height, the molecules passing downward through a horizontal plane at any instant carry more momentum than those which are passing upward through the plane. Thus, there is a net downward transport of x momentum. This downward momentum transport per unit time per unit area is simply the shearing stress.

In the simple two-dimensional steady-state motion example given above there is no net viscous force acting on the elements of fluid since the shearing stress acting across the top boundary of each fluid element is just equal and opposite to that acting across the lower boundary. For the more general case of nonsteady two-dimensional shear flow in an incompressible fluid, we may calculate the net viscous force by considering again a differential volume element of fluid centered at (x, y, z) with sides δx, δy, δz as shown in Fig. 1.4. If the shearing stress in the x direction acting through the center of the

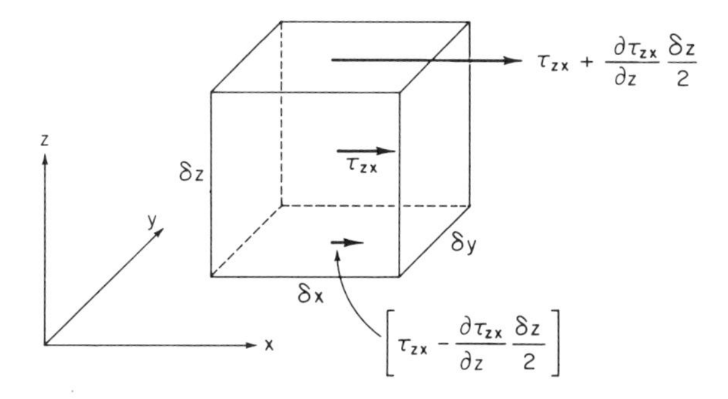

Fig. 1.4 The x component of the vertical shearing stress on a fluid element.

element is designated τ_{zx}, then the stress acting across the upper boundary may be written approximately as

$$\tau_{zx} + \frac{\partial \tau_{zx}}{\partial z}\frac{\delta z}{2}$$

while the stress acting across the lower boundary is

$$\tau_{zx} - \frac{\partial \tau_{zx}}{\partial z}\frac{\delta z}{2}$$

The net viscous force on the volume element acting in the x direction is then

$$\left(\tau_{zx} + \frac{\partial \tau_{zx}}{\partial z}\frac{\delta z}{2}\right)\delta y\,\delta x - \left(\tau_{zx} - \frac{\partial \tau_{zx}}{\partial z}\frac{\delta z}{2}\right)\delta y\,\delta x$$

so that the viscous force per unit mass due to vertical shear of the component of motion in the x direction is

$$\frac{1}{\rho}\frac{\partial \tau_{zx}}{\partial z} = \frac{1}{\rho}\frac{\partial}{\partial z}\left(\mu \frac{\partial u}{\partial z}\right) \tag{1.5}$$

For constant μ, the right-hand side of (1.5) may be simplified to $\nu\, \partial^2 u/\partial z^2$, where $\nu = \mu/\rho$ is the kinematic viscosity coefficient. For the atmosphere below 100 km, ν is so small that molecular viscosity is negligible except in a thin layer within a few centimeters of the earth's surface where the vertical shear is very large. Away from this surface molecular boundary layer, momentum is transferred primarily by turbulent eddy motions.

In a turbulent fluid such as the atmosphere it is often useful to picture the small-scale turbulent eddies as discrete "blobs" of fluid which move about bodily in the large-scale flow field and transfer momentum vertically in a manner analogous to the molecules in molecular viscosity. A mixing length may then be defined for the turbulent eddies, analogous to the mean free path of the molecules in molecular viscosity. By further analogy, the dissipative effects of the small-scale turbulent motions can be represented by defining an *eddy viscosity coefficient*. Thus, the simple formulation (1.5) can also be used for turbulent flow provided that the eddy viscosity coefficient is used instead of molecular viscosity. This approximation will be discussed further in Chapter 5.

1.5 Noninertial Reference Frames and "Apparent" Forces

In formulating the laws of atmospheric dynamics it is natural to use a *geocentric* reference frame which is fixed with respect to the rotating earth. Newton's first law of motion states that a mass in uniform motion relative

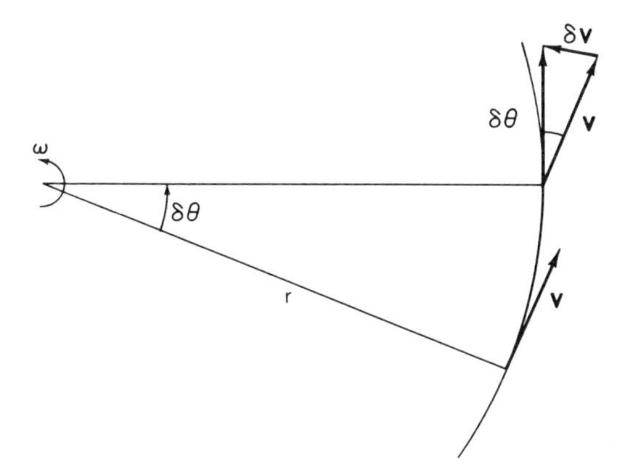

Fig. 1.5 Centripetal acceleration.

to a coordinate system fixed in space will remain in uniform motion in the absence of any forces. Such motion is referred to as *inertial motion*; and the fixed reference frame is an inertial, or absolute, frame of reference. It is clear, however, that an object at rest with respect to the rotating earth is not at rest or in uniform motion relative to a coordinate system fixed in space. Therefore, motion which appears to be inertial motion to an observer in the rotating earth reference frame is really accelerated motion. Hence, the rotating frame is a *noninertial* reference frame. Newton's laws of motion can only be applied in such a frame if the acceleration of the coordinates is taken into account. The most satisfactory way of including the effects of coordinate acceleration is to introduce "apparent" forces in the statement of Newton's second law. These apparent forces are the inertial reaction terms which arise because of the coordinate acceleration. For a coordinate system in uniform rotation, two such "apparent" forces are required, the centrifugal force and the Coriolis force.

1.5.1 The Centrifugal Force

We consider a ball of mass m which is attached to a string and whirled through a circle of radius r at a constant angular velocity ω. From the point of view of an observer in fixed space the speed of the ball is constant, but its direction of travel is continuously changing so that its velocity is not constant. To compute the acceleration we consider the change in velocity $\delta\mathbf{V}$ which occurs for a time increment δt during which the ball rotates through an angle $\delta\theta$ as shown in Fig. 1.5. Since $\delta\theta$ is also the angle between the vectors $\mathbf{V}$ and $\mathbf{V} + \delta\mathbf{V}$, the magnitude of $\delta\mathbf{V}$ is just

$$|\delta\mathbf{V}| = |\mathbf{V}|\,\delta\theta$$

If we divide by δt and note that in the limit $\delta t \to 0$, $\delta \mathbf{V}$ is directed toward the axis of rotation, we obtain

$$\frac{d\mathbf{V}}{dt} = |\mathbf{V}| \frac{d\theta}{dt} \left(-\frac{\mathbf{r}}{r} \right)$$

But, $|\mathbf{V}| = \omega r$ and $d\theta/dt = \omega$, so that

$$\frac{d\mathbf{V}}{dt} = -\omega^2 \mathbf{r} \tag{1.6}$$

Therefore, viewed from fixed coordinates the motion is one of uniform acceleration directed toward the axis of rotation, and equal to the square of the angular velocity times the distance from the axis of rotation. This acceleration is called the *centripetal acceleration*. It is caused by the force of the string pulling the ball.

Now suppose that we observe the motion in a coordinate system rotating with the ball. In this rotating system the ball is *stationary*, but there is still a force acting on the ball, namely the pull of the string. Therefore, in order to apply Newton's second law to describe the motion relative to this rotating coordinate system we must include an additional apparent force, the *centrifugal force*, which just balances the force of the string on the ball. Thus, the centrifugal force is equivalent to the inertial reaction of the ball on the string, and just equal and opposite to the centripetal acceleration.

To summarize: Observed from a fixed system the rotating ball undergoes a uniform centripetal acceleration in response to the force exerted by the string. Observed from a system rotating along with it, the ball is stationary and the force exerted by the string is balanced by a centrifugal force.

1.5.2 The Gravity Force

A particle of unit mass at rest on the surface of the earth, observed in a reference frame rotating with the earth, is subject to a centrifugal force $\Omega^2\mathbf{R}$, where Ω is the angular speed of rotation of the earth[3] and $\mathbf{R}$ the position vector from the axis of rotation to the particle.

Thus, the weight of a particle of mass m at rest on the earth's surface, which is just the reaction force of the earth on the particle, will generally be less than the gravitational force $m\mathbf{g}^*$ because the centrifugal force partly balances the gravitational force. It is, therefore, convenient to combine the effects of the gravitational force and centrifugal force by defining a gravity force $\mathbf{g}$ (often called simply *gravity*) such that

$$\mathbf{g} \equiv \mathbf{g}^* + \Omega^2\mathbf{R} \tag{1.7}$$

[3] The earth revolves around its axis once every 23 h 56 min 4 s (86,164 s). Thus, $\Omega = 2\pi/(86{,}164\ \mathrm{s}) = 7.292 \times 10^{-5}\ \mathrm{rad\ s^{-1}}$.

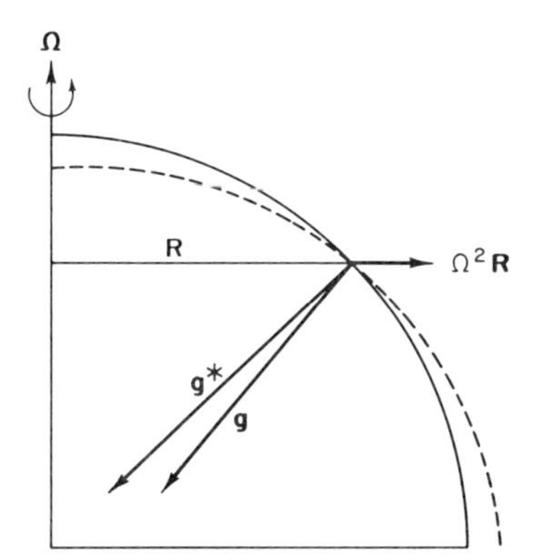

Fig. 1.6 The relationship between gravitation and gravity.

The gravitational force is directed toward the center of the earth whereas the centrifugal force is directed away from the axis of rotation. Therefore, except at the poles and the equator, gravity is not directed toward the center of the earth. As indicated by Fig. 1.6, if the earth were a perfect sphere, gravity would have an equatorward component parallel to the surface of the earth. The earth has adjusted to compensate for this equatorward force component by assuming the approximate shape of a spheroid with an equatorial "bulge" so that **g** is everywhere directed normal to the level surface. As a consequence, the equatorial radius of the earth is about 21 km larger than the polar radius. In addition, the local "vertical," which is taken to be parallel to **g**, does not pass through the center of the earth except at the equator and poles.

1.5.3 The Coriolis Force

We have previously seen that Newton's second law may be applied in rotating coordinates to describe an object at rest with respect to the rotating system provided that an apparent force, the centrifugal force, is included among the forces acting on the object. If the object is moving with respect to the rotating system, an additional apparent force, the Coriolis force, is required in order that Newton's second law remain valid.

Suppose that an object is set in uniform motion with respect to an inertial coordinate system. If the object is observed from a rotating system with the axis of rotation perpendicular to the plane of motion, the path will appear to be curved, as indicated in Fig. 1.7. Thus, as viewed in a rotating coordinate system there is an apparent force which deflects an object in inertial motion from a straight-line path. The resulting path is curved in a direction opposite to the direction of coordinate rotation. This deflection force is the Coriolis force. Viewed from the rotating system the relative motion is an accelerated motion, with the acceleration equal to the sum of the Coriolis force and the centrifugal force. The Coriolis force, which acts perpendicular to the velocity

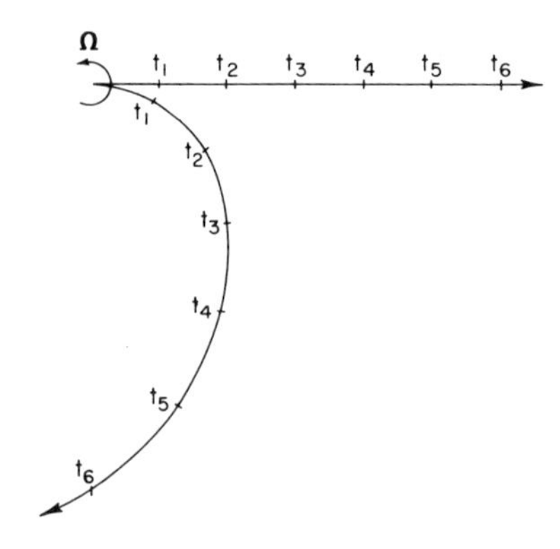

Fig. 1.7 Inertial motion as viewed from an inertial frame (straight line) and a rotating frame (curved line).

vector, can only change the direction of travel. However, the centrifugal force, which acts radially outward, has a component along the direction of motion which increases the speed of the particle relative to the rotating coordinates as the particle spirals outward. Thus, in this example of inertial motion viewed from a rotating system the effects of both the Coriolis force and the centrifugal force are included.

Although the above example is simple enough to be readily comprehended in terms of everyday experience, it gives little insight into the meteorological role of the Coriolis force. The mathematical form for the Coriolis force due to motion relative to the rotating earth can be obtained by considering the motion of a hypothetical particle of unit mass which is free to move on a frictionless horizontal surface on the rotating earth. If the particle is initially at rest with respect to the earth, then the only forces acting on it are the gravitational force and the apparent centrifugal force due to the rotation of the earth. As we saw in Section 1.5.2, the sum of these two forces defines gravity, which is directed perpendicular to the local horizontal. Suppose now that the particle is set in motion in the eastward direction by an impulsive force. Since the particle is now rotating faster than the earth, the centrifugal force on the particle will be increased. Letting Ω be the magnitude of the angular velocity of the earth, $\mathbf{R}$ the position vector from the axis of rotation to the particle, and u the eastward speed of the particle relative to the ground, we may write the total centrifugal force as

$$\left(\Omega + \frac{u}{R}\right)^2 \mathbf{R} = \Omega^2 \mathbf{R} + \frac{2\Omega u \mathbf{R}}{R} + \frac{u^2 \mathbf{R}}{R^2} \tag{1.8}$$

The first term on the right is just the centrifugal force due to the rotation of the earth. This is, of course, included in gravity. The other two terms represent *deflecting* forces which act outward along the vector $\mathbf{R}$ (that is, perpendicular to the axis of rotation). For synoptic scale motions $u \ll \Omega R$,

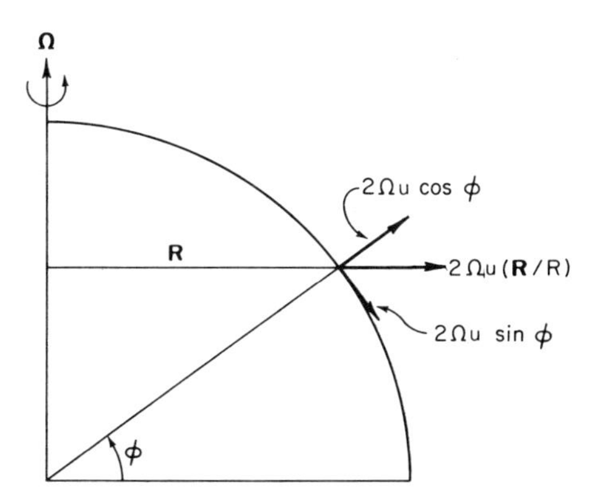

Fig. 1.8 Components of the Coriolis force due to relative motion along a latitude circle.

and the last term may be neglected, in a first approximation. The remaining term in (1.8), $2\Omega u(\mathbf{R}/R)$, is the *Coriolis force* due to relative motion parallel to a latitude circle. This Coriolis force can be divided into components in the vertical and meridional directions, respectively, as indicated in Fig. 1.8. Therefore, relative motion along the east–west coordinate will produce an acceleration in the north–south direction given by

$$\left(\frac{dv}{dt}\right)_{\mathrm{Co}} = -2\Omega u \sin\phi \tag{1.9}$$

and an acceleration in the vertical given by

$$\left(\frac{dw}{dt}\right)_{\mathrm{Co}} = 2\Omega u \cos\phi \tag{1.10}$$

where (u, v, w) designates the eastward, northward, and upward velocity components, respectively, ϕ is latitude, and the subscript Co indicates that this is the acceleration due only to the Coriolis force. A particle moving eastward in the horizontal plane in the Northern Hemisphere will be deflected southward by the Coriolis force, whereas a westward moving particle will be deflected northward. In either case the deflection is to the right of the direction of motion. The vertical component of the Coriolis force (1.10) is ordinarily much smaller than the gravitational force so that its only effect is to cause a very minor change in the apparent *weight* of an object depending on whether the object is moving eastward or westward.

So far we have considered only the Coriolis force due to relative motion parallel to latitude circles. Suppose now that a particle initially at rest on the earth is set in motion equatorward by an impulsive force. As the particle moves equatorward it will conserve its angular momentum in the absence of torques in the east–west direction. Since the distance to the axis of rotation R increases for a particle moving equatorward, a relative westward velocity must develop if the particle is to conserve its absolute angular momentum. Thus, letting δR designate the change in the distance to the axis of rotation for a southward displacement from a latitude ϕ_0 to latitude $\phi_0 + \delta\phi$ (note that $\delta\phi < 0$ for an equatorward displacement), we obtain by conservation of angular momentum

$$\Omega R^2 = \left(\Omega + \frac{\delta u}{R + \delta R}\right)(R + \delta R)^2$$

where δu is the eastward relative velocity when the particle has reached latitude $\phi_0 + \delta\phi$. Expanding the right-hand side, neglecting second-order differentials, and solving for δu, we obtain

$$\delta u = -2\Omega\, \delta R = +2\Omega a\, \delta\phi \sin \phi_0$$

where we have used $\delta R = -a\, \delta\phi \sin \phi_0$ (a is the radius of the earth). This relationship is illustrated in Fig. 1.9. Dividing through by the time increment δt and taking the limit as $\delta t \to 0$, we obtain from the above

$$\left(\frac{du}{dt}\right)_{\text{Co}} = 2\Omega a \frac{d\phi}{dt} \sin \phi_0 = 2\Omega v \sin \phi_0$$

where $v = a\, d\phi/dt$ is the northward velocity component.

Similarly, it is easy to show that if the particle is launched vertically at latitude ϕ_0, conservation of the absolute angular momentum will require an acceleration in the zonal direction equal to $-2\Omega w \cos \phi_0$, where w is the

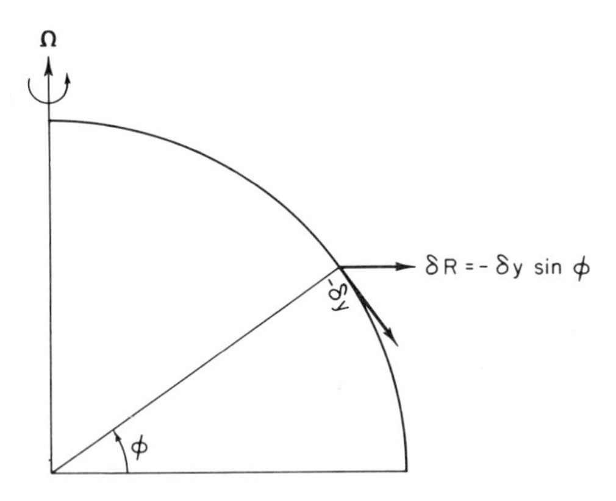

Fig. 1.9 Relationship of δR and $\delta y = a\delta\phi$ for an equatorward displacement.

vertical velocity. Thus, in the general case where both horizontal and vertical relative motions are included

$$\left(\frac{du}{dt}\right)_{Co} = 2\Omega v \sin\phi - 2\Omega w \cos\phi \qquad (1.11)$$

Again, the effect of the horizontal relative velocity is to deflect the particle to the right in the Northern Hemisphere. This deflection force is negligible for motions with time scales that are very short compared to the period of the earth's rotation (a point which is illustrated by several problems at the end of the chapter). Thus, the Coriolis force is not important for the dynamics of individual cumulus clouds, but is essential to the understanding of longer time scale phenomena such as synoptic scale systems. The Coriolis force must also be taken into account when computing long-range missile or artillery trajectories. As an example, suppose that a ballistic missile is fired due eastward at 43°N latitude ($2\Omega \sin\phi = 10^{-4}\ \mathrm{s}^{-1}$ at 43°N). If the missile travels 1000 km at a horizontal speed $u_0 = 1000\ \mathrm{m\ s}^{-1}$, by how much is the missile deflected from its eastward path by the Coriolis force? Integrating (1.9) with respect to time we find that

$$v = -2\Omega u_0 t \sin\phi \qquad (1.12)$$

where it is assumed that the deflection is sufficiently small so that we may let $u = u_0$ be constant. To find the total southward displacement we must integrate (1.12) with respect to time:

$$\int_0^{t_0} v\, dt = \int_{y_0}^{y_0+\delta y} dy = -2\Omega u_0 \int_0^{t_0} t\, dt \sin\phi$$

Thus the total displacement is

$$\delta y = -\Omega u_0 t_0^2 \sin\phi \approx -50\ \mathrm{km}$$

Therefore, the missile is deflected southward by 50 km due to the Coriolis effect. Further examples of the deflection of particles by the Coriolis force are given in some of the problems at the end of the chapter.

1.6 Structure of the Static Atmosphere

The thermodynamic state of the atmosphere at any point is determined by the values of pressure, temperature, and density (or specific volume) at that point. These field variables are related to each other by the equation of state for an ideal gas. Letting p, T, ρ, and α denote pressure, temperature,

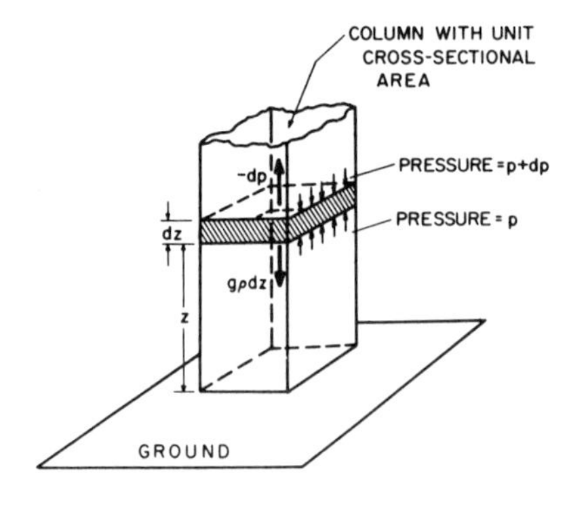

Fig. 1.10 Balance of forces for hydrostatic equilibrium. Note that dp is negative since pressure decreases with height. (After Wallace and Hobbs, 1977.)

density, and specific volume, respectively, we can express the equation of state for dry air as

$$p\alpha = RT \qquad \text{or} \qquad p = \rho RT \tag{1.13}$$

where R is the gas constant for dry air ($R = 287 \text{ J kg}^{-1} \text{ K}^{-1}$).

1.6.1 The Hydrostatic Equation

In the absence of atmospheric motions the gravity force must be exactly balanced by the vertical component of the pressure gradient force. Thus, as illustrated in Fig. 1.10,

$$dp/dz = -\rho g \tag{1.14}$$

This condition of *hydrostatic balance* provides an excellent approximation for the vertical dependence of the pressure field in the real atmosphere. Only for intense small-scale systems such as squall lines and tornadoes is it necessary to consider departures from hydrostatic balance. Integrating (1.14) from a height z to the top of the atmosphere we find that

$$p(z) = \int_z^\infty \rho g \, dz$$

so that the pressure at any point is simply equal to the weight of the unit cross section column of air overlying the point. Thus, mean sea level pressure $p(0) = 101.325$ kPa (1013.25 mb) is simply the average weight per square meter of the total atmospheric column.[4] It is often useful to express the hydrostatic equation in terms of the *geopotential* rather than the geometric height. The geopotential $\Phi(z)$ at height z is defined as the work required to

[4] For computational convenience we will generally let the mean surface pressure equal 100 kPa (1000 mb).

raise a unit mass to height z from mean sea level:

$$\Phi \equiv \int_0^z g\, dz \tag{1.15}$$

Noting from (1.15) that $d\Phi = g\, dz$ and from (1.13) that $\alpha = RT/p$ we can express the hydrostatic equation in the form

$$d\Phi = -(RT/p)\, dp = -RT\, d\ln p \tag{1.16}$$

Thus, the variation of geopotential with respect to pressure depends only on temperature. Integration of (1.16) in the vertical yields a form of the *hypsometric equation*:

$$\Phi(z_2) - \Phi(z_1) = R \int_{p_2}^{p_1} T\, d\ln p \tag{1.17}$$

Meteorologists often prefer to replace $\Phi(z)$ in (1.17) by a quantity called *geopotential height* which is defined as $Z \equiv \Phi(z)/g_0$, where $g_0 \equiv 9.80665$ m s^{-2} is the global average of gravity at mean sea level. Thus in the troposphere and lower stratosphere Z is numerically almost identical to the geometric height z. In terms of Z the hypsometric equation becomes

$$\Delta Z \equiv Z_2 - Z_1 = \frac{R}{g_0} \int_{p_2}^{p_1} T\, d\ln p \tag{1.18}$$

where ΔZ is the *thickness* of the atmospheric layer between the pressure surfaces p_2 and p_1. Defining a layer mean temperature

$$\overline{T} \equiv \int_{p_2}^{p_1} T\, d\ln p \bigg/ \int_{p_2}^{p_1} d\ln p$$

and a layer mean *scale height* $H \equiv R\overline{T}/g_0$ we have from (1.18)

$$\Delta Z = H \ln(p_1/p_2) \tag{1.19}$$

Thus the thickness of a layer is proportional to the mean temperature of the layer. Pressure decreases more rapidly with height in a cold layer than in a warm layer. It also follows immediately from (1.19) that in an isothermal atmosphere of temperature $\overline{T}$ the geopotential height is proportional to the natural logarithm of pressure normalized by the surface pressure;

$$Z = -H \ln(p/p_0),$$

where p_0 is the pressure at $Z = 0$. Thus, the pressure decreases exponentially with geopotential height by a factor of e^{-1} per scale height,

$$p(Z) = p(0)e^{-Z/H}$$

1.6.2 Pressure as a Vertical Coordinate

From the hydrostatic equation (1.14) it is clear that there exists a single-valued monotonic relationship between pressure and height in each vertical column of the atmosphere. Thus we may use pressure as the independent vertical coordinate and height (or geopotential) as a dependent variable. The thermodynamic state of the atmosphere is then specified by the fields of $\Phi(x, y, p, t)$ and $T(x, y, p, t)$.

Now the horizontal components of the pressure gradient force given by (1.1) are evaluated by partial differentiation holding z constant. However, when pressure is used as the vertical coordinate horizontal partial derivatives must be evaluated holding p constant. Transformation of the horizontal pressure gradient force from height to pressure coordinates may be carried out with the aid of Fig. 1.11. Considering only the x,z plane we see from the figure that

$$\left[\frac{(p_0 + \delta p) - p_0}{\delta x}\right]_z = \left[\frac{(p_0 + \delta p) - p_0}{\delta z}\right]_x \left(\frac{\delta z}{\delta x}\right)_p$$

where the subscripts indicate variables which remain constant in evaluating the differentials. Thus, for example, in the limit $\delta z \to 0$

$$\left[\frac{(p_0 + \delta p) - p_0}{\delta z}\right]_x \to \left(-\frac{\partial p}{\partial z}\right)_x$$

where the minus sign is included because $\delta z < 0$ for $\delta p > 0$.

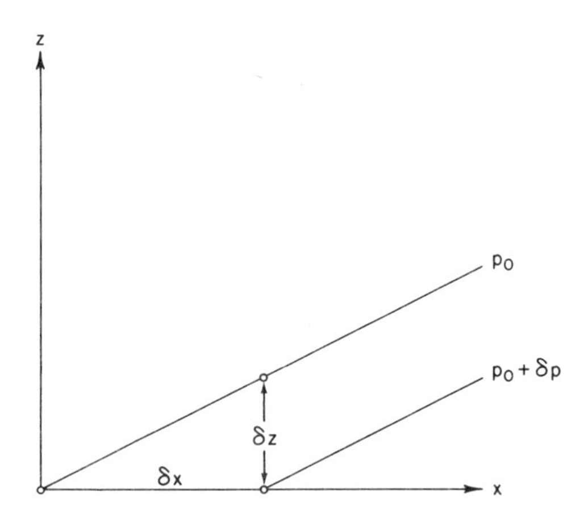

Fig. 1.11 Slope of pressure surfaces in the x,z plane.

Taking the limits δx, $\delta z \to 0$ we obtain[5]

$$\left(\frac{\partial p}{\partial x}\right)_z = -\left(\frac{\partial p}{\partial z}\right)_x \left(\frac{\partial z}{\partial x}\right)_p$$

which after substitution from the hydrostatic equation (1.14) yields

$$-\frac{1}{\rho}\left(\frac{\partial p}{\partial x}\right)_z = -g\left(\frac{\partial z}{\partial x}\right)_p = -\left(\frac{\partial \Phi}{\partial x}\right)_p \tag{1.20}$$

Similarly, it is easy to show that

$$-\frac{1}{\rho}\left(\frac{\partial p}{\partial y}\right)_z = -\left(\frac{\partial \Phi}{\partial y}\right)_p \tag{1.21}$$

Thus in the *isobaric* coordinate system the horizontal pressure gradient force is measured by the gradient of geopotential at constant pressure. Density no longer appears explicitly in the pressure gradient force—a distinct advantage of the isobaric system.

1.6.3 A Generalized Vertical Coordinate

Any single-valued monotonic function of pressure or height may be used as the independent vertical coordinate. For example, in many numerical weather prediction models pressure normalized by the pressure at the ground [$\sigma \equiv p(x, y, z, t)/p_s(x, y, t)$] is used as a vertical coordinate. This choice guarantees that the ground will be a coordinate surface ($\sigma = 1$) even in the presence of spatial and temporal surface pressure variations. Thus this so-called "σ-coordinate" system is particularly useful in regions of strong topographic variations.

We now obtain a general expression for the horizontal pressure gradient which is applicable to any vertical coordinate $s = s(x, y, z, t)$ which is a single-valued monotonic function of height. Referring to Fig. 1.12 we see that for a horizontal distance δx the pressure difference evaluated along the surface $s =$ const. is related to that evaluated at $z =$ const. as

$$\frac{p_C - p_A}{\delta x} = \frac{p_C - p_B}{\delta z}\frac{\delta z}{\delta x} + \frac{p_B - p_A}{\delta x}$$

Taking the limit as δx, $\delta z \to 0$ we obtain

$$\left(\frac{\partial p}{\partial x}\right)_s = \frac{\partial p}{\partial z}\left(\frac{\partial z}{\partial x}\right)_s + \left(\frac{\partial p}{\partial x}\right)_z \tag{1.22}$$

[5] It is important to note the minus sign on the right in this expression!

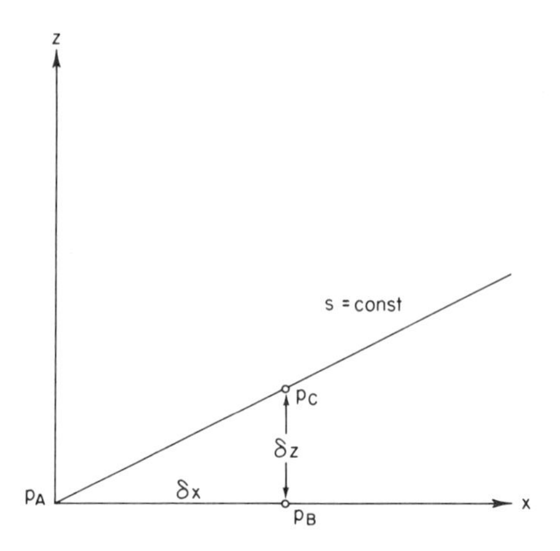

Fig. 1.12 Transformation of the pressure gradient force to s coordinates.

Using the identity

$$\frac{\partial p}{\partial z} \equiv \frac{\partial s}{\partial z}\frac{\partial p}{\partial s}$$

we can express (1.22) in the alternate form

$$\left(\frac{\partial p}{\partial x}\right)_s = \left(\frac{\partial p}{\partial x}\right)_z + \frac{\partial s}{\partial z}\left(\frac{\partial z}{\partial x}\right)_s \frac{\partial p}{\partial s} \tag{1.23}$$

In later chapters we will apply (1.22) or (1.23) and similar expressions for other fields to transform the dynamical equations to several different vertical coordinate systems.

Problems

The more difficult problems are marked by an asterisk.

1. Neglecting the latitudinal variation in the radius of the earth, calculate the angle between the gravitational force and gravity vectors at the surface of the earth as a function of latitude.
2. Calculate the altitude at which an artificial satellite orbiting in the equatorial plane can be a synchronous satellite (i.e., can remain above the same spot on the surface of the earth).

3. An artificial satellite is placed into a natural synchronous orbit above the equator and is attached to the earth below by a wire. A second satellite is attached to the first by a wire of the same length and is placed in orbit directly above the first at the same angular velocity. Assuming that the wires have zero mass, calculate the tension in the wires per unit mass of satellite. Could this tension be used to lift objects into orbit with no additional expenditure of energy?

4. A train is running smoothly along a curved track at the rate of 50 m s^{-1}. A passenger standing on a set of scales observes that his weight is 10% greater than when the train is at rest. The track is banked so that the force acting on the passenger is normal to the floor of the train. What is the radius of curvature of the track?

5. If a baseball player throws a ball a horizontal distance of 100 m at 30° latitude in 4 s, by how much is it deflected horizontally as a result of the rotation of the earth?

6. Two balls 4 cm in diameter are placed 100 m apart on a frictionless horizontal plane at 43°N latitude. If the balls are impulsively propelled directly at each other with equal speeds, at what speed must they travel so that they just miss each other?

7. A locomotive of 2×10^5-kg mass travels 50 m s^{-1} along a straight horizontal track at 43°N. What lateral force is exerted on the rails? Compare upward reaction force exerted by the rails for cases where the locomotive is traveling eastward and westward, respectively.

8. Find the horizontal displacement of a body dropped from a fixed platform at a height h at the equator neglecting the effects of air resistance. What is the numerical value of the displacement for $h = 5$ km?

9. A bullet is fired vertically upward with initial speed w_0 at latitude ϕ. Neglecting air resistance, by what distance will it be displaced horizontally when it returns to the ground? (Neglect $2\Omega u \cos \phi$ compared to g in the vertical equation.)

***10.** A block of mass $M = 1$ kg is suspended from the end of a weightless string. The other end of the string is passed through a small hole in a horizontal platform and a ball of mass $m = 10$ kg is attached. At what angular velocity must the ball rotate on the horizontal platform to balance the weight of the block if the horizontal distance of the ball from the hole is 1 m? While the ball is rotating, the block is pulled down 10 cm. What is the new angular velocity of the ball? How much work is done in pulling down the block?

***11.** A particle is free to slide on a horizontal frictionless plane located at a latitude ϕ on the earth. Find the equation governing the path of the particle if it is given an impulsive eastward velocity $u = u_0$ at $t = 0$. Give the solution for the position of the particle as a function of time.

12. Calculate the 100–50-kPa (1000–500-mb) thicknesses for isothermal conditions with temperatures of 273 K and 250 K, respectively.

13. Isolines of 100–50-kPa (1000–500-mb) thickness are drawn on a weather map using a contour interval of 60 m. What is the corresponding layer mean temperature interval?

14. Show that a homogeneous atmosphere (density independent of height) has a finite height which depends only on the temperature at the lower boundary. Compute the height of a homogeneous atmosphere with surface temperature $T_0 = 273$ K and surface pressure 100 kPa. (Use the ideal gas law and hydrostatic balance.)

15. For the conditions of Problem 14 compute the variation of the temperature with respect to height.

16. Show that for an atmosphere with uniform *lapse rate* γ [$\gamma \equiv -dT/dz$], the geopotential height at pressure level p_1 is given by

$$Z = \frac{T_0}{\gamma}\left[1 - \left(\frac{p_0}{p_1}\right)^{-R\gamma/g}\right]$$

where T_0 and p_0 are the sea level temperature and pressure, respectively.

17. Calculate the 100–50-kPa (1000–500-mb) thickness for a constant lapse rate atmosphere with $\gamma = 6.5$ K km^{-1} and $T_0 = 273$ K. Compare your results with the results in Problem 12.

18. Derive an expression for the variation of density with respect to height in a constant lapse rate atmosphere.

***19.** Derive an expression for the altitude variation of the pressure change δp which occurs when an atmosphere with constant lapse rate is subjected to a height independent temperature change δT while surface pressure remains constant. At what height is the magnitude of the pressure change a maximum?

Suggested References

Complete reference information is provided in the Bibliography at the end of the book.

Kittel *et al.*, *Mechanics* (Vol. 1 of the Berkeley physics course) has excellent discussions of gravitation and noninertial reference frames at the introductory level.

Wallace and Hobbs, *Atmospheric Science: An Introductory Survey*, discusses much of the material in this chapter at an elementary level.

Iribarne and Godson, *Atmospheric Thermodynamics*, contains a more advanced discussion of atmospheric statics.

Chapter

2 The Basic Conservation Laws

Atmospheric motions are governed by three fundamental physical principles: conservation of mass, conservation of momentum, and conservation of energy. The mathematical relations which express these laws may be derived by considering the budgets of mass, momentum, and energy for an infinitesimal *control volume* in the fluid. Two types of control volumes are commonly used in fluid dynamics. In the *eulerian* frame of reference the control volume consists of a parallelepiped of sides δx, δy, δz whose position is fixed relative to the coordinate axes. Mass, momentum, and energy budgets will depend on fluxes due to the flow of fluid through the boundaries of the control volume. (This type of control volume was used in Section 1.4.1.) In the *lagrangian* frame, however, the control volume consists of an infinitesimal mass of "tagged" fluid particles; thus, the control volume moves about following the motion of the fluid, always containing the same fluid particles.

The lagrangian frame is particularly useful for deriving conservation laws since such laws may be stated most simply in terms of a particular mass element of the fluid. The eulerian system is, however, more convenient for solving most problems because in that system the field variables are related by a set of partial differential equations in which the independent variables are the coordinates x, y, z, t. In the lagrangian system, on the other hand, it is necessary to follow the time evolution of the fields for various individual fluid parcels. Thus the independent variables are x_0, y_0, z_0, and t, where x_0, y_0, z_0 designate the position which a particular parcel passed through at a reference time t_0.

2.1 Total Differentiation

The conservation laws to be derived in this chapter contain expressions for the rates of change per unit volume of mass, momentum, and thermodynamic energy following the motion of particular fluid parcels. In order to apply these laws in the eulerian frame it is necessary to derive a relationship between the rate of change of a field variable following the motion and its rate of change at a fixed point. The former is called the *substantial* or *total* derivative, while the latter is called the *local* derivative (it is merely the partial derivative with respect to time).

To derive a relationship between the total derivative and the local derivative it is convenient to refer to a particular field variable, temperature, for example. Suppose that the temperature measured on a balloon which moves with the wind is T_0 at the point x_0, y_0, z_0 and time t_0. If the balloon moves to the point $x_0 + \delta x$, $y_0 + \delta y$, $z_0 + \delta z$ in a time increment δt, then the temperature change recorded on the balloon δT may be expressed in a Taylor series expansion as

$$\delta T = \left(\frac{\partial T}{\partial t}\right)\delta t + \left(\frac{\partial T}{\partial x}\right)\delta x + \left(\frac{\partial T}{\partial y}\right)\delta y + \left(\frac{\partial T}{\partial z}\right)\delta z + \text{(higher order terms)}$$

Dividing through by δt and taking the limit $\delta t \to 0$ we obtain

$$\frac{dT}{dt} = \frac{\partial T}{\partial t} + \frac{\partial T}{\partial x}\frac{dx}{dt} + \frac{\partial T}{\partial y}\frac{dy}{dt} + \frac{\partial T}{\partial z}\frac{dz}{dt}$$

where

$$\frac{dT}{dt} \equiv \lim_{\delta t \to 0} \frac{\delta T}{\delta t}$$

is the rate of change of T following the motion. If we now let

$$\frac{dx}{dt} \equiv u, \qquad \frac{dy}{dt} \equiv v, \qquad \frac{dz}{dt} \equiv w$$

then u, v, w are the velocity components in the x, y, z directions, respectively, and we have

$$\frac{dT}{dt} = \frac{\partial T}{\partial t} + \left(u\frac{\partial T}{\partial x} + v\frac{\partial T}{\partial y} + w\frac{\partial T}{\partial z}\right) \tag{2.1}$$

Using vector notation this expression may be rewritten as

$$\frac{\partial T}{\partial t} = \frac{dT}{dt} - \mathbf{U} \cdot \nabla T$$

where $\mathbf{U} = \mathbf{i}u + \mathbf{j}v + \mathbf{k}w$ is the velocity vector. The term $-\mathbf{U} \cdot \nabla T$ is called the temperature *advection.* It gives the contribution to the local temperature change due to the air motion. For example, if the wind is blowing from a cold region toward a warm region $-\mathbf{U} \cdot \nabla T$ will be negative (cold advection) and the advection term will contribute negatively to the local temperature change. Thus, the local rate of change of temperature equals the rate of change of temperature following the motion (that is, the heating or cooling of individual air parcels) plus the advective rate of change of temperature.

The relationship given for temperature in (2.1) holds for any of the field variables. Furthermore, the total derivative can be defined following a motion field other than the actual wind field. For example, we may wish to relate the pressure change measured by a barometer on a moving ship to the local pressure change.

Example. The surface pressure decreases by 0.3 kPa/180 km in the eastward direction. A ship steaming eastward at 10 km/h measures a pressure fall of 0.1 kPa/3 h. What is the pressure change on an island which the ship is passing? If we take the x axis oriented eastward, then the local rate of change of pressure on the island is

$$\frac{\partial p}{\partial t} = \frac{dp}{dt} - u\frac{\partial p}{\partial x}$$

where dp/dt is the pressure change observed by the ship and u the velocity of the ship. Thus,

$$\frac{\partial p}{\partial t} = \frac{-0.1 \text{ kPa}}{3 \text{ h}} - \left(10 \frac{\text{km}}{\text{h}}\right)\left(\frac{-0.3 \text{ kPa}}{180 \text{ km}}\right) = -\frac{0.1 \text{ kPa}}{6 \text{ h}}$$

Thus, the rate of pressure fall on the island is only half the rate measured on the moving ship.

If the total derivative of a field variable is zero, then that variable is a conservative quantity following the motion. The local change is then entirely due to advection. As we shall see later, field variables that are approximately conserved following the motion play an important role in dynamic meteorology.

2.1.1 Total Differentiation of a Vector in a Rotating System

The conservation law for momentum (Newton's second law of motion) relates the rate of change of the absolute momentum following the motion in an inertial reference frame to the sum of the forces acting on the fluid. For most applications in meteorology it is desirable to refer the motion to a reference frame rotating with the earth. The transformation of the momentum equation to a rotating coordinate system requires a relationship between the total derivative of a vector in an inertial reference frame and the corresponding total derivative in a rotating system.

To derive this relationship we let **A** be an arbitrary vector whose cartesian components in an inertial frame are given by

$$\mathbf{A} = \mathbf{i}A_x + \mathbf{j}A_y + \mathbf{k}A_z$$

and whose components in a frame rotating with an angular velocity $\mathbf{\Omega}$ are

$$\mathbf{A} = \mathbf{i}'A'_x + \mathbf{j}'A'_y + \mathbf{k}'A'_z$$

Letting $d_a\mathbf{A}/dt$ be the total derivative of **A** in the inertial frame we can write

$$\begin{aligned}\frac{d_a\mathbf{A}}{dt} &= \mathbf{i}\frac{dA_x}{dt} + \mathbf{j}\frac{dA_y}{dt} + \mathbf{k}\frac{dA_z}{dt} \\ &= \mathbf{i}'\frac{dA'_x}{dt} + \mathbf{j}'\frac{dA'_y}{dt} + \mathbf{k}'\frac{dA'_z}{dt} + \frac{d\mathbf{i}'}{dt}A'_x + \frac{d\mathbf{j}'}{dt}A'_y + \frac{d\mathbf{k}'}{dt}A'_z\end{aligned}$$

Now,

$$\mathbf{i}'\frac{dA'_x}{dt} + \mathbf{j}'\frac{dA'_y}{dt} + \mathbf{k}'\frac{dA'_z}{dt} \equiv \frac{d\mathbf{A}}{dt}$$

is just the total derivative of **A** as viewed in the rotating coordinates (that is, the rate of change of **A** following the relative motion). Furthermore, since $\mathbf{i}'$ may be regarded as a position vector of unit length, $d\mathbf{i}'/dt$ is the velocity of $\mathbf{i}'$ due to its rotation. Thus $d\mathbf{i}'/dt = \mathbf{\Omega} \times \mathbf{i}'$ and similarly $d\mathbf{j}'/dt = \mathbf{\Omega} \times \mathbf{j}'$ and $d\mathbf{k}'/dt = \mathbf{\Omega} \times \mathbf{k}'$. Therefore the above can be rewritten as

$$\frac{d_a\mathbf{A}}{dt} = \frac{d\mathbf{A}}{dt} + \mathbf{\Omega} \times \mathbf{A} \tag{2.2}$$

which is the desired relationship.

2.2 The Vectorial Form of the Momentum Equation in Rotating Coordinates

In an inertial reference frame Newton's second law of motion may be written symbolically as

$$\frac{d_a\mathbf{U}_a}{dt} = \sum \mathbf{F} \tag{2.3}$$

The left-hand side represents the rate of change of the absolute velocity $\mathbf{U}_a$, following the motion as viewed in an inertial system. The right-hand side represents the sum of the real forces acting *per unit mass.* In Section 1.5 we found through simple physical reasoning that when the motion is viewed in a rotating coordinate system certain additional apparent forces must be included if Newton's second law is to be valid. The same result may be obtained by a formal transformation of coordinates in (2.3).

In order to transform this expression to rotating coordinates we must first find a relationship between $\mathbf{U}_a$ and the velocity relative to the rotating system, which we will designate by $\mathbf{U}$. This relationship is obtained by applying (2.2) to the position vector $\mathbf{r}$ for an air parcel on the rotating earth:

$$\frac{d_a\mathbf{r}}{dt} = \frac{d\mathbf{r}}{dt} + \boldsymbol{\Omega} \times \mathbf{r} \tag{2.4}$$

But $d_a\mathbf{r}/dt \equiv \mathbf{U}_a$ and $d\mathbf{r}/dt \equiv \mathbf{U}$; therefore (2.4) may be written as

$$\mathbf{U}_a = \mathbf{U} + \boldsymbol{\Omega} \times \mathbf{r} \tag{2.5}$$

which states simply that the absolute velocity of an object on the rotating earth is equal to its velocity relative to the earth plus the velocity due to the rotation of the earth.

Next we apply (2.2) to the velocity vector $\mathbf{U}_a$ and obtain

$$\frac{d_a\mathbf{U}_a}{dt} = \frac{d\mathbf{U}_a}{dt} + \boldsymbol{\Omega} \times \mathbf{U}_a \tag{2.6}$$

Substituting from (2.5) into the right-hand side of (2.6) gives

$$\begin{aligned} \frac{d_a\mathbf{U}_a}{dt} &= \frac{d}{dt}(\mathbf{U} + \boldsymbol{\Omega} \times \mathbf{r}) + \boldsymbol{\Omega} \times (\mathbf{U} + \boldsymbol{\Omega} \times \mathbf{r}) \\ &= \frac{d\mathbf{U}}{dt} + 2\boldsymbol{\Omega} \times \mathbf{U} - \Omega^2\mathbf{R} \end{aligned} \tag{2.7}$$

where $\boldsymbol{\Omega}$ is assumed to be constant. Here $\mathbf{R}$ is a vector perpendicular to the axis of rotation, with magnitude equal to the distance to the axis of rotation, so that with the aid of a vector identity,

$$\boldsymbol{\Omega} \times (\boldsymbol{\Omega} \times \mathbf{r}) = \boldsymbol{\Omega} \times (\boldsymbol{\Omega} \times \mathbf{R}) \equiv -\Omega^2\mathbf{R}$$

Equation (2.7) states that the acceleration following the motion in an inertial system equals the acceleration following the relative motion in a rotating system plus the Coriolis acceleration plus the centripetal acceleration.

If we assume that the only real forces acting on the atmosphere are the pressure gradient force, gravitation, and friction, we can rewrite Newton's second law (2.3) with the aid of (2.7) as

$$\frac{d\mathbf{U}}{dt} = -2\boldsymbol{\Omega} \times \mathbf{U} - \frac{1}{\rho}\nabla p + \mathbf{g} + \mathbf{F}_r \tag{2.8}$$

where $\mathbf{F}_r$ designates the friction force, and the centrifugal force has been combined with gravitation in the gravity term $\mathbf{g}$ (see Section 1.5.2). Equation (2.8) is the statement of Newton's second law for motion relative to a rotating coordinate frame. It states that the acceleration following the relative motion in the rotating frame equals the sum of the Coriolis force, the pressure gradient force, effective gravity, and friction. It is this form of the momentum equation which is basic to most work in dynamic meteorology.

2.3 The Component Equations in Spherical Coordinates

For purposes of theoretical analysis and numerical prediction, it is necessary to expand the vectorial momentum equation (2.8) into its scalar components. Since the departure of the shape of the earth from sphericity is entirely negligible for meteorological purposes, it is convenient to expand (2.8) in spherical coordinates so that the (level) surface of the earth corresponds to a coordinate surface. The coordinate axes are then (λ, ϕ, z), where λ is longitude, ϕ is latitude, and z is the vertical distance above the surface of the earth. If the unit vectors $\mathbf{i}, \mathbf{j}, \mathbf{k}$ are now taken to be directed eastward, northward, and upward, respectively, the relative velocity becomes

$$\mathbf{U} \equiv \mathbf{i}u + \mathbf{j}v + \mathbf{k}w$$

where the components u, v, and w are defined as follows:

$$u \equiv r\cos\phi\,\frac{d\lambda}{dt}, \qquad v \equiv r\,\frac{d\phi}{dt}, \qquad w \equiv \frac{dz}{dt} \tag{2.9}$$

Here, r is the distance to the center of the earth, which is related to z by $r = a + z$, where a is the radius of the earth. Traditionally, the variable r in (2.9) is replaced by the constant a. This is a very good approximation since $z \ll a$ for the regions of the atmosphere with which the meteorologist is concerned. For notational simplicity, it is conventional to define x and y as eastward and northward distance, such that $dx = a\cos\phi\, d\lambda$ and $dy = a\, d\phi$. Thus, the horizontal velocity components are $u \equiv dx/dt$ and $v \equiv dy/dt$ in the eastward and northward directions, respectively. The (x, y, z)-coordinate system defined in this way is not, however, a cartesian coordinate system

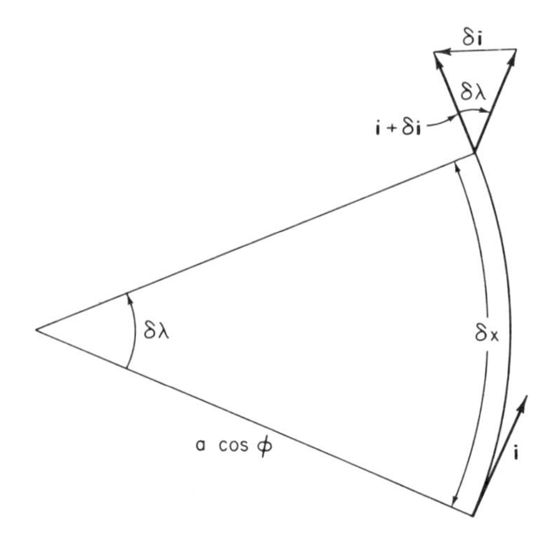

Fig. 2.1 The longitudinal dependence of the unit vector **i**.

because the *directions* of the **i, j, k** unit vectors are not constant, but are functions of position on the spherical earth. This position dependence of the unit vectors must be taken into account when the acceleration vector is expanded into its components on the sphere. Thus, we write

$$\frac{d\mathbf{U}}{dt} = \mathbf{i}\frac{du}{dt} + \mathbf{j}\frac{dv}{dt} + \mathbf{k}\frac{dw}{dt} + u\frac{d\mathbf{i}}{dt} + v\frac{d\mathbf{j}}{dt} + w\frac{d\mathbf{k}}{dt} \tag{2.10}$$

In order to obtain the component equations, it is necessary first to evaluate the rates of change of the unit vectors following the motion.

We first consider $d\mathbf{i}/dt$. Expanding the total derivative as in (2.1) and noting that **i** is a function only of x (that is, an eastward-directed vector does not change its orientation if the motion is in the north–south or vertical directions), we get

$$\frac{d\mathbf{i}}{dt} = u\frac{\partial \mathbf{i}}{\partial x}$$

From Fig. 2.1 we see that

$$\lim_{\delta x \to 0} \frac{|\delta \mathbf{i}|}{\delta x} = \left|\frac{\partial \mathbf{i}}{\partial x}\right| = \frac{1}{a\cos\phi}$$

and that the vector $\partial\mathbf{i}/\partial x$ is directed toward the axis of rotation. Thus, as illustrated in Fig. 2.2

$$\frac{\partial \mathbf{i}}{\partial x} = \frac{1}{a\cos\phi}(\mathbf{j}\sin\phi - \mathbf{k}\cos\phi)$$

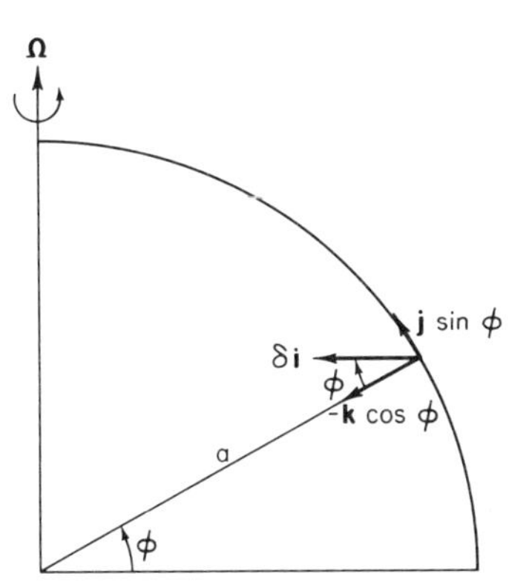

Fig. 2.2 Resolution of $\delta\mathbf{i}$ into northward and vertical components.

Therefore

$$\frac{d\mathbf{i}}{dt} = \frac{u}{a\cos\phi}(\mathbf{j}\sin\phi - \mathbf{k}\cos\phi) \tag{2.11}$$

Considering now $d\mathbf{j}/dt$, we note that $\mathbf{j}$ is a function only of x and y. Thus, with the aid of Fig. 2.3 we see that for eastward motion $|\delta\mathbf{j}| = \delta x/(a/\tan\phi)$. Since the vector $\partial\mathbf{j}/\partial x$ is directed in the negative x direction, we have then

$$\frac{\partial\mathbf{j}}{\partial x} = -\frac{\tan\phi}{a}\mathbf{i}$$

From Fig. 2.4 it is clear that for northward motion $|\delta\mathbf{j}| = \delta\phi$. But $\delta y = a\,\delta\phi$, and $\delta\mathbf{j}$ is directed downward, so that

$$\frac{\partial\mathbf{j}}{\partial y} = -\frac{\mathbf{k}}{a}$$

Hence,

$$\frac{d\mathbf{j}}{dt} = -\frac{u\tan\phi}{a}\mathbf{i} - \frac{v}{a}\mathbf{k} \tag{2.12}$$

Finally, by similar arguments it can be shown that

$$\frac{d\mathbf{k}}{dt} = \mathbf{i}\frac{u}{a} + \mathbf{j}\frac{v}{a} \tag{2.13}$$

Substituting (2.11)–(2.13) into (2.10) and rearranging the terms, we obtain the spherical polar coordinate expansion of the acceleration following the relative motion

$$\frac{d\mathbf{U}}{dt} = \left(\frac{du}{dt} - \frac{uv\tan\phi}{a} + \frac{uw}{a}\right)\mathbf{i} + \left(\frac{dv}{dt} + \frac{u^2\tan\phi}{a} + \frac{wv}{a}\right)\mathbf{j} + \left(\frac{dw}{dt} - \frac{u^2+v^2}{a}\right)\mathbf{k} \tag{2.14}$$

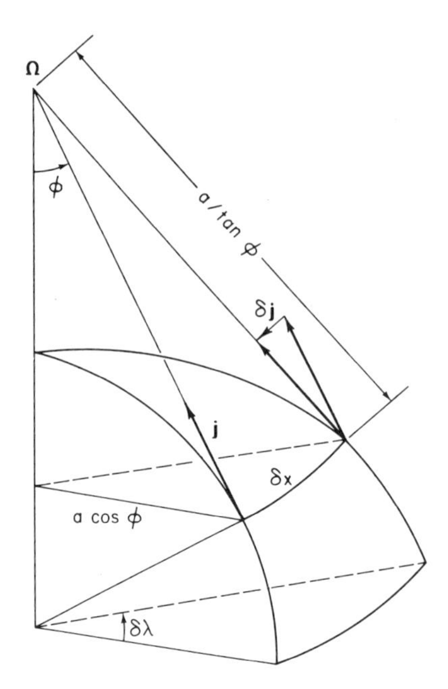

Fig. 2.3 The dependence of unit vector **j** on longitude.

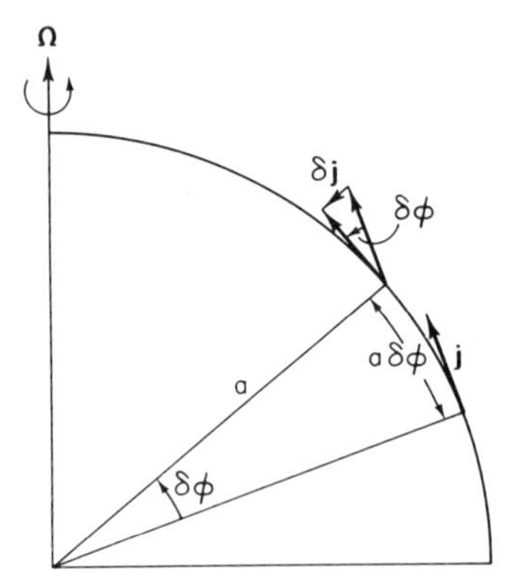

Fig. 2.4 The dependence of unit vector **j** on latitude.

We next turn to the component expansion of the force terms in (2.8). The Coriolis force is expanded by noting that $\mathbf{\Omega}$ has no component parallel to **i**, and that its components parallel to **j** and **k** are $2\Omega\cos\phi$ and $2\Omega\sin\phi$, respectively. Thus, using the definition of the vector cross product,

$$-2\mathbf{\Omega}\times\mathbf{U} = -2\Omega\begin{vmatrix}\mathbf{i} & \mathbf{j} & \mathbf{k}\\ 0 & \cos\phi & \sin\phi\\ u & v & w\end{vmatrix}$$
$$= -(2\Omega w\cos\phi - 2\Omega v\sin\phi)\mathbf{i} - 2\Omega u\sin\phi\mathbf{j} + 2\Omega u\cos\phi\mathbf{k} \tag{2.15}$$

The pressure gradient force may be expressed as

$$\nabla p = \mathbf{i}\frac{\partial p}{\partial x} + \mathbf{j}\frac{\partial p}{\partial y} + \mathbf{k}\frac{\partial p}{\partial z} \tag{2.16}$$

and gravity is conveniently represented as

$$\mathbf{g} = -g\mathbf{k} \tag{2.17}$$

where g is a positive scalar ($g \simeq 9.8 \text{ m s}^{-2}$ at the earth's surface). Finally, friction is expanded in components as

$$\mathbf{F}_r = \mathbf{i}F_x + \mathbf{j}F_y + \mathbf{k}F_z \tag{2.18}$$

Substituting (2.14)–(2.18) into the equation of motion (2.8) and equating all terms in the **i**, **j**, and **k** directions, respectively, we obtain

$$\frac{du}{dt} - \frac{uv \tan\phi}{a} + \frac{uw}{a} = -\frac{1}{\rho}\frac{\partial p}{\partial x} + 2\Omega v \sin\phi - 2\Omega w \cos\phi + F_x \tag{2.19}$$

$$\frac{dv}{dt} + \frac{u^2 \tan\phi}{a} + \frac{vw}{a} = -\frac{1}{\rho}\frac{\partial p}{\partial y} - 2\Omega u \sin\phi + F_y \tag{2.20}$$

$$\frac{dw}{dt} - \frac{u^2 + v^2}{a} = -\frac{1}{\rho}\frac{\partial p}{\partial z} - g + 2\Omega u \cos\phi + F_z \tag{2.21}$$

which are the eastward, northward, and vertical component momentum equations, respectively. The terms proportional to $1/a$ on the left-hand sides in (2.19)–(2.21) are called the *curvature* terms because they arise due to the curvature of the earth.[1] Because they are nonlinear terms (that is, they are quadratic in the dependent variable), they are difficult to handle in theoretical analyses. Fortunately, as will be shown in the next section, the curvature terms are unimportant for midlatitude synoptic scale motions. However, even when the curvature terms are neglected (2.19)–(2.21) are still nonlinear partial differential equations as can be seen by expanding the total derivatives into their local and advective parts:

$$\frac{du}{dt} = \frac{\partial u}{\partial t} + u\frac{\partial u}{\partial x} + v\frac{\partial u}{\partial y} + w\frac{\partial u}{\partial z}$$

with similar expressions for dv/dt and dw/dt. In general the advective acceleration terms are comparable in magnitude to the local acceleration. It is primarily the presence of nonlinear advection processes that makes dynamic meteorology an interesting and challenging subject.

2.4 Scale Analysis of the Equations of Motion

In Section 1.3 we discussed the basic notion of scaling the equations of motion in order to determine whether some terms in the equations are

[1] It can be shown that when r is replaced by a as done here (the traditional approximation) the Coriolis terms proportional to $\cos\phi$ in (2.19) and (2.21) must be neglected if the equations are to satisfy angular momentum conservation.

negligible for motions of meteorological concern. Elimination of terms on scaling considerations not only has the advantage of simplifying the mathematics, but as we will show in later chapters the elimination of small terms in some cases has the very important property of completely eliminating, or *filtering*, an unwanted type of motion. The complete equations of motion (2.19)–(2.21) describe all types and scales of atmospheric motions. Sound waves, for example, are a perfectly valid solution to these equations. However, sound waves are of negligible importance in meteorological problems. Therefore, it will be a distinct advantage if, as turns out to be true, we can neglect the terms which lead to sound wave-type solutions and filter out this unwanted class of motions.

In order to simplify (2.19)–(2.21) for synoptic scale motions we define the following characteristic scales of the field variables based on observed values for midlatitude synoptic systems.

$U \sim 10 \text{ m s}^{-1}$	horizontal velocity scale
$W \sim 1 \text{ cm s}^{-1}$	vertical velocity scale
$L \sim 10^6 \text{ m}$	length scale [$\sim 1/(2\pi)$ wavelength]
$D \sim 10^4 \text{ m}$	depth scale
$\Delta P/\rho \sim 10^3 \text{ m}^2 \text{ s}^{-2}$	horizontal pressure fluctuation scale
$L/U \sim 10^5 \text{ s}$	time scale

The horizontal pressure fluctuation ΔP is normalized by the density ρ in order to produce a scale estimate which is valid at all heights in the troposphere despite the approximate exponential decrease with height of both ΔP and ρ. Note that $\Delta P/\rho$ has units of geopotential. Referring back to (1.21) we see that indeed the magnitude of the fluctuation of $\Delta P/\rho$ on a surface of constant height must equal the magnitude of the fluctuation of geopotential on an isobaric surface. The time scale here is an advective time scale which is appropriate for pressure systems which move at approximately the speed of the horizontal wind, as is observed for synoptic scale motions. Thus L/U is the time required to travel a distance L at a speed U.

It should be pointed out here that the synoptic scale vertical velocity is not a directly measurable quantity. However, as we will show in Chapter 3, the magnitude of w can be deduced from knowledge of the horizontal velocity field.

We can now estimate the magnitude of each term in (2.19) and (2.20) for synoptic scale motions at a given latitude. It is convenient to consider disturbances centered at latitude $\phi_0 = 45°$, and introduce the notation

$$f_0 = 2\Omega \sin \phi_0 = 2\Omega \cos \phi_0 \simeq 10^{-4} \quad \text{s}^{-1}$$

Table 2.1 *Scale Analysis of the Horizontal Momentum Equations*

	A	B	C	D	E	F
x-component momentum equation	$\frac{du}{dt}$	$-2\Omega v \sin\phi$	$+2\Omega w \cos\phi$	$+\frac{uw}{a}$	$-\frac{uv\tan\phi}{a}$	$= -\frac{1}{\rho}\frac{\partial p}{\partial x}$
y-component momentum equation	$\frac{dv}{dt}$	$+2\Omega u \sin\phi$		$+\frac{vw}{a}$	$+\frac{u^2\tan\phi}{a}$	$= -\frac{1}{\rho}\frac{\partial p}{\partial y}$
Scales of individual terms	$\frac{U^2}{L}$	$f_0 U$	$f_0 W$	$\frac{UW}{a}$	$\frac{U^2}{a}$	$\frac{\Delta P}{\rho L}$
Magnitudes of the terms ($\mathrm{m\,s^{-2}}$)	10^{-4}	10^{-3}	10^{-6}	10^{-8}	10^{-5}	10^{-3}

Table 2.1 shows the characteristic magnitude of each term in (2.19) and (2.20) based on the above scaling considerations. Frictional terms are not included because for the synoptic time scale frictional dissipation is thought to be of secondary importance above the lowest kilometer. The role of friction in the boundary layer near the ground will be discussed in Chapter 5.

2.4.1 The Geostrophic Approximation and the Geostrophic Wind

It is apparent from Table 2.1 that for midlatitude synoptic scale disturbances the Coriolis force (term B) and the pressure gradient force (term F) are in approximate balance. Therefore, retaining only these two terms in (2.19) and (2.20), we obtain as a first approximation the *geostrophic* relationship

$$-fv \simeq -\frac{1}{\rho}\frac{\partial p}{\partial x}, \qquad fu \simeq -\frac{1}{\rho}\frac{\partial p}{\partial y} \tag{2.22}$$

where $f \equiv 2\Omega \sin\phi$ is called the *Coriolis parameter*. The geostrophic balance is a *diagnostic* expression which gives the approximate relationship between the pressure field and horizontal velocity in synoptic scale systems. The approximation (2.22) contains no reference to time, and therefore cannot be used to predict the evolution of the velocity field. It is for this reason that the geostrophic relationship is called a diagnostic relationship.

By analogy to the geostrophic approximation (2.22) it is possible to define a horizontal velocity field, $\mathbf{V}_g \equiv \mathbf{i}u_g + \mathbf{j}v_g$, called the *geostrophic wind*, which satisfies (2.22) identically. Thus in vectorial form

$$\mathbf{V}_g \equiv \mathbf{k} \times \frac{1}{\rho f}\nabla p \tag{2.23}$$

Thus, knowledge of the pressure distribution at any time determines the geostrophic wind. It should be kept clearly in mind that (2.23) always *defines* the geostrophic wind; but only for large-scale motions should the geostrophic wind be used as an approximation to the actual horizontal wind field. For the scales used in Table 2.1 the geostrophic wind approximates the true horizontal velocity to within 10–15% in midlatitudes.

2.4.2 Approximate Prognostic Equations; The Rossby Number

To obtain prediction equations it is necessary to retain the acceleration (term A) in (2.19) and (2.20). The resulting approximate horizontal momentum equations are

$$\frac{du}{dt} - fv = -\frac{1}{\rho}\frac{\partial p}{\partial x} \tag{2.24}$$

$$\frac{dv}{dt} + fu = -\frac{1}{\rho}\frac{\partial p}{\partial y} \tag{2.25}$$

Our scale analysis showed that the acceleration terms in (2.24) and (2.25) are about an order of magnitude smaller than the Coriolis force and pressure gradient force. The fact that the horizontal flow is in approximate geostrophic balance is helpful for diagnostic analysis. However, it makes actual applications of these equations in weather prognosis difficult because acceleration (which must be measured accurately) is given by the small difference between two large terms. Thus, a small error in measurement of either velocity or pressure gradient will lead to very large errors in estimating the acceleration. This problem will be discussed in some detail in Chapter 8.

A convenient measure of the magnitude of the acceleration compared to the Coriolis force may be obtained by forming the ratio of the characteristic scales for the acceleration and the Coriolis force terms

$$\frac{U^2/L}{f_0 U}$$

This ratio is a nondimensional number called the *Rossby number* after the Swedish meteorologist C. G. Rossby [1898–1957], and is designated by

$$Ro \equiv \frac{U}{f_0 L}$$

Thus, the smallness of the Rossby number is a measure of the validity of the geostrophic approximation.

2.4.3 The Hydrostatic Approximation

A similar scale analysis can be applied to the vertical component of the momentum equation (2.21). Since pressure decreases by about an order of magnitude from the ground to the tropopause, the vertical pressure gradient may be scaled by P_0/H where P_0 is the surface pressure and H is the depth of the troposphere. The terms in (2.21) may then be estimated for synoptic scale motions and are shown in Table 2.2. As with the horizontal component equations we consider motions centered at 45° latitude and neglect friction. The scaling indicates that to a high degree of accuracy the pressure field is in *hydrostatic* equilibrium, that is, the pressure at any point is simply equal to the weight of a unit cross-section column of air above that point.

The above analysis of the vertical momentum equation is, however, somewhat misleading. It is not sufficient to show merely that the vertical acceleration is small compared to g. Since only that part of the pressure field which varies horizontally is directly coupled to the horizontal velocity field, it is actually necessary to show that the horizontally varying pressure component is itself in hydrostatic equilibrium with the horizontally varying density field. To do this it is convenient to first define a standard pressure $p_0(z)$, which is the horizontally averaged pressure at each height, and a corresponding standard density $\rho_0(z)$, defined so that $p_0(z)$ and $\rho_0(z)$ are in *exact* hydrostatic balance:

$$\frac{1}{\rho_0}\frac{dp_0}{dz} \equiv -g \tag{2.26}$$

We may then write the total pressure and density fields as

$$\begin{aligned} p(x, y, z, t) &= p_0(z) + p'(x, y, z, t) \\ \rho(x, y, z, t) &= \rho_0(z) + \rho'(x, y, z, t) \end{aligned} \tag{2.27}$$

where p' and ρ' are perturbations from the standard values of pressure and density. For an atmosphere at rest, p' and ρ' would thus be zero. Using the

Table 2.2 *Scale Analysis of the Vertical Momentum Equation*

z-component momentum equation	$\dfrac{dw}{dt}$	$-2\Omega u \cos\phi$	$-\dfrac{u^2+v^2}{a}$	$= -\dfrac{1}{\rho}\dfrac{\partial p}{\partial z}$	$-g$
Scales of individual terms	$\dfrac{UW}{L}$	$f_0 U$	$\dfrac{U^2}{a}$	$\dfrac{P_0}{\rho H}$	g
Magnitudes of the terms (m s^{-2})	10^{-7}	10^{-3}	10^{-5}	10	10

definitions (2.26) and (2.27) and assuming that ρ'/ρ_0 is much less than unity in magnitude so that $(\rho_0 + \rho')^{-1} \simeq \rho_0^{-1}(1 - \rho'/\rho_0)$ we find that

$$-\frac{1}{\rho}\frac{\partial p}{\partial z} - g = -\frac{1}{(\rho_0 + \rho')}\frac{\partial}{\partial z}(p_0 + p') - g$$

$$\simeq \frac{1}{\rho_0}\left[\frac{\rho'}{\rho_0}\frac{dp_0}{dz} - \frac{\partial p'}{\partial z}\right] = -\frac{1}{\rho_0}\left[\rho' g + \frac{\partial p'}{\partial z}\right] \tag{2.28}$$

For synoptic scale motions, the terms in (2.28) have the magnitudes

$$\frac{1}{\rho_0}\frac{\partial p'}{\partial z} \sim \left[\frac{\Delta P}{\rho_0 H}\right] \sim 10^{-1} \quad \mathrm{m\ s^{-2}}, \qquad \frac{\rho' g}{\rho_0} \sim 10^{-1} \quad \mathrm{m\ s^{-2}}$$

Comparing these with the magnitudes of other terms in the vertical momentum equation (Table 2.2), we see that to a very good approximation the perturbation pressure field is in hydrostatic equilibrium with the perturbation density field so that

$$\frac{\partial p'}{\partial z} + \rho' g = 0 \tag{2.29}$$

Therefore, for synoptic scale motions, vertical accelerations are negligible and the vertical velocity cannot be determined from the vertical momentum equation. However, we will show in Chapter 3 that it is, nevertheless, possible to deduce the vertical motion field indirectly.

2.5 The Continuity Equation

We turn now to the second of the three fundamental conservation principles, conservation of mass. The mathematical relationship which expresses conservation of mass for a fluid is called the *continuity equation.* In this section we develop the continuity equation using two alternative methods. The first method is based on an eulerian control volume, while the second is based on a lagrangian control volume.

2.5.1 An Eulerian Derivation

We consider a volume element $\delta x\, \delta y\, \delta z$ which is fixed in a cartesian coordinate frame as shown in Fig. 2.5. For such a *fixed* control volume the net rate of mass inflow through the sides must equal the rate of accumulation of mass within the volume. The rate of inflow of mass through the left-hand face per unit area is

$$\left[\rho u - \frac{\partial}{\partial x}(\rho u)\frac{\delta x}{2}\right]$$

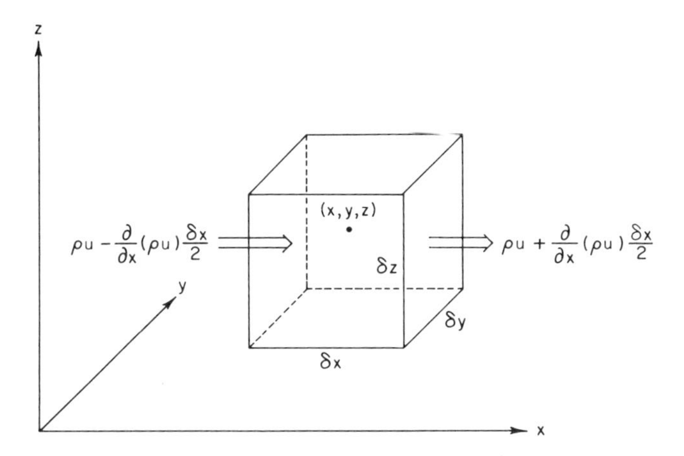

Fig. 2.5 Mass inflow into a fixed (eulerian) control volume, due to motion parallel to the x axis.

whereas the rate of outflow per unit area through the right-hand face is

$$\left[\rho u + \frac{\partial}{\partial x}(\rho u)\frac{\delta x}{2}\right]$$

Since the area of each of these faces is $\delta y\,\delta z$, the net rate of flow into the volume due to the x velocity component is

$$\left[\rho u - \frac{\partial}{\partial x}(\rho u)\frac{\delta x}{2}\right]\delta y\,\delta z - \left[\rho u + \frac{\partial}{\partial x}(\rho u)\frac{\delta x}{2}\right]\delta y\,\delta z = -\frac{\partial}{\partial x}(\rho u)\,\delta x\,\delta y\,\delta z$$

Similar expressions obviously hold for the y and z directions. Thus, the net rate of mass inflow is

$$-\left[\frac{\partial}{\partial x}(\rho u) + \frac{\partial}{\partial y}(\rho v) + \frac{\partial}{\partial z}(\rho w)\right]\delta x\,\delta y\,\delta z$$

and the mass inflow per unit volume is just $-\nabla\cdot(\rho\mathbf{U})$, which must equal the rate of mass increase per unit volume. Now the increase of mass per unit volume is just the local density change $\partial\rho/\partial t$. Therefore,

$$\frac{\partial\rho}{\partial t} + \nabla\cdot(\rho\mathbf{U}) = 0 \tag{2.30}$$

Equation (2.30) is the mass divergence form of the continuity equation.

An alternative form of the continuity equation is obtained by applying the vector identity

$$\nabla\cdot(\rho\mathbf{U}) \equiv \rho\nabla\cdot\mathbf{U} + \mathbf{U}\cdot\nabla\rho$$

and the relationship

$$\frac{d}{dt} \equiv \frac{\partial}{\partial t} + \mathbf{U} \cdot \mathbf{\nabla}$$

to get

$$\frac{1}{\rho}\frac{d\rho}{dt} + \mathbf{\nabla} \cdot \mathbf{U} = 0 \tag{2.31}$$

Equation (2.31) is the velocity divergence form of the continuity equation. It states that the fractional rate of increase of the density *following the motion* of an air parcel is equal to minus the velocity divergence. This should be clearly distinguished from (2.30) which states that the *local* rate of change of density is equal to minus the mass divergence.

2.5.2 A Lagrangian Derivation

The physical meaning of divergence can be illustrated by the following alternative derivation of (2.31). Consider a control volume of fixed mass δM which moves with the fluid. Letting $\delta V = \delta x\, \delta y\, \delta z$ be the volume we find that, since $\delta M = \rho\, \delta V = \rho\, \delta x\, \delta y\, \delta z$ is conserved following the motion, we can write

$$\frac{1}{\delta M}\frac{d}{dt}(\delta M) = \frac{1}{\rho}\frac{d\rho}{dt} + \frac{1}{\delta V}\frac{d}{dt}(\delta V) = 0 \tag{2.32}$$

But

$$\frac{1}{\delta V}\frac{d}{dt}(\delta V) = \frac{1}{\delta x}\frac{d}{dt}(\delta x) + \frac{1}{\delta y}\frac{d}{dt}(\delta y) + \frac{1}{\delta z}\frac{d}{dt}(\delta z)$$

Referring to Fig. 2.6 we see that the faces of our control volume in the y, z plane (designated A and B) are advected with the flow in the x direction at

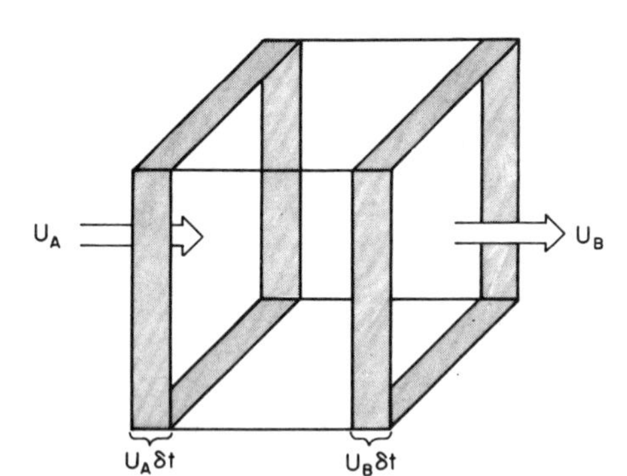

Fig. 2.6 Change in lagrangian control volume due to fluid motion parallel to the x axis.

the speeds $u_A = dx/dt$ and $u_B = d(x + \delta x)/dt$, respectively. Thus the difference in speeds of the two faces is $\delta u = u_B - u_A = d(x + \delta x)/dt - dx/dt$; or $\delta u = d(\delta x)/dt$. Similarly, $\delta v = d(\delta y)/dt$ and $\delta w = d(\delta z)/dt$. Therefore,

$$\lim_{\delta x,\ \delta y,\ \delta z \to 0} \left[\frac{1}{\delta V} \frac{d}{dt} (\delta V) \right] = \frac{\partial u}{\partial x} + \frac{\partial v}{\partial y} + \frac{\partial w}{\partial z} = \nabla \cdot \mathbf{U}$$

so that in the limit $\delta V \to 0$ (2.32) reduces to the continuity equation (2.31). It is thus clear that the divergence of the three dimensional velocity field is equal to the fractional rate of change of volume of a fluid parcel in the limit $\delta V \to 0$. It is left as a problem for the student to show that the divergence of the *horizontal* velocity field is equal to the fractional rate of change of the horizontal area δA of a fluid parcel in the limit $\delta A \to 0$.

2.5.3 Scale Analysis of the Continuity Equation

Following the technique developed in Section 2.4.3 we can rewrite the continuity equation (2.31) as

$$\underbrace{\frac{1}{\rho_0}\left(\frac{\partial \rho'}{\partial t} + \mathbf{U} \cdot \nabla \rho'\right)}_{\text{A}} + \underbrace{\frac{w}{\rho_0}\frac{d\rho_0}{dz}}_{\text{B}} + \underbrace{\nabla \cdot \mathbf{U}}_{\text{C}} \simeq 0 \tag{2.33}$$

where ρ' designates the local deviation of density from its horizontally averaged value, $\rho_0(z)$. For synoptic scale motions $\rho'/\rho_0 \sim 10^{-2}$ so that using the characteristic scales given in Section 2.4 we find that term A has magnitude

$$\frac{1}{\rho_0}\left(\frac{\partial \rho'}{\partial t} + \mathbf{U} \cdot \nabla \rho'\right) \sim \frac{\rho'}{\rho_0}\frac{U}{L} \simeq 10^{-7} \quad \text{s}^{-1}$$

For motions in which the depth scale D is comparable to the density scale height H, $d \ln \rho_0/dz \sim D^{-1}$, so that term B scales as

$$\frac{w}{\rho_0}\frac{d\rho_0}{dz} \sim \frac{W}{D} \simeq 10^{-6} \quad \text{s}^{-1}$$

Expanding Term C in cartesian coordinates we have

$$\nabla \cdot \mathbf{U} = \frac{\partial u}{\partial x} + \frac{\partial v}{\partial y} + \frac{\partial w}{\partial z}$$

For synoptic scale motions the terms $\partial u/\partial x$ and $\partial v/\partial y$ tend to be of equal magnitude but opposite sign. Thus, they tend to balance so that

$$\left(\frac{\partial u}{\partial x} + \frac{\partial v}{\partial y}\right) \sim 10^{-1}\frac{U}{L} \simeq 10^{-6} \quad \text{s}^{-1}$$

and in addition

$$\frac{\partial w}{\partial z} \sim \frac{W}{D} \simeq 10^{-6} \quad \mathrm{s}^{-1}$$

Thus, terms B and C are each an order of magnitude greater than term A, and to a first approximation terms B and C balance in the continuity equation so that we have

$$\frac{\partial u}{\partial x} + \frac{\partial v}{\partial y} + \frac{\partial w}{\partial z} + w\frac{d \ln \rho_0}{dz} = 0$$

or alternatively in vector form

$$\nabla \cdot (\rho_0 \mathbf{U}) = 0 \tag{2.34}$$

Thus for synoptic scale motions the mass flux computed using the basic state density ρ_0 is nondivergent. This approximation is similar to the idealization of incompressibility which is often used in fluid mechanics. However, an *incompressible* fluid has density constant following the motion

$$\frac{d\rho}{dt} = 0$$

Thus by (2.31) the velocity divergence vanishes ($\nabla \cdot \mathbf{U} = 0$) in an incompressible fluid, which is not the same as (2.34). Our approximation (2.34) shows that for purely *horizontal* flow the atmosphere behaves as though it were an incompressible fluid. However, when there is vertical motion the compressibility associated with the height dependence of ρ_0 must be taken into account.

2.6 The Thermodynamic Energy Equation

We now turn to the third fundamental conservation principle, the conservation of thermodynamic energy as applied to a moving fluid element. The first law of thermodynamics is usually derived by considering a system in thermodynamic equilibrium, i.e., a system which is initially at rest and after exchanging heat with its surroundings and doing work on the surroundings is again at rest. For such a system the first law states that *the change in internal energy of the system is equal to the difference between the heat added to the system and the work done by the system.* A lagrangian control volume consisting of a specified mass of fluid may be regarded as a thermo-

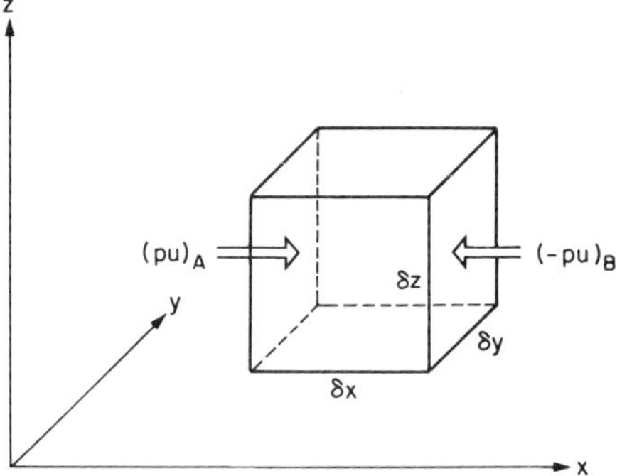

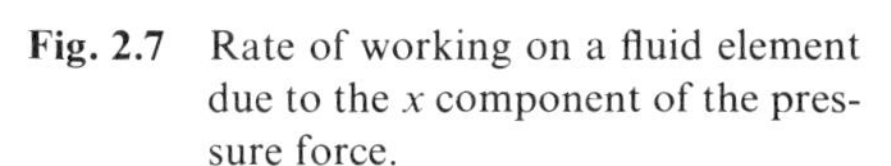
Fig. 2.7 Rate of working on a fluid element due to the x component of the pressure force.

dynamic system. However, unless the fluid is at rest it will not be in thermodynamic equilibrium. Nevertheless, the first law of thermodynamics can still be applied to a fluid system provided that the instantaneous energy of the control volume is considered to consist of the sum of the internal energy (due to the kinetic energy of the individual molecules) and the kinetic energy due to the macroscopic motion of the fluid. The modified form of the first law of thermodynamics, or energy equation, which must be applied to a fluid element thus states that the rate of change of total thermodynamic energy (internal plus kinetic) is equal to the rate of heating plus the rate at which work is done on the element by external forces.

If we let e designate the internal energy per unit mass, then the total thermodynamic energy contained in a lagrangian fluid element of density ρ and volume δV is $\rho(e + \frac{1}{2}\mathbf{U} \cdot \mathbf{U})\,\delta V$. The external forces which act on a fluid element may be divided into surface forces, such as pressure and viscosity, and body forces, such as gravity or the Coriolis force. The rate at which work is done on the fluid element by the x component of the pressure force is illustrated in Fig. 2.7. Recalling that pressure is a force per unit area, and that the rate at which a force does work is given by the dot product of the force and velocity vectors, we see that the rate at which the surrounding fluid does work on the element due to the pressure force on the two boundary surfaces in the y,z plane is given by

$$(pu)_A\,\delta y\,\delta z - (pu)_B\,\delta y\,\delta z$$

(The negative sign is needed before the second term because the work done *on* the fluid element is positive if u is negative across face B.) Now by expanding in a Taylor series we can write

$$(pu)_B = (pu)_A + \left[\frac{\partial}{\partial x}(pu)\right]_A \delta x + \cdots$$

Thus the net rate of working of the pressure force due to the x component of motion is

$$[(pu)_A - (pu)_B]\,\delta y\,\delta z = -\left[\frac{\partial}{\partial x}(pu)\right]_A \delta V$$

where $\delta V = \delta x\,\delta y\,\delta z$.

Similarly, we can show that the rates of working by the pressure due to the y and z components of motion are

$$-[\partial(pv)/\partial y]\,\delta V \qquad \text{and} \qquad -[\partial(pw)/\partial z]\,\delta V,$$

respectively. Thus the total rate of working by the pressure force is simply

$$-\mathbf{\nabla}\cdot(p\mathbf{U})\,\delta V$$

The only body forces of meteorological significance which act on an element of mass in the atmosphere are the Coriolis force and gravity. However, since the Coriolis force, $-2\mathbf{\Omega}\times\mathbf{U}$, is perpendicular to the velocity vector it can do no work. Thus the rate at which body forces do work on the mass element is just $\rho\mathbf{g}\cdot\mathbf{U}\,\delta V$.

Applying the principle of energy conservation to our lagrangian control volume (neglecting effects of molecular viscosity) we thus obtain

$$\frac{d}{dt}\left[\rho\left(e + \frac{1}{2}\mathbf{U}\cdot\mathbf{U}\right)\delta V\right] = -\mathbf{\nabla}\cdot(p\mathbf{U})\,\delta V + \rho\mathbf{g}\cdot\mathbf{U}\,\delta V + \rho\dot{q}\,\delta V \quad (2.35)$$

Here $\dot{q}$ is the rate of heating per unit mass due to radiation, conduction, and latent heat release. With the aid of the chain rule of differentiation we can rewrite (2.35) as

$$\rho\,\delta V\frac{d}{dt}\left(e + \frac{1}{2}\mathbf{U}\cdot\mathbf{U}\right) + \left(e + \frac{1}{2}\mathbf{U}\cdot\mathbf{U}\right)\frac{d(\rho\,\delta V)}{dt}$$
$$= -\mathbf{U}\cdot\mathbf{\nabla}p\,\delta V - p\mathbf{\nabla}\cdot\mathbf{U}\,\delta V - \rho g w\,\delta V + \rho\dot{q}\,\delta V \quad (2.36)$$

where we have used $\mathbf{g} = -g\mathbf{k}$. Now from (2.32) we see that the second term on the left in (2.36) vanishes so that we have

$$\rho\frac{de}{dt} + \rho\frac{d}{dt}\left(\frac{1}{2}\mathbf{U}\cdot\mathbf{U}\right) = -\mathbf{U}\cdot\mathbf{\nabla}p - p\mathbf{\nabla}\cdot\mathbf{U} - \rho g w + \rho\dot{q} \quad (2.37)$$

This equation can be further simplified by noting that if we take the dot product of $\mathbf{U}$ with the momentum equation (2.8) we obtain (neglecting friction)

$$\rho\frac{d}{dt}\left(\frac{1}{2}\mathbf{U}\cdot\mathbf{U}\right) = -\mathbf{U}\cdot\mathbf{\nabla}p - \rho g w \quad (2.38)$$

Subtracting (2.38) from (2.37) we obtain

$$\rho \frac{de}{dt} = -p \, \nabla \cdot \mathbf{U} + \rho \dot{q} \tag{2.39}$$

The terms in (2.37) which were eliminated by subtracting (2.38) represent the balance of mechanical energy due to the motion of the fluid element; the remaining terms represent the thermal energy balance.

Using the definition of geopotential (1.15) we have

$$gw = g \frac{dz}{dt} = \frac{d\Phi}{dt}$$

so that (2.38) can be rewritten as

$$\rho \frac{d}{dt} \left(\frac{1}{2} \mathbf{U} \cdot \mathbf{U} + \Phi \right) = -\mathbf{U} \cdot \nabla p \tag{2.40}$$

which is referred to as the *mechanical energy equation*. The sum of the kinetic energy plus the gravitational potential energy is called the *mechanical energy*. Thus (2.40) states that following the motion, the rate of change of mechanical energy per unit volume equals the rate at which work is done by the pressure gradient force.

The thermal energy equation (2.39) can be written in more familiar form by noting from (2.31) that

$$\frac{1}{\rho} \nabla \cdot \mathbf{U} = -\frac{1}{\rho^2} \frac{d\rho}{dt} = \frac{d\alpha}{dt}$$

and that for dry air the internal energy per unit mass is given by $e = c_v T$, where c_v [$= 717 \text{ J kg}^{-1} \text{ K}^{-1}$] is the specific heat at constant volume. We then obtain

$$c_v \frac{dT}{dt} + p \frac{d\alpha}{dt} = \dot{q} \tag{2.41}$$

which is the usual form of the energy equation. Thus the first law of thermodynamics indeed is applicable to a fluid in motion. The second term on the left, representing the rate of working by the fluid system (per unit mass), represents a conversion between thermal and mechanical energy. It is this conversion process which enables the solar heat energy to drive the motions of the atmosphere.

2.7 Thermodynamics of the Dry Atmosphere

Taking the total derivative of the equation of state (1.13) we obtain

$$p \frac{d\alpha}{dt} + \alpha \frac{dp}{dt} = R \frac{dT}{dt}$$

Substituting for $p\,d\alpha/dt$ in (2.41) and using $c_p = c_v + R$, where c_p [$= 1004$ J kg^{-1} K^{-1}] is the specific heat at constant pressure, we can rewrite the first law of thermodynamics as

$$c_p \frac{dT}{dt} - \alpha \frac{dp}{dt} = \dot{q} \tag{2.42}$$

Dividing through by T and again using the equation of state we obtain the entropy form of the first law of thermodynamics:

$$c_p \frac{d \ln T}{dt} - R \frac{d \ln p}{dt} = \frac{\dot{q}}{T} \equiv \frac{ds}{dt} \tag{2.43}$$

Equation (2.43) gives the rate of change of entropy per unit mass following the motion for a thermodynamically reversible process. The entropy s defined by (2.43) is a field variable which depends only on the state of the fluid. Thus ds is a perfect differential, and ds/dt is to be regarded as a total derivative. However, "heat" is not a field variable, so that the heating rate $\dot{q}$ is not a total derivative.[2]

2.7.1 Potential Temperature

For an ideal gas undergoing an *adiabatic process* (i.e., a process in which no heat is exchanged with the surroundings) the first law of thermodynamics can be written in the form

$$c_p\, d \ln T - R\, d \ln p = 0$$

Integrating this expression from a state at pressure p and temperature T to a state in which the pressure is p_s and the temperature is θ, we obtain after taking the antilogarithm

$$\theta = T(p_s/p)^{R/c_p} \tag{2.44}$$

This relationship is referred to as *Poisson's equation*, and the temperature θ defined by (2.44) is called the *potential temperature*. θ is simply the temperature which a parcel of dry air at pressure p and temperature T would have if it were expanded or compressed adiabatically to a standard pressure p_s [usually taken to be 100 kPa (1000 mb)]. Thus, a parcel of dry air moving adiabatically will conserve its potential temperature.

Taking the logarithm of (2.44) and differentiating, we find that

$$c_p \frac{d \ln \theta}{dt} = c_p \frac{d \ln T}{dt} - R \frac{d \ln p}{dt} \tag{2.45}$$

[2] For a discussion of entropy and its role in the second law of thermodynamics see Wallace and Hobbs (1977), for example.

Comparing (2.43) and (2.45) we obtain

$$c_p \frac{d \ln \theta}{dt} = \frac{ds}{dt} \tag{2.46}$$

Thus, for reversible dry adiabatic processes, fractional potential temperature changes are indeed proportional to entropy changes. A parcel which conserves entropy following the motion must move along an *isentropic* (constant θ) surface.

2.7.2 The Adiabatic Lapse Rate

A relationship between the *lapse rate* of temperature (i.e., the rate of *decrease* of temperature with respect to height) and the rate of change of potential temperature with respect to height can be obtained by taking the logarithm of (2.44) and differentiating with respect to height. Using the hydrostatic equation and the ideal gas law to simplify the result we obtain

$$\frac{T}{\theta} \frac{\partial \theta}{\partial z} = \frac{\partial T}{\partial z} + \frac{g}{c_p} \tag{2.47}$$

For an atmosphere in which the potential temperature is constant with respect to height the lapse rate is thus

$$-\frac{dT}{dz} = \frac{g}{c_p} \equiv \Gamma_{\mathrm{d}} \tag{2.48}$$

Hence, the dry adiabatic lapse rate is approximately constant throughout the lower atmosphere.

2.7.3 Static Stability

If potential temperature is a function of height the actual lapse rate $\Gamma \equiv -\partial T/\partial z$ will differ from the adiabatic lapse rate and

$$\frac{T}{\theta} \frac{\partial \theta}{\partial z} = \Gamma_{\mathrm{d}} - \Gamma \tag{2.49}$$

If $\Gamma < \Gamma_{\mathrm{d}}$ so that θ increases with height, an air parcel which undergoes an adiabatic displacement from its equilibrium level will be positively (negatively) buoyant when displaced vertically downward (upward) so that it will tend to return to its equilibrium level and the atmosphere is said to be statically stable or *stably stratified.*

Adiabatic oscillations of a fluid parcel about its equilibrium level in a stably stratified atmosphere are referred to as *buoyancy oscillations.* The characteristic frequency of such oscillations can be derived by considering a parcel which is displaced vertically a small distance δz without disturbing its environment. If the environment is in hydrostatic balance we have $\bar{\rho} g = -d\bar{p}/dz$, where $\bar{p}$ and $\bar{\rho}$ are the pressure and density of the environment. The vertical acceleration of the parcel will be

$$\frac{dw}{dt} = \frac{d^2}{dt^2}(\delta z) = -g - \frac{1}{\rho}\frac{\partial p}{\partial z} \tag{2.50}$$

where p and ρ are the pressure and density of the parcel, respectively. In the parcel method it is assumed that the pressure of the parcel instantaneously adjusts to the environmental pressure during the displacement: $p = \bar{p}$. This condition must be true if the parcel is to leave the environment undisturbed. Thus with the aid of the hydrostatic relationship pressure can be eliminated in (2.50) to give

$$\frac{d^2}{dt^2}(\delta z) = g\left(\frac{\bar{\rho} - \rho}{\rho}\right) = g\left(\frac{\theta - \bar{\theta}}{\bar{\theta}}\right) \tag{2.51}$$

where (2.44) and the ideal gas law have been used to express the buoyancy force in terms of potential temperature. If the parcel is initially at level $z = 0$ where the potential temperature is θ_0, then for a small displacement δz we can represent the environmental potential temperature as

$$\bar{\theta}(\delta z) \simeq \theta_0 + \frac{d\bar{\theta}}{dz}\delta z$$

If the parcel displacement is adiabatic, the potential temperature of the parcel is conserved: $\theta(\delta z) = \theta_0$. Thus (2.51) becomes

$$\frac{d^2(\delta z)}{dt^2} = -N^2\,\delta z \tag{2.52}$$

where

$$N^2 = \frac{g}{\bar{\theta}}\frac{d\bar{\theta}}{dz}$$

is a measure of the static stability of the environment. Equation (2.52) has a general solution of the form

$$\delta z = Ae^{iNt}$$

Therefore, if $N > 0$ the parcel will oscillate about its initial level with a period

$$\tau = 2\pi/N$$

The corresponding frequency N is the buoyancy frequency.[3] For average tropospheric conditions $N \simeq 1.2 \times 10^{-2}\ \mathrm{s}^{-1}$, so that the period of a buoyancy oscillation is about 8 min.

In the case of $N = 0$, examination of (2.52) indicates that no accelerating force will exist and the parcel will be in neutral equilibrium at its new level. On the other hand, if $N^2 < 0$ (potential temperature decreasing with height) the displacement will increase exponentially in time. We thus arrive at the familiar gravitational or static stability criteria for dry air:

$$\frac{d\theta}{dz}\begin{cases} >0 & \text{stable} \\ =0 & \text{neutral} \\ <0 & \text{unstable} \end{cases}$$

On the synoptic scale the atmosphere is always stably stratified because any unstable regions which develop are quickly stabilized by convective overturning. For a moist atmosphere, the situation is more complicated and discussion of that situation will be deferred until Chapter 12.

2.7.4 Scale Analysis of the Thermodynamic Energy Equation

If potential temperature is divided into basic state and perturbation parts by letting $\theta = \theta_0(z) + \theta'(x, y, z, t)$, the first law of thermodynamics (2.46) can be written approximately as

$$\frac{1}{\theta_0}\left(\frac{\partial \theta'}{\partial t} + \mathbf{U} \cdot \nabla \theta'\right) + w\,\frac{d \ln \theta_0}{dz} = \frac{\dot{q}}{c_p T} \tag{2.53}$$

where we have used the fact that for $|\theta'/\theta_0| \ll 1$

$$\ln \theta \simeq \ln \theta_0 + \theta'/\theta_0$$

Outside regions of active precipitation the diabatic heating $\dot{q}$ is due primarily to the net radiative heating. In the troposphere radiative heating is quite weak so that typically $|\dot{q}/c_p| \lesssim 1^\circ\mathrm{C\,d^{-1}}$ (except near cloud tops where substantially larger cooling can occur). The typical amplitude of horizontal potential temperature fluctuations in a midlatitude synoptic system (above the boundary layer) is $\theta' \sim 4^\circ\mathrm{C}$. Thus,

$$\frac{T}{\theta_0}\left(\frac{\partial \theta'}{\partial t} + \mathbf{U} \cdot \nabla \theta'\right) \sim \frac{\theta' U}{L} \simeq 4^\circ\mathrm{C\,d^{-1}}$$

[3] N is often referred to as the Brunt–Väisällä frequency.

The cooling due to vertical advection of the basic state potential temperature (usually called the *adiabatic cooling*) has a typical magnitude of

$$w\left(\frac{T}{\theta_0}\frac{d\theta_0}{dz}\right) = w(\Gamma_d - \Gamma) \sim 4°\mathrm{C\ d}^{-1}$$

where $w \sim 1\ \mathrm{cm\ s}^{-1}$ and $\Gamma_d - \Gamma$, the difference between the dry adiabatic and actual lapse rates, is $\sim 4°\mathrm{C\ km}^{-1}$.

Thus to a first approximation we find that, in the absence of strong diabatic heating, the rate of change of the perturbation potential temperature is equal to the adiabatic heating or cooling due to vertical motion in the statically stable basic state:

$$\frac{d\theta'}{dt} + w\frac{d\theta_0}{dz} \simeq 0 \tag{2.54}$$

Problems

1. A ship is steaming northward at a rate of $10\ \mathrm{km\ h}^{-1}$. The surface pressure increases towards the northwest at the rate of $5\ \mathrm{Pa\ km}^{-1}$. What is the pressure tendency recorded at a nearby island station if the pressure aboard the ship decreases at a rate of 100 Pa/3 h?

2. The temperature at a point 50 km north of a station is 3°C cooler than at the station. If the wind is blowing from the northeast at $20\ \mathrm{m\ s}^{-1}$ and the air is being heated by radiation at the rate of $1°\mathrm{C\ h}^{-1}$, what is the local temperature change at the station?

3. Derive the relationship

$$\mathbf{\Omega} \times (\mathbf{\Omega} \times \mathbf{r}) = -\Omega^2\mathbf{R}$$

which was used in Eq. (2.7).

4. Derive the expression given in Eq. (2.13) for the rate of change of **k** following the motion.

5. Compare the magnitudes of the curvature term $u^2 \tan\phi/a$ and the Coriolis force for a ballistic missile fired eastward with a velocity of $1000\ \mathrm{m\ s}^{-1}$ at 45° latitude. If the missile travels 1000 km by how much is it deflected from its eastward path due to both these terms? Can the curvature term be neglected in this case?

6. Suppose a 1-kg parcel of dry air is rising at a constant vertical velocity. If the parcel is being heated by radiation at the rate of $10^{-1}\ \mathrm{W\ kg}^{-1}$, what must the speed of rise be to maintain the parcel at a constant temperature?

7. Suppose an air parcel starts from rest at the 80-kPa (800-mb) level and rises vertically to 50 kPa (500 mb) while maintaining a constant 1°C temperature excess over the environment. Assuming that the mean temperature of the 80–50-kPa layer is 260 K, compute the energy released due to the work of the buoyancy force. Assuming that all the released energy is realized as kinetic energy of the parcel what will the vertical velocity of the parcel be at 50 kPa?

8. Show that for an atmosphere with an adiabatic lapse rate (i.e., constant potential temperature) the geopotential height is given by

$$Z = H_\theta[1 - (p/p_0)^{R/c_p}]$$

where p_0 is the pressure at $Z = 0$ and $H_\theta \equiv c_p\theta/g_0$ is the total geopotential height of the atmosphere.

***9.** In the so-called *isentropic coordinate* system potential temperature is used as the vertical coordinate. Since in adiabatic flow potential temperature is conserved following the motion, isentropic coordinates are useful for tracing the actual paths of travel of individual air parcels. Show that the transformation of the horizontal pressure gradient force from z to θ coordinates is given by

$$\frac{1}{\rho}\mathbf{\nabla}_z p = \mathbf{\nabla}_\theta \Psi$$

where $\Psi \equiv c_p T + \Phi$ is the *Montgomery streamfunction.*

Suggested References

Haltiner, *Numerical Weather Prediction*, contains a more complete discussion of scale analysis for synoptic scale motions.

Iribarne and Godson, *Atmospheric Thermodynamics*, contains a thorough discussion of the thermodynamics of both dry and moist atmospheres.

Chapter

3 Elementary Applications of the Basic Equations

3.1 The Basic Equations in Isobaric Coordinates

3.1.1 The Horizontal Momentum Equation

The approximate horizontal momentum equations (2.24) and (2.25) may be written in vectorial form as

$$\frac{d\mathbf{V}}{dt} + f\mathbf{k} \times \mathbf{V} = -\frac{1}{\rho}\nabla p \tag{3.1}$$

where $\mathbf{V} = \mathbf{i}u + \mathbf{j}v$ is the *horizontal* velocity vector. In order to express (3.1) in isobaric coordinate form we transform the pressure gradient force using (1.20) and (1.21) to obtain

$$\frac{d\mathbf{V}}{dt} + f\mathbf{k} \times \mathbf{V} = -\nabla_p \Phi \tag{3.2}$$

where ∇_p is the horizontal gradient operator applied with pressure held constant.

Since p is the independent vertical coordinate we must expand the total derivative as follows:

$$\begin{aligned}\frac{d}{dt} &\equiv \frac{\partial}{\partial t} + \frac{dx}{dt}\frac{\partial}{\partial x} + \frac{dy}{dt}\frac{\partial}{\partial y} + \frac{dp}{dt}\frac{\partial}{\partial p} \\ &= \frac{\partial}{\partial t} + u\frac{\partial}{\partial x} + v\frac{\partial}{\partial y} + \omega\frac{\partial}{\partial p}\end{aligned} \tag{3.3}$$

Here $\omega \equiv dp/dt$ (usually called the "omega" vertical motion) is the pressure change following the motion, which plays the same role in the isobaric coordinate system as $w \equiv dz/dt$ plays in height coordinates.

From (3.2) we see that the isobaric coordinate form of the geostrophic relationship is

$$f\mathbf{V}_g = \mathbf{k} \times \nabla_p \Phi \tag{3.4}$$

One advantage of isobaric coordinates is easily seen by comparing (2.23) and (3.4). In the latter equation density does not appear. Thus, a given geopotential gradient implies the same geostrophic wind at any height, whereas a given horizontal pressure gradient implies different values of the geostrophic wind depending on the density. Furthermore, if f is regarded as a constant, the horizontal divergence of the geostrophic wind at constant pressure is zero,

$$\nabla_p \cdot \mathbf{V}_g = 0$$

3.1.2 The Continuity Equation

It is possible to transform the continuity equation (2.31) from height coordinates to pressure coordinates. However, it is simpler to directly derive the isobaric form by considering a lagrangian control volume $\delta V = \delta x\, \delta y\, \delta z$ and applying the hydrostatic equation $\delta p = -\rho g\, \delta z$ to express the volume element as $\delta V = -\delta x\, \delta y\, \delta p/(\rho g)$. The mass of this fluid element is then $\delta M = -\delta x\, \delta y\, \delta p/g$. (Note that $\delta p < 0$.) Since the mass of the fluid element is conserved following the motion,

$$\frac{1}{\delta M}\frac{d}{dt}\delta M = \frac{g}{\delta x\, \delta y\, \delta p}\frac{d}{dt}\left(\frac{\delta x\, \delta y\, \delta p}{g}\right) = 0$$

After differentiating, using the chain rule, and changing the order of the differential operators we obtain[1]

$$\frac{1}{\delta x}\delta\left(\frac{dx}{dt}\right) + \frac{1}{\delta y}\delta\left(\frac{dy}{dt}\right) + \frac{1}{\delta p}\delta\left(\frac{dp}{dt}\right) = 0$$

[1] From now on g will be regarded as a constant.

or

$$\frac{\delta u}{\delta x} + \frac{\delta v}{\delta y} + \frac{\delta \omega}{\delta p} = 0$$

Taking the limit δx, δy, $\delta p \to 0$ we obtain the continuity equation in the isobaric system

$$\left(\frac{\partial u}{\partial x} + \frac{\partial v}{\partial y}\right)_p + \frac{\partial \omega}{\partial p} = 0 \tag{3.5}$$

This form of the continuity equation contains no reference to the density field, and does not involve time derivatives. The simplicity of (3.5) is one of the chief advantages of the isobaric coordinate system.

3.1.3 The Thermodynamic Energy Equation

The first law of thermodynamics (2.42) can be expressed in the isobaric system by letting $dp/dt = \omega$ and expanding dT/dt by using (3.3):

$$c_p\left(\frac{\partial T}{\partial t} + u\frac{\partial T}{\partial x} + v\frac{\partial T}{\partial y} + \omega\frac{\partial T}{\partial p}\right) - \alpha\omega = \dot{q}$$

This may be rewritten as

$$\left(\frac{\partial T}{\partial t} + u\frac{\partial T}{\partial x} + v\frac{\partial T}{\partial y}\right) - S_p\omega = \dot{q}/c_p \tag{3.6}$$

where, with the aid of the equation of state and Poisson's equation (2.44), we have

$$S_p \equiv \frac{RT}{c_p p} - \frac{\partial T}{\partial p} = \frac{-T}{\theta}\frac{\partial \theta}{\partial p} \tag{3.7}$$

which is the static stability parameter for the isobaric system. Comparing with (2.49) we see with the aid of the hydrostatic equation

$$S_p = (\Gamma_d - \Gamma)/\rho g$$

Thus, S_p is positive provided that the lapse rate is less than dry adiabatic. However, since density decreases approximately exponentially with height, S_p increases rapidly with height. This strong height dependence of the stability measure S_p is one disadvantage of isobaric coordinates.

3.2 Balanced Flow

Despite the apparent complexity of atmospheric motion systems as depicted on synoptic weather charts, the pressure and velocity distributions of weather disturbances are actually related by rather simple approximate force balances. In order to gain a qualitative understanding of the horizontal balance of forces in atmospheric motions it is useful to idealize by considering flows which are steady state (i.e., time independent) and have no vertical component of velocity. Furthermore it is useful to describe the flow field by expanding the horizontal momentum equation (3.1) into its components in a so-called *natural* coordinate system.

3.2.1 NATURAL COORDINATES

The directions of the coordinates (s, n, z) in the natural coordinate system are defined by unit vectors **t**, **n**, and **k**, respectively; **t** is oriented parallel to the direction of flow at each point, **n** is a normal vector which is positive to the left of the flow direction, and **k** is directed vertically upward. In this system the horizontal velocity may be written $\mathbf{V} = V\mathbf{t}$ where $V = ds/dt$, the horizontal speed, is a nonnegative scalar. The acceleration following the motion is thus

$$\frac{d\mathbf{V}}{dt} = \mathbf{t}\frac{dV}{dt} + V\frac{d\mathbf{t}}{dt}$$

The rate of change of **t** following the motion may be derived from geometrical considerations with the aid of Fig. 3.1. Recalling that $|\mathbf{t}| = 1$, we see that

$$\delta\psi = \frac{\delta s}{R} = |\delta\mathbf{t}|$$

Here R is the *radius of curvature* following the parcel motion. R is taken to be positive when the center of curvature is in the positive **n** direction. Thus,

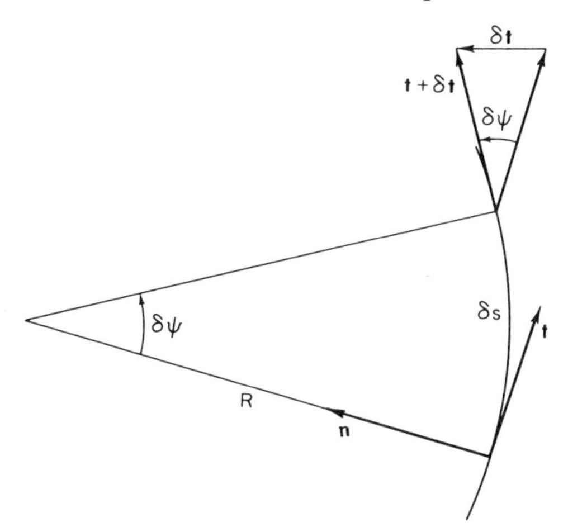

Fig. 3.1 Rate of change of the unit tangent vector **t** following the motion.

for $R > 0$ the air parcels turn toward the left following the motion, and for $R < 0$ the air parcels turn toward the right following the motion.

Noting that $\delta\mathbf{t}$ is directed parallel to $\mathbf{n}$, we find that in the limit $\delta s \to 0$

$$\frac{d\mathbf{t}}{ds} = \frac{\mathbf{n}}{R}$$

Thus,

$$\frac{d\mathbf{t}}{dt} = \frac{d\mathbf{t}}{ds}\frac{ds}{dt} = \frac{\mathbf{n}}{R}V$$

and

$$\frac{d\mathbf{V}}{dt} = \mathbf{t}\frac{dV}{dt} + \mathbf{n}\frac{V^2}{R} \tag{3.8}$$

Therefore, the acceleration following the motion is the sum of the rate of change of speed of the air parcel and its centripetal acceleration due to the curvature of the trajectory. Since the Coriolis force always acts normal to the direction of motion, we can write

$$f\mathbf{k} \times \mathbf{V} = fV\mathbf{n}$$

The horizontal momentum equation may thus be expanded into the following component equations in the natural coordinate system:

$$\frac{dV}{dt} = -\frac{1}{\rho}\frac{\partial p}{\partial s} \tag{3.9}$$

$$\frac{V^2}{R} + fV = -\frac{1}{\rho}\frac{\partial p}{\partial n} \tag{3.10}$$

Equations (3.9) and (3.10) express the force balances parallel to and normal to the direction of flow, respectively. For motion parallel to the isobars, $\partial p/\partial s = 0$ and the speed is constant following the motion.

3.2.2 Geostrophic Flow

If the flow is parallel to the isobars so that $\partial p/\partial s = 0$, and in addition the isobars are straight ($|R| \to \infty$) the flow is said to be in *geostrophic motion.* In geostrophic motion the horizontal components of the Coriolis force and pressure gradient force are in exact balance. This balance is indicated schematically in Fig. 3.2. It should be clearly understood that the actual

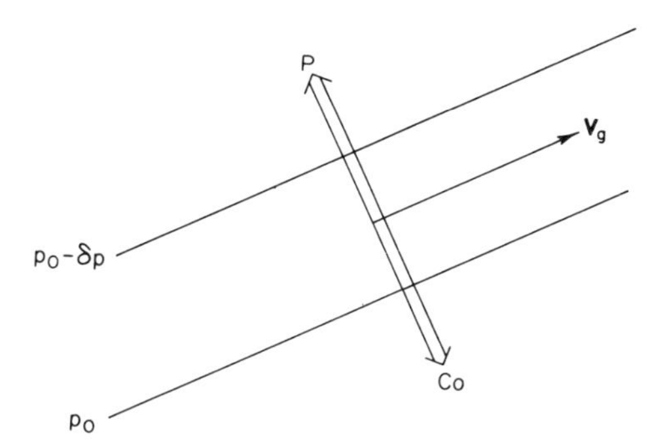

Fig. 3.2 The balance of forces for geostrophic equilibrium. The pressure gradient force is designated by P and the Coriolis force by Co.

wind is in geostrophic motion with

$$fV = -\frac{1}{\rho}\frac{\partial p}{\partial n} \tag{3.11}$$

only for the very restrictive conditions specified above. It is, however, always possible to define a vector called the geostrophic wind which exactly satisfies (3.4).

3.2.3 Inertial Flow

If the pressure field is horizontally uniform so that the horizontal pressure gradient vanishes, (3.10) reduces to a balance between the Coriolis force and centrifugal force:

$$\frac{V^2}{R} + fV = 0 \tag{3.12}$$

Equation (3.12) may be solved for the radius of curvature

$$R = -\frac{V}{f}$$

Since from (3.9), the speed must be constant in this case, the radius of curvature is also constant (neglecting the latitudinal dependence of f). Thus the air parcels follow circular paths in an anticyclonic sense.[2] The period of this oscillation is

$$P = -\frac{2\pi R}{V} = \frac{2\pi}{|f|} = \frac{\frac{1}{2}\text{ day}}{|\sin\phi|} \tag{3.13}$$

[2] *Anticyclonic* flow is a clockwise rotation in the Northern Hemisphere and counterclockwise in the Southern Hemisphere. *Cyclonic* flow has the opposite sense of rotation in each hemisphere.

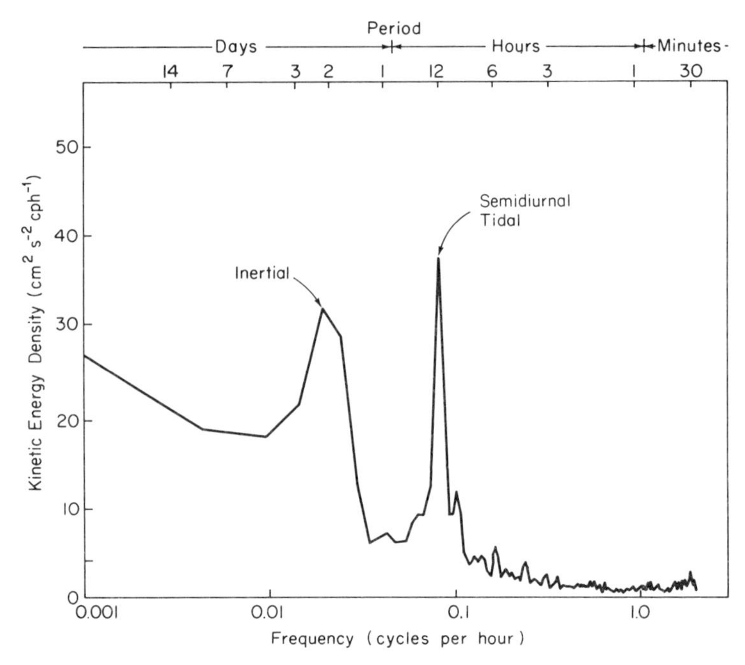

Fig. 3.3 Power spectrum of kinetic energy at 30-m depth in the ocean near Barbados (13°N). Ordinate shows kinetic energy density per unit frequency interval (cph^{-1} designates cycles per hour). This type of plot indicates the manner in which the total kinetic energy is partitioned among oscillations of different periods. Note the strong peak at 53 h, which is the period of an inertial oscillation at 13° latitude. (After Warsh *et al.*, 1971. Reproduced with permission of the American Meteorological Society.)

P is equivalent to the time which is required for a Foucault pendulum to turn through an angle of 180°. Hence, P is often referred to as one-half *pendulum day*. Since both the Coriolis force and the centrifugal force due to the relative motion are caused by the inertia of the fluid, this type of motion is referred to as an inertial oscillation, and the circle of radius $|R|$ is called the *inertia circle*. Pure inertial oscillations are apparently not of importance in the earth's atmosphere. However, in the oceans where velocities (and hence the radius of inertia circles) are much smaller than in the atmosphere, significant amounts of energy have been detected in currents which oscillate with the inertial period. An example of inertial oscillations in the oceans recorded by a current meter near the island of Barbados is shown in Fig. 3.3.

3.2.4 Cyclostrophic Flow

If the horizontal scale of a disturbance is small enough, the Coriolis force may be neglected in (3.10) compared to the pressure gradient force and the centrifugal force. The force balance normal to the direction of flow is then

$$\frac{V^2}{R} = -\frac{1}{\rho}\frac{\partial p}{\partial n}$$

If this equation is solved for V, we obtain the speed of the *cyclostrophic wind*

$$V = \left(-\frac{R}{\rho}\frac{\partial p}{\partial n}\right)^{1/2} \tag{3.14}$$

As indicated in Fig. 3.4, cyclostrophic flow may be either cyclonic or anticyclonic. In both cases the pressure gradient force is directed toward the center of curvature, and the centrifugal force away from the center of curvature.

The cyclostrophic balance approximation is valid provided that the ratio of the centrifugal force to the Coriolis force is large. This ratio V/fR is equivalent to the Rossby number discussed in Section 2.4.2. As an example of cyclostrophic scale motion we consider a typical tornado. Suppose that the tangential velocity is 30 m s^{-1} at a distance of 300 m from the center of the vortex. Assuming that $f = 10^{-4}$ s^{-1}, we obtain a Rossby number of

$$Ro = \frac{V}{fR} \simeq 10^3$$

Therefore, the Coriolis force can be neglected in computing the balance of forces for a mature tornado. However, the majority of tornadoes in the

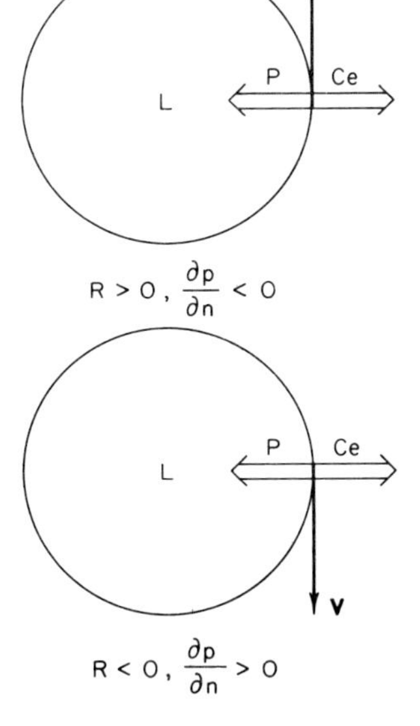

Fig. 3.4 The force balance in cyclostrophic flow, P designates the pressure gradient, Ce the centrifugal force.

Northern Hemisphere are observed to rotate in a cyclonic (counterclockwise) sense. Thus, in the initial stages of development of a tornado the Coriolis force serves to deflect air parcels accelerating toward the center of low pressure to the right so that a cyclonic circulation is established about the pressure minimum. Smaller scale vortices, on the other hand, such as dust devils and water spouts, do not have a preferred direction of rotation. According to data collected by Sinclair (1965), they are observed to be anticyclonic as often as cyclonic.

3.2.5 Gradient Flow and the Gradient Wind Approximation

Horizontal frictionless flow which is parallel to the isobars so that the tangential acceleration vanishes ($dV/dt = 0$) is called *gradient flow*. Gradient flow is a three-way balance between the Coriolis force, the centrifugal force, and the horizontal pressure gradient force. Like geostrophic flow, pure gradient flow can exist only under very special circumstances. However, it is always possible to define a *gradient wind* which at any point is just the wind component parallel to the isobars which satisfies (3.10), the so-called gradient wind equation. Because (3.10) takes account of the centrifugal force due to the curvature of parcel trajectories, the gradient wind is often a better approximation to the actual wind than is the geostrophic wind.

The gradient wind speed is obtained by solving (3.10) for V to yield

$$V = -\frac{fR}{2} \pm \left(\frac{f^2R^2}{4} - \frac{R}{\rho}\frac{\partial p}{\partial n}\right)^{1/2} \tag{3.15}$$

Not all the mathematically possible roots of (3.15) correspond to physically possible solutions since it is required that V be real and nonnegative. In Table 3.1 the various roots of (3.15) are classified according to the signs of R

Table 3.1 *Classification of Roots of the Gradient Wind Equation in the Northern Hemisphere*

	$R > 0$	$R < 0$
$\frac{\partial p}{\partial n} > 0$	Two − roots: not permitted	+ root: antibaric flow (anomalous low) − root: not permitted
$\frac{\partial p}{\partial n} < 0$	+ root: cyclonic flow (regular low) − root: not permitted	Two + roots: larger ($V > -fR/2$), anticyclonic flow (anomalous high) smaller ($V < -fR/2$), anticyclonic flow (regular high)

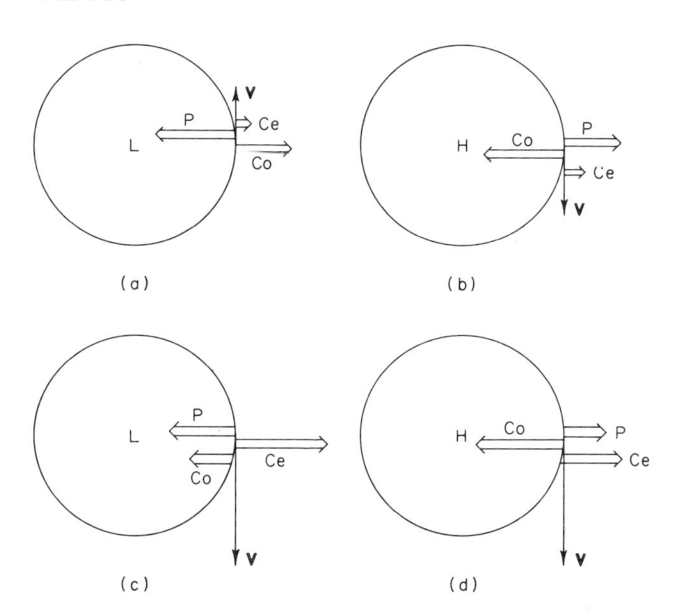

Fig. 3.5 Force balances in the Northern Hemisphere for the four types of gradient flow: (a) regular low; (b) regular high; (c) anomalous low; (d) anomalous high.

and $\partial p/\partial n$ in order to isolate the physically meaningful solutions. The force balances for the four permitted solutions are illustrated in Fig. 3.5. Returning to Eq. (3.15), we note that in the cases of both the regular and anomalous highs the pressure gradient is limited by the requirement that the quantity under the radical be nonnegative; that is,

$$\left|\frac{\partial p}{\partial n}\right| < \frac{\rho |R| f^2}{4} \tag{3.16}$$

Thus the pressure gradient in a high must decrease linearly toward the center. It is for this reason that the pressure field near the center of a high is always flat and the wind gentle compared to the region near the center of a low.

The so-called "anomalous" gradient wind balance cases illustrated in Fig. 3.5 are unlikely to occur because as the wind field accelerates in response to an imposed pressure distribution the deflection by the Coriolis force will tend to produce a normal gradient balance. Thus, as previously argued in Section 3.2.4, the Coriolis force will tend to establish a normal cyclonic circulation about a low-pressure center. Conversely, as air parcels are accelerated outward from the center of a developing high-pressure system, the Coriolis force will gradually deflect the parcels so that they will circulate anticyclonically about the high. When the motion reaches gradient equilibrium the speed will be that of a normal anticyclone.

In all cases except the anomalous low the flow is *baric* (i.e., the horizontal components of the Coriolis and pressure gradient forces are oppositely directed). Furthermore, the flow is cyclonic when the centrifugal force and the horizontal component of the Coriolis force have the same sense ($Rf > 0$) and anticyclonic when these forces have the opposite sense ($Rf < 0$). This definition is independent of the hemisphere considered.

Geostrophic versus Gradient Winds. The geostrophic wind speed in natural coordinates is given by the relationship

$$V_g \equiv -\frac{1}{\rho f}\frac{\partial p}{\partial n}$$

Using this definition, (3.10) can be rewritten as

$$\frac{V^2}{R} + fV - fV_g = 0 \tag{3.17}$$

Thus the ratio of the geostrophic wind to the gradient wind is

$$\frac{V_g}{V} = 1 + \frac{V}{fR}$$

For normal cyclonic flow ($fR > 0$) V_g is larger than V, while for anticyclonic flow ($fR < 0$) V_g is smaller than V. Therefore, the geostrophic wind is an overestimate of the balanced wind in a region of cyclonic curvature, and an underestimate in a region of anticyclonic curvature. For midlatitude synoptic systems, the difference between the gradient and geostrophic wind speeds generally does not exceed 10–20%. (Note that the magnitude of V/fR is just the Rossby number.) However, for the tropical cyclone scale, the Rossby number is of order unity and the gradient wind formula must be applied rather than the geostrophic wind.

3.3 Trajectories and Streamlines

In the natural coordinate system used in the previous section to discuss balanced flow the s coordinate was defined as the curve in the horizontal plane traced out by the path of an air parcel. The path followed by a particular air parcel over a finite period of time is called the *trajectory* of the parcel. Thus, the radius of curvature R of the path s referred to in the gradient wind equation is the radius of curvature for a parcel trajectory. In practice R is often estimated by using the radius of curvature of the isobars, since the latter can easily be measured at any time from a synoptic map. However, the isobars are actually *streamlines* of the gradient wind (that is, lines which are everywhere parallel to the instantaneous wind velocity).

It is important to distinguish clearly between streamlines, which give a "snapshot" of the velocity field at any instant, and trajectories, which trace the motion of individual fluid parcels over a finite time interval. Mathematically, the horizontal trajectory is given by the integration of

$$\frac{ds}{dt} = V(x, y, t) \tag{3.18}$$

over a finite time span, whereas the streamline is given by the integration of

$$\frac{dy}{dx} = \frac{v(x, y, t_0)}{u(x, y, t_0)} \tag{3.19}$$

with respect to x for a constant time. Only for steady-state systems (in which the local rate of change of velocity is zero) will the streamlines and trajectories coincide. In order to gain an appreciation for the possible errors involved in using the curvature of the streamlines instead of the curvature of the trajectories in the gradient wind equation, it is necessary to investigate the relationship between the curvature of the trajectories and the curvature of the streamlines for a moving pressure system.

If β is the angular direction of the wind and R_t and R_s designate the radii of curvature of the trajectories and streamlines, respectively, we see with the aid of Fig. 3.6 that $\delta s = R\delta\beta$ so that in the limit $\delta s \to 0$

$$\frac{d\beta}{ds} = \frac{1}{R_t} \quad \text{and} \quad \frac{\partial\beta}{\partial s} = \frac{1}{R_s} \tag{3.20}$$

where $d\beta/ds$ means the rate of change of wind direction along a trajectory (positive for counterclockwise turning) and $\partial\beta/\partial s$ is the rate of change of

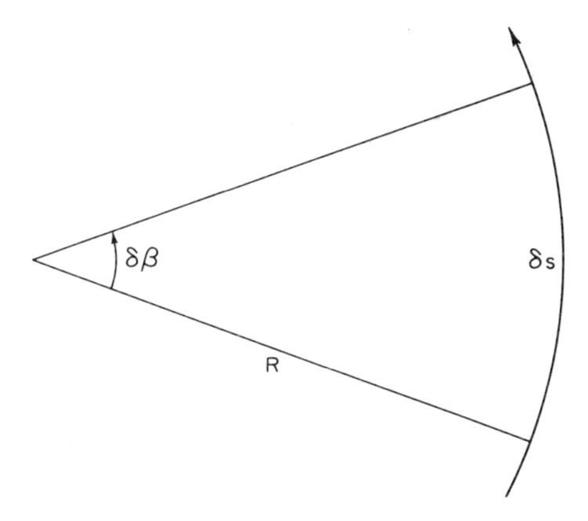

Fig. 3.6 The relationship between the change in angular direction of the wind $\delta\beta$ and the radius of curvature R.

wind direction along a streamline at any instant. Thus, the rate of change of wind direction following the motion is

$$\frac{d\beta}{dt} = \frac{d\beta}{ds}\frac{ds}{dt} = \frac{V}{R_t} \tag{3.21}$$

or,

$$\frac{d\beta}{dt} = \frac{\partial\beta}{\partial t} + V\frac{\partial\beta}{\partial s} = \frac{\partial\beta}{\partial t} + \frac{V}{R_s} \tag{3.22}$$

Combining (3.21) and (3.22) we obtain a formula for the local turning of the wind:

$$\frac{\partial\beta}{\partial t} = V\left(\frac{1}{R_t} - \frac{1}{R_s}\right) \tag{3.23}$$

Equation (3.23) indicates that only if the trajectories and streamlines coincide will the wind direction remain constant in time.

In general, midlatitude synoptic pressure patterns move eastward as a result of advection by the upper level westerly winds. In such cases there will be local turning of the wind due to the motion of the system even if the shape of the pressure pattern does not change following its motion. For simplicity, we now consider a circular pattern of isobars moving at a constant velocity **C** without change of shape. It must be kept clearly in mind here that **C** designates the velocity at which the pressure pattern is moving and should not be confused with the wind velocity, which in this case (neglecting friction) is given by the gradient wind equation. Thus, the isobars are streamlines and the local turning of the wind is entirely due to the motion of the streamline pattern, so that

$$\frac{\partial\beta}{\partial t} = -\mathbf{C}\cdot\nabla\beta = -C\cos\gamma\frac{\partial\beta}{\partial s}$$

where γ is the angle between the streamlines (isobars) and direction of motion of the system. Substituting the above into (3.23) and solving for R_s we obtain the desired relationship between the curvature of the streamlines and the curvature of trajectories:

$$R_s = R_t\left(1 - \frac{C\cos\gamma}{V}\right) \tag{3.24}$$

Equation (3.24) can be used to compute the curvature of the trajectory anywhere on a moving pattern of streamlines. In Fig. 3.7 the curvatures of the

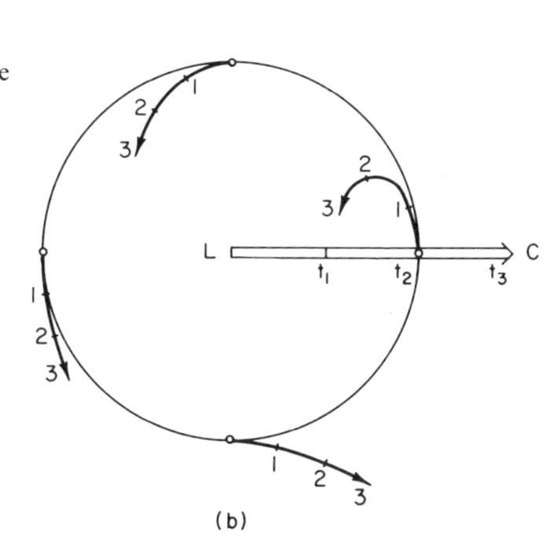

Fig. 3.7 Trajectories for moving circular cyclonic circulation systems in the Northern Hemisphere with (a) $V = 2C$ and (b) $2V = C$. Numbers indicate positions at successive times. The "L" designates a pressure minimum.

trajectories for parcels initially located due north, east, south, and west of the center of a cyclonic system are shown both for the case of a wind speed greater than the speed of movement of the isobars, and the case of a wind speed less than the speed of movement of the isobars. In these examples the plotted trajectories are based on a geostrophic balance so that the isobars are equivalent to streamlines. It is also assumed for simplicity that the wind speed does not depend on the distance from the center of the system. In the case shown in Fig. 3.7b there is actually a region where the curvature of the trajectories is opposite that of the streamlines. Since synoptic scale pressure systems usually move at a speed comparable to the wind speed, the gradient wind speed computed on the basis of the curvature of the isobars is often no better an approximation to the actual wind speed than is the geostrophic wind. In fact the actual gradient wind speed will vary along an isobar with the variation of the trajectory curvature.

3.4 Vertical Shear of the Geostrophic Wind: The Thermal Wind

The geostrophic wind must have vertical shear in the presence of a horizontal temperature gradient, as can easily be shown from simple physical considerations based on hydrostatic equilibrium. Recalling that the geostrophic wind is proportional to the horizontal pressure gradient, we see that for the case shown in Fig. 3.8 there is a geostrophic wind directed along the y axis which increases in magnitude with height as the slope of the isobars increases. From the hydrostatic approximation (1.16) it can be shown that the height increment δz corresponding to a given pressure increment δp is

$$\delta z \approx -\frac{\delta p}{p}\frac{RT}{g} \tag{3.25}$$

Therefore, referring to Fig. 3.8, since $(\delta z)_{x_1} < (\delta z)_{x_2}$ we see that $T_{x_2} > T_{x_1}$. Thus, the increase with height of the positive x-directed pressure gradient must be associated with a positive x-directed temperature gradient. The air in a vertical column at x_2 since it is warmer (less dense) must occupy a greater depth for a given pressure drop than the air at x_1.

The equations for the rate of change of the geostrophic wind components with height are most easily derived using the isobaric coordinate system. Recall that in isobaric coordinates the geostrophic wind components are

$$v_g = \frac{1}{f}\frac{\partial \Phi}{\partial x} \quad \text{and} \quad u_g = -\frac{1}{f}\frac{\partial \Phi}{\partial y} \tag{3.26}$$

where the derivatives are evaluated with pressure held constant. Also, with the aid of the ideal gas law we can write the hydrostatic equation as

$$\frac{\partial \Phi}{\partial p} = -\alpha = -\frac{RT}{p} \tag{3.27}$$

Differentiating (3.26) with respect to pressure and applying (3.27) we obtain

$$p\frac{\partial v_g}{\partial p} \equiv \frac{\partial v_g}{\partial \ln p} = -\frac{R}{f}\left(\frac{\partial T}{\partial x}\right)_p \tag{3.28}$$

$$p\frac{\partial u_g}{\partial p} \equiv \frac{\partial u_g}{\partial \ln p} = \frac{R}{f}\left(\frac{\partial T}{\partial y}\right)_p \tag{3.29}$$

or in vectorial form

$$\frac{\partial \mathbf{V}_g}{\partial \ln p} = -\frac{R}{f}\mathbf{k} \times (\nabla T)_p \tag{3.30}$$

Equation (3.30) is often referred to as the *thermal wind* equation. However, it is actually a relationship for the vertical wind *shear* (that is, the rate of

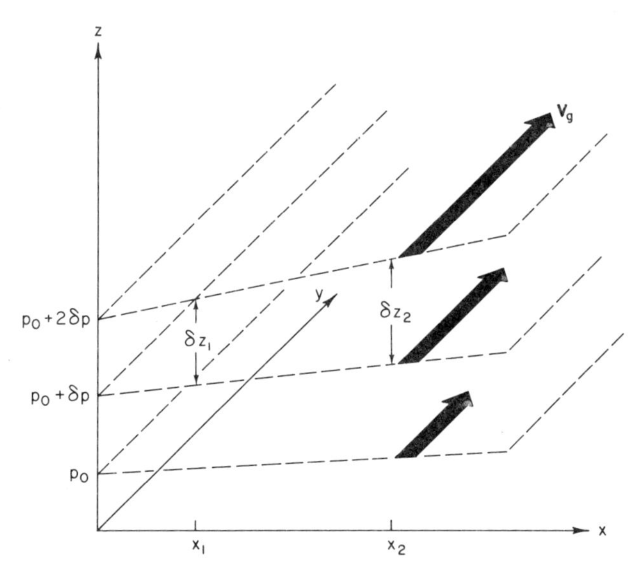

Fig. 3.8 Relationship between vertical shear of the geostrophic wind and horizontal temperature gradients. (Note $\delta p < 0$.)

change of the geostrophic wind with respect to ln p). Strictly speaking, the term thermal wind refers to the vector difference between the geostrophic winds at two levels. Designating the thermal wind vector by $\mathbf{V}_T$ we may integrate (3.30) from pressure level p_0 to level p_1 $(p_1 < p_0)$ to get

$$\mathbf{V}_T = \mathbf{V}_g(p_1) - \mathbf{V}_g(p_0) = -\frac{R}{f}\int_{p_0}^{p_1} (\mathbf{k} \times \nabla T)\, d \ln p \tag{3.31}$$

Letting $\overline{T}$ denote the mean temperature in the layer between pressure p_0 and p_1, and x and y components of the thermal wind are thus given by

$$\begin{aligned} u_T &= -\frac{R}{f}\left(\frac{\partial \overline{T}}{\partial y}\right)_p \ln\left(\frac{p_0}{p_1}\right) \\ v_T &= \frac{R}{f}\left(\frac{\partial \overline{T}}{\partial x}\right)_p \ln\left(\frac{p_0}{p_1}\right) \end{aligned} \tag{3.32}$$

Alternatively, we may express the thermal wind for a given layer in terms of the horizontal gradient of the geopotential difference between the top and bottom of the layer:

$$\begin{aligned} u_T &= u_g(p_1) - u_g(p_0) = -\frac{1}{f}\frac{\partial}{\partial y}(\Phi_1 - \Phi_0) \\ v_T &= v_g(p_1) - v_g(p_0) = \frac{1}{f}\frac{\partial}{\partial x}(\Phi_1 - \Phi_0) \end{aligned} \tag{3.33}$$

The equivalence of (3.32) and (3.33) can be readily verified by integrating the hydrostatic equation (3.27) vertically from p_0 to p_1 after replacing T by the mean $\overline{T}$. The result is the hypsometric equation (1.17):

$$\delta\Phi \equiv \Phi_1 - \Phi_0 = R \ln\left(\frac{p_0}{p_1}\right)\overline{T} \tag{3.34}$$

The quantity $\delta\Phi$ is the *thickness* of the layer between p_0 and p_1 measured in units of geopotential. From (3.34) we see that the thickness is proportional to the mean temperature in the layer. Hence, lines of equal $\delta\Phi$ (isolines of thickness) are equivalent to the isotherms of mean temperature in the layer. Note also that $\delta z \equiv \delta\Phi/g_0$ gives the approximate thickness in geopotential height.

The thermal wind equation is an extremely useful diagnostic tool which is often used to check analyses of the observed wind and temperature fields for consistency. It can also be used to estimate the mean horizontal temperature advection in a layer as shown in Fig. 3.9. It is clear from the vector form of (3.33),

$$\mathbf{V}_T = \frac{1}{f}\mathbf{k} \times \nabla(\Phi_1 - \Phi_0)$$

that the thermal wind blows parallel to the isotherms (lines of constant thickness) with the warm air to the right facing downstream in the Northern Hemisphere. Thus, as is illustrated in Fig. 3.9a, a geostrophic wind which turns counterclockwise with height (backs) is associated with cold-air advection. Conversely, as shown in Fig. 3.9b, clockwise turning (veering) of the geostrophic wind with height implies warm advection by the geostrophic wind in the layer. It is therefore possible to obtain a reasonable estimate of the horizontal temperature advection and its vertical dependence at a given location solely from data on the vertical profile of the wind given by a single sounding. Alternatively, the geostrophic wind at any level can be estimated from the mean temperature field provided that the geostrophic velocity is known at a single level. Thus, for example, if the geostrophic wind at 850 mb is known and the mean horizontal temperature gradient in the layer 850–500 mb is also known, the thermal wind equation can be applied to obtain the geostrophic wind at 500 mb.

The Barotropic Atmosphere. A barotropic atmosphere is one in which the density depends only on the pressure, $\rho \equiv \rho(p)$, so that isobaric surfaces are also surfaces of constant density. For an ideal gas, the isobaric surfaces will also be isothermal if the atmosphere is barotropic. Thus, $(\nabla T)_p = 0$ in a barotropic atmosphere, and the thermal wind equation (3.30) becomes

$$\frac{\partial \mathbf{V}_g}{\partial \ln p} = 0$$

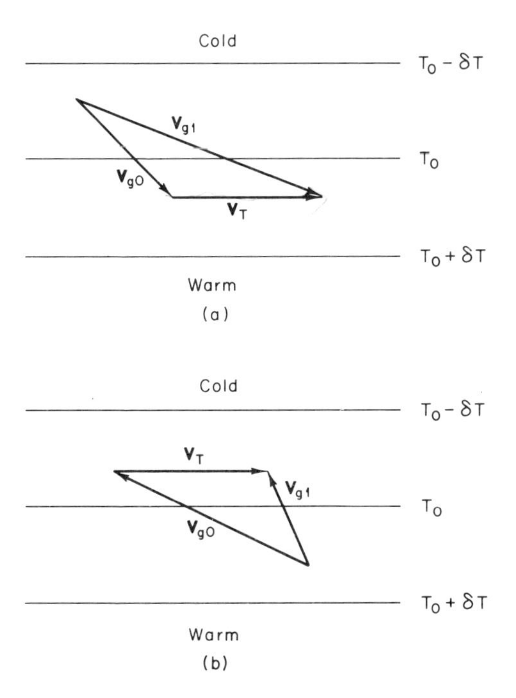

Fig. 3.9 The relationship between turning of the geostrophic wind and temperature advection: (a) backing of the wind with height; (b) veering of the wind with height.

which states that the geostrophic wind is *independent of height* in a barotropic atmosphere. Thus, barotropy provides a very strong constraint on the types of motion possible in a rotating fluid.

The Baroclinic Atmosphere. An atmosphere in which density depends on both the temperature and the pressure, $\rho = \rho(p, T)$, is referred to as a baroclinic atmosphere. In a baroclinic atmosphere the geostrophic wind generally has vertical shear, and this shear is related to the horizontal temperature gradient by the thermal wind equation. Obviously, the baroclinic atmosphere is of primary importance in dynamic meteorology. However, as will be seen in later chapters, much can be learned by study of the simpler barotropic atmosphere.

3.5 Vertical Motion

As previously mentioned, for synoptic scale motions the vertical velocity component is typically of the order of a few centimeters per second. Routine meteorological soundings, however, only give the wind speed to an accuracy of about a meter per second. Thus, in general the vertical velocity is not measured directly but must be inferred from the fields which are directly measured.

Two commonly used methods for inferring the vertical motion field are the kinematic method, based on the equation of continuity, and the adiabatic method, based on the thermodynamic energy equation. Both methods are usually applied using the isobaric coordinate system so that $\omega(p)$ is inferred rather than $w(z)$.

3.5.1 The Relationship between ω and w

Expanding dp/dt in the x, y, z system we have

$$\omega \equiv \frac{dp}{dt} = \frac{\partial p}{\partial t} + \mathbf{V} \cdot \nabla p + w \frac{\partial p}{\partial z} \tag{3.35}$$

Now, for synoptic scale motions, the horizontal velocity is geostrophic to a first approximation. Therefore we can write $\mathbf{V} = \mathbf{V}_g + \mathbf{V}'$, where $|\mathbf{V}'| \ll |\mathbf{V}_g|$. But $\mathbf{V}_g = (1/\rho f)\mathbf{k} \times \nabla p$ so that $\mathbf{V}_g \cdot \nabla p = 0$. Using this result plus the hydrostatic approximation, (3.35) may be rewritten as

$$\omega = \frac{\partial p}{\partial t} + \mathbf{V}' \cdot \nabla p - g\rho w \tag{3.36}$$

Comparing the magnitudes of the three terms on the right in (3.36) we find that for synoptic scale motions

$$\frac{\partial p}{\partial t} \sim 1 \quad \text{kPa d}^{-1}$$

$$\mathbf{V}' \cdot \nabla p \sim (1 \text{ m s}^{-1})(1 \text{ Pa km}^{-1}) \sim 0.1 \quad \text{kPa d}^{-1}$$

$$g\rho w \sim 10 \quad \text{kPa d}^{-1}$$

Thus, in the first approximation,

$$\omega = -\rho g w \tag{3.37}$$

3.5.2 The Kinematic Method

One method of deducing the vertical velocity is based on integrating the continuity equation in the vertical. Integration of (3.5) with respect to pressure from a reference level p_0 to any level p yields

$$\begin{aligned}\omega(p) &= \omega(p_0) - \int_{p_0}^{p} \left(\frac{\partial u}{\partial x} + \frac{\partial v}{\partial y}\right)_p dp \\ &= \omega(p_0) + (p_0 - p)\left(\frac{\partial \langle u \rangle}{\partial x} + \frac{\partial \langle v \rangle}{\partial y}\right)_p \end{aligned} \tag{3.38}$$

Here the angle brackets denote a pressure weighted vertical average.

$$\langle\ \rangle \equiv \frac{1}{(p - p_0)} \int_{p_0}^{p} (\ )\, dp$$

With the aid of (3.37) and the hydrostatic equation, (3.38) can be rewritten as

$$w(z) = \frac{\rho(z_0)w(z_0)}{\rho(z)} - \frac{p_0 - p}{\rho(z)g}\left(\frac{\partial}{\partial x}\langle u\rangle + \frac{\partial}{\partial y}\langle v\rangle\right) \tag{3.39}$$

where z and z_0 are the heights of pressure levels p and p_0, respectively.

Applicationof (3.39) to infer the vertical velocity field requires knowledge of the horizontal divergence. In order to determine the horizontal divergence the partial derivatives $\partial u/\partial x$ and $\partial v/\partial y$ are generally estimated from the fields of u and v by using *finite difference* approximations. For example, to determine the horizontal divergence at the point x_0, y_0 in Fig. 3.10 we write

$$\frac{\partial u}{\partial x} + \frac{\partial v}{\partial y} \approx \frac{u(x_0 + d) - u(x_0 - d)}{2d} + \frac{v(y_0 + d) - v(y_0 - d)}{2d} \tag{3.40}$$

However, for synoptic scale motions in midlatitudes the horizontal velocity is nearly in geostrophic equilibrium. Except for the small effect due to the variation of the Coriolis parameter (see Problem 19) the geostrophic wind is nondivergent, that is, $\partial u/\partial x$ and $\partial v/\partial y$ are nearly equal in magnitude but opposite in sign. Thus, the horizontal divergence is due primarily to the small departures of the wind from geostrophic balance. A 10% error in evaluating one of the wind components in (3.40) can easily cause the estimated divergence to be in error by 100%. For this reason, the continuity equation method is not recommended for estimating the vertical motion field from observed horizontal winds. In Chapter 6 we will develop an alternative method, the so-called omega equation, which is not nearly so sensitive to observational errors.

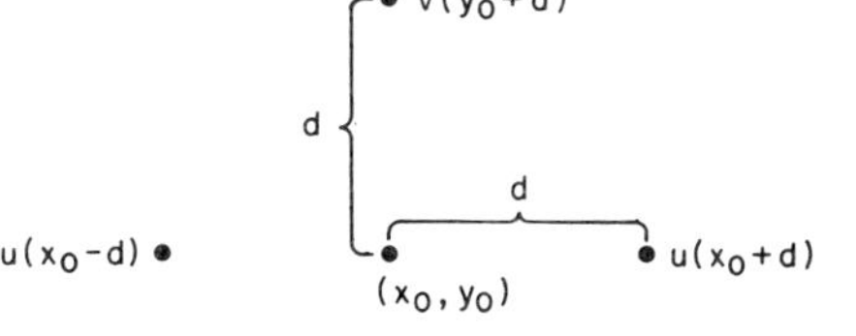

Fig. 3.10 Grid for estimation of the horizontal divergence.

3.5.3 THE ADIABATIC METHOD

A second method for inferring vertical velocities, which is not so sensitive to errors in the measured horizontal velocities, is based on the thermodynamic energy equation. If we assume that the diabatic heating $\dot{q}$ is small compared to the other terms in the heat balance we can solve (3.6) for ω to yield

$$\omega = S_p^{-1}\left(\frac{\partial T}{\partial t} + u\frac{\partial T}{\partial x} + v\frac{\partial T}{\partial y}\right) \tag{3.41}$$

Since temperature advection can usually be estimated quite accurately in midlatitudes by using the geostrophic winds, the adiabatic method can be applied when only geopotential and temperature data are available. A disadvantage of the adiabatic method is that the local rate of change of temperature is required. Unless observations are taken at close intervals in time it may be difficult to accurately estimate $\partial T/\partial t$ over a wide area.

Problems

1. An aircraft flying a heading of 60° at air speed 200 m s^{-1} moves relative to the ground due east (90°) at 225 m s^{-1}. If the plane is flying at constant pressure, what is its rate of change in altitude assuming a steady pressure field and $f = 10^{-4}\ s^{-1}$?

2. The actual wind is directed 30° to the right of the geostrophic wind. If the geostrophic wind is 20 m s^{-1}, what is the rate of change of wind speed? Let $f = 10^{-4}\ s^{-1}$.

3. A tornado rotates with constant angular velocity ω. Show that the surface pressure at the center of the tornado is given by

$$p = p_0 \exp\left(\frac{-\omega^2 r_0^2}{2RT}\right)$$

where p_0 is the surface pressure at a distance r_0 from the center and T is the temperature (assumed constant). If the temperature is 288 K, pressure at 100 m from the center is 10^2 kPa, and wind speed at 100 m from center is 100 m s^{-1}, what is the central pressure?

4. Calculate the geostrophic wind speed in meters per second for a pressure gradient of 1 kPa/10^3 km and compare with *all* possible gradient wind speeds for the same pressure gradient and a radius of curvature of ± 500 km. Let $\rho = 1$ kg m^{-3} and $f = 10^{-4}\ s^{-1}$.

5. Determine the maximum possible ratio of the normal anticyclonic gradient wind speed to the geostrophic wind speed for the same pressure gradient.

6. Show that the geostrophic balance in isothermal coordinates may be written

$$f\mathbf{V}_g = \mathbf{k} \times \nabla_T(RT \ln p + \Phi)$$

7. Determine the radii of curvature for the trajectories of air parcels located 500 km to the east, north, south, and west of the center of a circular low-pressure system, respectively. The system is moving eastward at 15 m s^{-1}. Assume geostrophic flow with a uniform tangential wind speed of 15 m s^{-1}

8. Determine the gradient wind speeds for the four air parcels in Problem 7 and compare these speeds with the geostrophic speed. (Let $f = 10^{-4}\ s^{-1}$.)

9. Show that as the pressure gradient approaches zero the gradient wind reduces to the geostrophic wind for a normal anticyclone and to the inertia circle for an anomalous anticyclone.

10. The mean temperature in the layer between 75 and 50 kPa decreases eastward by 3°C per 100 km. If the 75-kPa geostrophic wind is from the southeast at 20 m s^{-1}, what is the geostrophic wind speed and direction at 50 kPa? Let $f = 10^{-4}\ s^{-1}$.

11. What is the mean temperature advection in the 75–50-kPa layer in Problem 10?

12. Suppose that a vertical column of the atmosphere is initially isothermal from 90 to 50 kPa. The geostrophic wind is 10 m s^{-1} from the south at 90 kPa, 10 m s^{-1} from the west at 70 kPa, and 20 m s^{-1} from the west at 50 kPa. Calculate the mean horizontal temperature gradients in the two layers 90–70 kPa and 70–50 kPa. Compute the rate of advective temperature change in each layer. How long would this advection pattern have to persist in order to establish a dry adiabatic lapse rate between 60 and 80 kPa? (Assume that the lapse rate is linear from 90–50 kPa.

***13.** An airplane pilot crossing the ocean at 45°N latitude has both a pressure altimeter and a radar altimeter, the latter measuring his absolute height above the sea. Flying at an air speed of 100 m s^{-1} he maintains altitude by referring to his pressure altimeter set at an altimeter setting of 101.3 kPa. He holds an indicated 6000-m altitude.

At the beginning of a one-hour period he notes that his radar altimeter reads 5700 m and at the end of the hour he notes that it reads 5950 m. In what direction and approximately how far has he drifted from his heading?

14. Work out a gradient wind classification scheme equivalent to Table 3.1 for the Southern Hemisphere ($f < 0$) case.

15. In the so-called *geostrophic momentum* approximation (Hoskins, 1975) the gradient wind formula for steady circular flow (3.17) is replaced by the approximation

$$\frac{VV_g}{R} + fV = fV_g$$

Compare the speeds V computed using this approximation with those obtained in Problem 8 using the gradient wind formula.

16. How large can the ratio V_g/fR be before the geostrophic momentum approximation differs from the gradient wind approximation by 10% for cyclonic flow?

***17.** The planet Venus rotates about its axis so slowly that to a reasonable approximation the Coriolis parameter may be set equal to zero. For steady, frictionless motion parallel to latitude circles the momentum equation (2.20) then reduces to a type of cyclostrophic balance:

$$\frac{u^2 \tan \phi}{a} = -\frac{1}{\rho}\frac{\partial p}{\partial y}$$

By transforming this expression to isobaric coordinates show that the "thermal wind" equation in this case can be expressed in the form

$$\omega_r^2(p_1) - \omega_r^2(p_0) = \frac{-R \ln(p_0/p_1)}{(a \sin \phi \cos \phi)} \frac{\partial \langle T \rangle}{\partial y}$$

where R is the gas constant, a the radius of the planet, and $\omega_r \equiv u/(a \cos \phi)$ is the relative angular velocity. How must $\langle T \rangle$ (the vertically averaged temperature) vary with respect to latitude in order that ω_r be a function only of height? If the zonal velocity at $\simeq$60-km height above the equator ($p_1 = 2.9 \times 10^5$ Pa) is 100 m s^{-1} and the zonal velocity vanishes at the surface of the planet ($p_0 = 9.5 \times 10^6$ Pa) what is the vertically averaged temperature difference between the equator and pole assuming that ω_r depends only on height? The planetary radius is $a = 6100$ km, and the gas constant is $R = 187$ J kg^{-1} K^{-1}.

18. Suppose that during the passage of a cyclonic storm the radius of curvature of the isobars is observed to be +800 km at a station where the wind is veering (turning clockwise) at a rate of 10° per hour. What is the radius of curvature of the trajectory for an air parcel which is passing over the station? (The wind speed is 20 m s^{-1}.)

19. Show that the divergence of the geostrophic wind in isobaric coordinates on the spherical earth is given by

$$\nabla \cdot \mathbf{V}_g = -\frac{1}{fa}\frac{\partial \Phi}{\partial x}\left(\frac{\cos \phi}{\sin \phi}\right)$$

(Use the spherical coordinate expression for the divergence operator given in Appendix C.)

20. The following wind data were received from 50 km to the east, north, west, and south of a station, respectively: 90°, 10 m s^{-1}; 120°, 4 m s^{-1}; 90°, 8 m s^{-1}; 60°, 4 m s^{-1}. Calculate the approximate horizontal divergence at the station.

21. Suppose that the wind speeds given in Problem 20 are each in error by ±10%. What would be the percent error in the calculated horizontal divergence in the worst case?

22. The divergence of the horizontal wind at various pressure levels above a given station is shown in the following table.

Pressure (kPa)	$\nabla \cdot \mathbf{V}$ ($\times 10^{-5}$ s^{-1})
100	+0.9
85	+0.6
70	+0.3
50	0.0
30	−0.6
10	−1.0

Compute the vertical velocity at each level assuming an isothermal atmosphere with temperature 260 K and letting $w = 0$ at 100 kPa (1000 mb).

23. Suppose that the lapse rate at the 85-kPa (850-mb) level is 4 K km^{-1}. If the temperature at a given location is decreasing at a rate of 2 K h^{-1}, the wind is westerly at 10 m s^{-1}, and the temperature decreases towards the west at a rate of 5 K/100 km, compute the vertical velocity at the 85-kPa level using the adiabatic method.

Chapter

4 Circulation and Vorticity

In classical mechanics the principle of conservation of angular momentum is often invoked in the analysis of motions that involve rotation. This principle provides a powerful constraint on the behavior of rotating objects. Analogous conservation laws also apply to the rotational field of a fluid. However, it should be obvious that in a continuous medium such as the atmosphere the definition of "rotation" is more difficult than for a solid object.

Circulation and vorticity are the two primary measures of rotation in a fluid. Circulation, which is a scalar integral quantity, is a *macroscopic* measure of rotation for a finite area of the fluid. Vorticity, however, is a vector field which gives a *microscopic* measure of the rotation at any point in the fluid.

4.1 The Circulation Theorem

The *circulation* about a closed contour in a fluid is defined as the line integral about the contour of the component of the velocity vector which is locally tangent to the contour. Thus, for a contour in the horizontal plane the circulation C is given by

$$C \equiv \oint \mathbf{V} \cdot d\mathbf{l} = \oint |\mathbf{V}| \cos \alpha \, dl$$

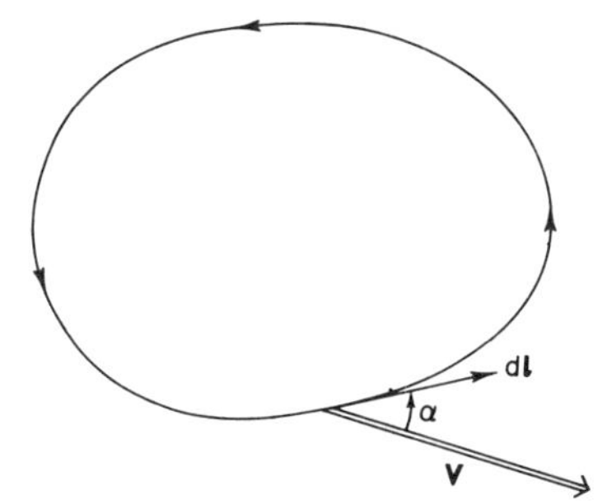

Fig. 4.1 Circulation about a closed contour.

where, as indicated in Fig. 4.1, $d\mathbf{l}$ represents an element of the contour. By convention the circulation is taken to be positive if $C > 0$ for counterclockwise integration around the contour.

That circulation is a measure of rotation can be seen from the following example. Suppose that a circular disk of fluid of radius r is in solid-body rotation at angular velocity $\mathbf{\Omega}$ about the z axis. In this case $\mathbf{V} = \mathbf{\Omega} \times \mathbf{r}$ where $\mathbf{r}$ is the distance to the axis of rotation. Thus the circulation about the edge of the disk is given by

$$C = \oint \mathbf{V} \cdot d\mathbf{l} = \int_0^{2\pi} \Omega r^2 \, d\lambda = 2\Omega \pi r^2$$

or

$$\frac{C}{\pi r^2} = 2\Omega$$

Hence, in the case of solid-body rotation the circulation divided by the area enclosed by the loop is just twice the angular speed of rotation.

The circulation theorem is obtained by taking the line integral of Newton's second law for a closed chain of fluid particles. In the absolute coordinate system the result neglecting viscous forces is

$$\oint \frac{d_a \mathbf{U}_a}{dt} \cdot d\mathbf{l} = - \oint \frac{\nabla p \cdot d\mathbf{l}}{\rho} - \oint \nabla \Phi \cdot d\mathbf{l} \tag{4.1}$$

where the gravitational force $\mathbf{g}$ is represented as the gradient of the geopotential Φ $(-\nabla\Phi = \mathbf{g} = -g\mathbf{k})$. The integrand on the left-hand side can be rewritten as[1]

$$\frac{d_a \mathbf{U}_a}{dt} \cdot d\mathbf{l} = \frac{d}{dt}(\mathbf{U}_a \cdot d\mathbf{l}) - \mathbf{U}_a \cdot \frac{d_a}{dt}(d\mathbf{l})$$

[1] Note that for a scalar $d_a/dt = d/dt$ (i.e., the rate of change following the motion does not depend on the reference system). For a vector, however, this is not the case, as was shown in Section 2.1.1.

or after observing that since $\mathbf{l}$ is a position vector, $d_a\mathbf{l}/dt \equiv \mathbf{U}_a$,

$$\frac{d_a\mathbf{U}_a}{dt}\cdot d\mathbf{l} = \frac{d}{dt}(\mathbf{U}_a \cdot d\mathbf{l}) - \mathbf{U}_a \cdot d\mathbf{U}_a \tag{4.2}$$

Substituting (4.2) into (4.1) and using the fact that the line integral of a perfect differential is zero so that

$$\oint \nabla\Phi \cdot d\mathbf{l} = \oint d\Phi = 0$$

and

$$\oint \mathbf{U}_a \cdot d\mathbf{U}_a = \tfrac{1}{2}\oint d(\mathbf{U}_a \cdot \mathbf{U}_a) = 0$$

we obtain the circulation theorem:

$$\frac{dC_a}{dt} = \frac{d}{dt}\oint \mathbf{U}_a \cdot d\mathbf{l} = -\oint \frac{dp}{\rho} \tag{4.3}$$

The term on the right-hand side in (4.3) is called the *solenoidal* term. For a barotropic fluid, the density is a function only of pressure and the solenoidal term is zero. Thus, in a barotropic fluid the *absolute* circulation is conserved following the motion. This result, called Kelvin's circulation theorem, is a fluid analog of angular momentum conservation in solid-body mechanics.

For meteorological analysis, it is more convenient to work with the relative circulation C rather than the absolute circulation since a portion of the absolute circulation, C_e, is due to the rotation of the earth about its axis. To compute C_e we apply Stokes' theorem to the vector $\mathbf{U}_e$, where $\mathbf{U}_e = \boldsymbol{\Omega} \times \mathbf{r}$ is the velocity of the earth at the position $\mathbf{r}$. Thus,

$$C_e = \oint \mathbf{U}_e \cdot d\mathbf{l} = \int_A\int (\nabla \times \mathbf{U}_e)\cdot \mathbf{n}\, dA$$

where A is the area enclosed by the contour, and the unit normal $\mathbf{n}$ is defined by the counterclockwise sense of the line integration using the "right-hand screw" rule. Thus, for the contour of Fig. 4.1 $\mathbf{n}$ would be directed out of the page. But if the line integral is computed in the horizontal plane $\mathbf{n}$ is directed along the local vertical and $(\nabla \times \mathbf{U}_e)\cdot\mathbf{n} = 2\Omega\sin\phi \equiv f$ is just the Coriolis parameter. Hence the circulation due to the rotation of the earth is $C_e = 2\Omega \sin\bar{\phi}\, A$, where $\bar{\phi}$ denotes the average value of latitude over the area element A. Finally, we may write

$$C = C_a - C_e = C_a - 2\Omega A_e \tag{4.4}$$

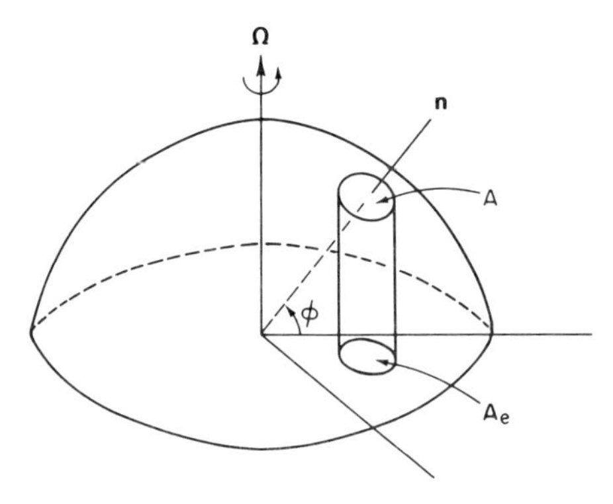

Fig. 4.2 Area A_e subtended on the equatorial plane by horizontal area A centered at latitude ϕ.

where A is the area enclosed by the contour, and the unit normal $\mathbf{n}$ is defined as shown in Fig. 4.2. Differentiating (4.4) following the motion, and substituting from (4.3) we obtain the Bjerknes circulation theorem:

$$\frac{dC}{dt} = -\oint \frac{dp}{\rho} - 2\Omega \frac{dA_e}{dt} \tag{4.5}$$

For a barotropic fluid, (4.5) can be integrated following the motion from an initial state designated by subscript 1 to a final state designated by subscript 2, yielding

$$C_2 - C_1 = -2\Omega(A_2 \sin \phi_2 - A_1 \sin \phi_1) \tag{4.6}$$

Equation (4.6) indicates that in a barotropic fluid the relative circulation for a closed chain of fluid particles will be changed if either the horizontal area enclosed by the loop changes or the average latitude changes.

Example. Suppose that the air within a circular region of radius 100 km centered at the equator is initially motionless with respect to the earth. If this circular air mass were moved to the North Pole along an isobaric surface preserving its area, the circulation about the circumference would be

$$C = -2\Omega\pi r^2 (\sin \pi/2 - \sin 0)$$

Thus the mean tangential velocity at the radius $r = 100$ km would be

$$\overline{V} = \frac{C}{2\pi r} = -\Omega r \approx -7 \quad \text{m s}^{-1}$$

The negative sign here indicates that the air has acquired anticyclonic relative circulation.

We have yet to consider the role of the solenoidal term in the circulation theorem. The generation of circulation by pressure–density solenoids can be

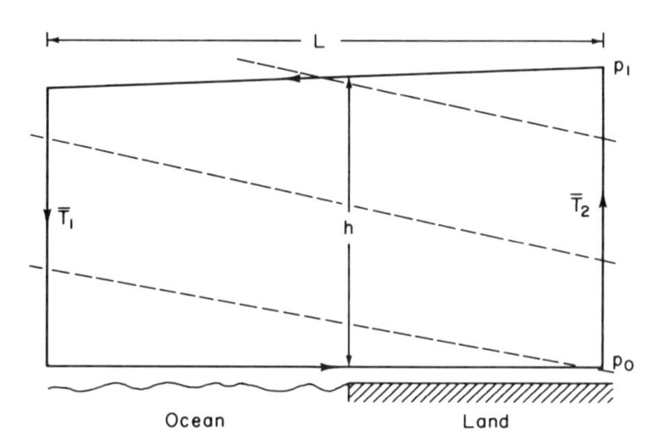

Fig. 4.3 Application of the circulation theorem to the seabreeze problem. The closed heavy solid line is the loop about which the circulation is to be evaluated. Dashed lines indicate isosteric surfaces.

effectively illustrated by considering the development of a seabreeze circulation. The situation is shown in Fig. 4.3. The mean temperature in the air over the ocean is colder than the mean temperature over the adjoining land. Thus, if the pressure is uniform at ground level, the isobaric surfaces above the ground will slope downward toward the ocean while the isosteric surfaces (isolines of specific volume) slope downward toward the land. To compute the acceleration as a result of the intersection of the pressure–density surfaces we apply the circulation theorem on a circuit in a vertical plane perpendicular to the coastline. Substituting the ideal gas law into (4.3) we obtain

$$\frac{dC_{\mathrm{a}}}{dt} = -\oint RT\, d\ln p$$

Evaluating the line integral for the circuit shown in Fig. 4.3 we see that there is a contribution only for the vertical segments of the loop since the horizontal segments are taken at constant pressure (the slope of the isobaric surface for the upper segment is neglected here since it is small compared to the slopes of the isosteric surfaces). The resulting rate of increase in the circulation is

$$\frac{dC_{\mathrm{a}}}{dt} = R\ln\left(\frac{p_0}{p_1}\right)(\bar{T}_2 - \bar{T}_1) > 0$$

Letting $\bar{v}$ be the mean tangential velocity along the circuit, we find that

$$\frac{d\bar{v}}{dt} = \frac{R\ln(p_0/p_1)}{2(h+L)}(\bar{T}_2 - \bar{T}_1) \tag{4.7}$$

As a specific example we let $p_0 = 100$ kPa, $p_1 = 90$ kPa, $\bar{T}_2 - \bar{T}_1 = 10°$C, $L = 20$ km, and $h = 1$ km. The acceleration is then

$$\frac{d\bar{v}}{dt} \approx 7.1 \times 10^{-3} \quad \text{m s}^{-2}$$

so that in the absence of frictional retarding forces the wind would reach a speed of 25 m s^{-1} in about one hour. In reality, however, frictional drag (which is roughly proportional to the square of the wind speed) would quickly retard the acceleration and at the same time temperature advection would reduce the temperature difference between land and sea, so that eventually a balance would be reached between the generation of kinetic energy by the pressure–density solenoids and dissipation by friction.

4.2 Vorticity

Vorticity, the microscopic measure of rotation in a fluid, is a vector field defined as the curl of velocity.

The *absolute vorticity* $\boldsymbol{\omega}_a$ is given by the curl of the absolute velocity, while the *relative vorticity* $\boldsymbol{\omega}$ is given by the curl of the relative velocity:

$$\boldsymbol{\omega}_a \equiv \nabla \times \mathbf{U}_a, \qquad \boldsymbol{\omega} \equiv \nabla \times \mathbf{U}$$

However, in dynamic meteorology we are in general concerned only with the vertical components of absolute and relative vorticity:

$$\eta \equiv \mathbf{k} \cdot (\nabla \times \mathbf{U}_a), \qquad \zeta \equiv \mathbf{k} \cdot (\nabla \times \mathbf{U})$$

In particular, the vertical component of relative vorticity ζ is highly correlated with synoptic scale weather disturbances. (Large positive ζ tends to occur in association with cyclonic storms in the Northern Hemisphere.) Furthermore, η tends to be conserved following the motion in the middle troposphere. Thus, analysis of the η field and its evolution due to advection forms the basis for the simplest dynamical forecast scheme to be discussed in Chapter 8.

In the remainder of this book we generally will refer to η and ζ as the absolute and relative vorticities, respectively, without adding the explicit modifier "vertical component of." The difference between absolute and relative vorticity is given by the vertical component of the vorticity of the earth due to its rotation; or $\mathbf{k} \cdot \nabla \times \mathbf{U}_e = 2\Omega \sin \phi \equiv f$. Thus we have $\eta = \zeta + f$, or using cartesian coordinates

$$\zeta = \frac{\partial v}{\partial x} - \frac{\partial u}{\partial y}, \qquad \eta = \frac{\partial v}{\partial x} - \frac{\partial u}{\partial y} + f$$

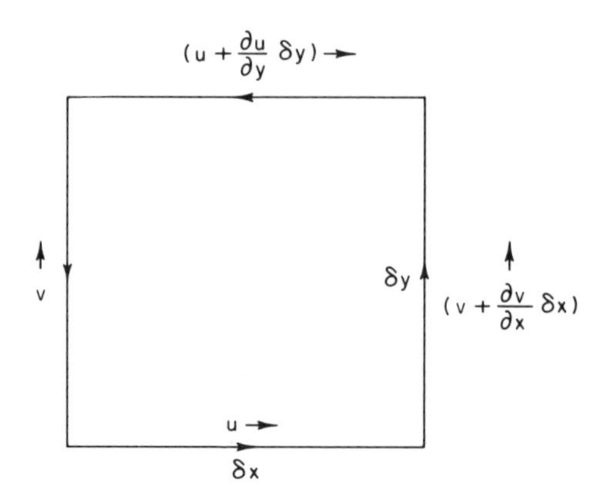

Fig. 4.4 The relationship between circulation and vorticity for an area element in the horizontal plane.

The relationship between relative vorticity and the relative circulation C discussed in the previous section can be clearly seen by considering an alternative approach in which the vertical component of vorticity is defined as the circulation about a closed contour in the horizontal plane divided by the area enclosed, in the limit where the area approaches zero:

$$\zeta \equiv \lim_{A \to 0} \frac{\oint \mathbf{U} \cdot d\mathbf{l}}{A} \tag{4.8}$$

This latter definition makes explicit the relationship between circulation and vorticity discussed in the introduction to this chapter. The equivalence of these two definitions of ζ may be easily shown by considering the circulation about a rectangular element of area in the x,y plane as shown in Fig. 4.4. With the aid of the figure we see that the circulation is

$$\delta C = u\,\delta x + \left(v + \frac{\partial v}{\partial x}\,\delta x\right)\delta y - \left(u + \frac{\partial u}{\partial y}\,\delta y\right)\delta x - v\,\delta y$$

$$= \left(\frac{\partial v}{\partial x} - \frac{\partial u}{\partial y}\right)\delta x\,\delta y$$

Dividing through by the area $\delta A = \delta x\,\delta y$, we get

$$\frac{\delta C}{\delta A} = \frac{\partial v}{\partial x} - \frac{\partial u}{\partial y} \equiv \zeta$$

In more general terms the relationship between vorticity and circulation is given simply by Stokes' theorem applied to the velocity vector:

$$\oint \mathbf{U} \cdot d\mathbf{l} = \int_A \int (\boldsymbol{\nabla} \times \mathbf{U}) \cdot \mathbf{n}\, dA$$

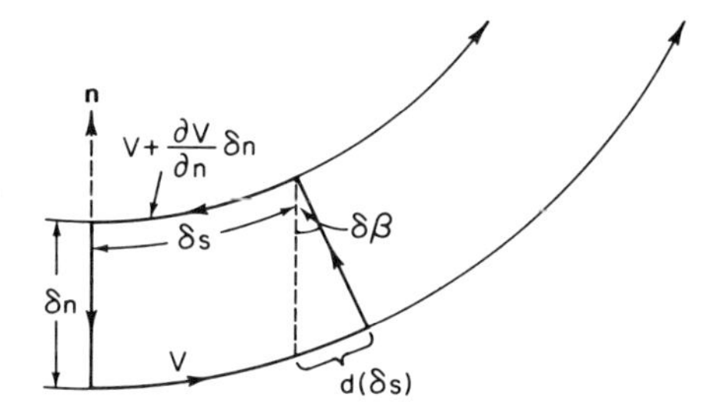

Fig. 4.5 Circulation for an infinitesimal loop in the natural coordinate system.

Here A is the area enclosed by the contour, and $\mathbf{n}$ is a unit normal to the area element dA (positive in the right-hand sense). Thus, Stokes' theorem states that the circulation about any closed loop is equal to the integral of the normal component of vorticity over the area enclosed by the contour. Hence, for a finite area, the circulation divided by the area gives the *average* normal component of vorticity in the region. As a consequence, the vorticity of a fluid in solid-body rotation is just twice the angular velocity of rotation. Vorticity may thus be regarded as a measure of the local angular velocity of the fluid.

Physical interpretation of vorticity may be facilitated by considering the vertical component of vorticity in the natural coordinate system (see Section 3.2.1). If we compute the circulation about the infinitesimal contour shown in Fig. 4.5 we obtain[2]

$$\delta C = V[\delta s + d(\delta s)] - \left(V + \frac{\partial V}{\partial n}\delta n\right)\delta s$$

But from Fig. 4.5 we see that

$$d(\delta s) = \delta\beta\,\delta n$$

where $\delta\beta$ is the angular change in the wind direction in the distance δs. Hence

$$\delta C = \left(-\frac{\partial V}{\partial n} + V\frac{\delta\beta}{\delta s}\right)\delta n\,\delta s$$

or in the limit $\delta n, \delta s \to 0$

$$\zeta = \lim_{\delta n,\,\delta s \to 0}\frac{\delta C}{(\delta n\,\delta s)} = -\frac{\partial V}{\partial n} + \frac{V}{R_s} \tag{4.9}$$

[2] Recall that $\mathbf{n}$ is here a unit vector in the horizontal plane directed to the left of the local flow direction.

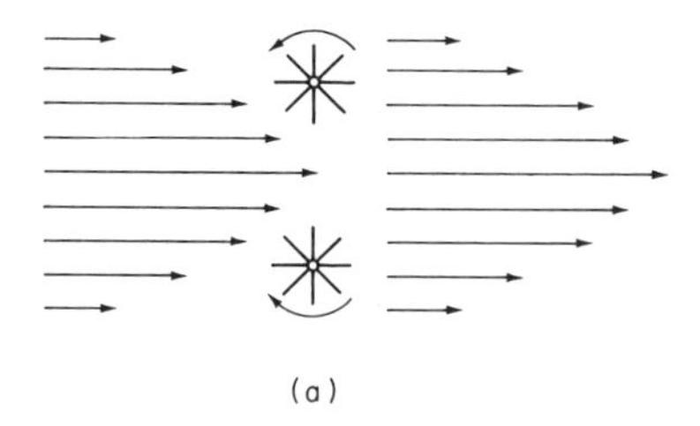

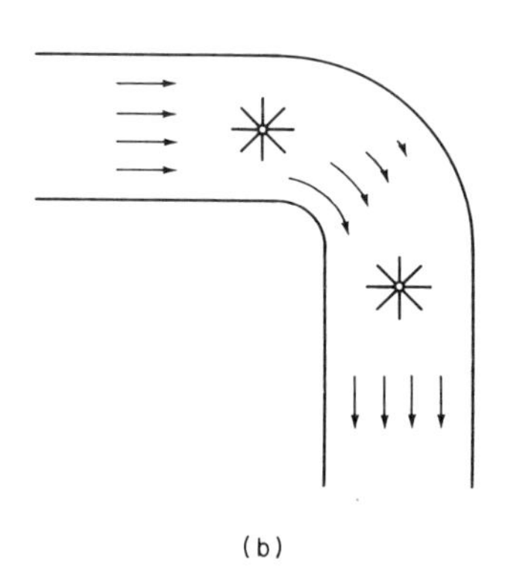

Fig. 4.6 Two types of two-dimensional flow: (a) linear shear flow with vorticity; and (b) curved flow with zero vorticity.

where R_s is the radius of curvature of the streamlines [Eq. (3.20)]. It is now apparent that the net vertical vorticity component is the result of the sum of two parts: (1) the rate of change of wind speed normal to the direction of flow $-\partial V/\partial n$, called the *shear vorticity*; and (2) the turning of the wind along a streamline V/R_s, called the *curvature vorticity*. Thus, even straight-line motion may have vorticity if the speed changes normal to the flow axis. For example, in the jet stream shown schematically in Fig. 4.6 there will be cyclonic vorticity north of the velocity maximum and anticyclonic vorticity to the south as is easily recognized when the motion of a small exploring paddle wheel is considered. The lower paddle wheel shown in the figure will be turned in a clockwise direction since the wind force on its northern blades is stronger than the force on the blades to the south of the axis of the wheel.

On the other hand, curved flow may have zero vorticity provided that the shear vorticity is equal and opposite to the curvature vorticity. This will be the case in the example shown in Fig. 4.6b where a frictionless fluid with zero relative vorticity upstream flows around a bend in a canal. The water along the inner margin of the channel flows faster in just the right proportion so that the paddle wheel does not turn.

4.3 Potential Vorticity

Since potential temperature θ is conserved following the motion in adiabatic flow, a parcel of air which moves adiabatically will remain on the same potential temperature surface. Now from the definition of potential temperature it can be shown that for constant θ

$$\rho \propto p^{c_v/c_p}$$

Therefore, density is a function of pressure alone on an adiabatic surface and the solenoidal term in the circulation theorem will vanish:

$$\oint \frac{dp}{\rho} \propto \oint dp^{(1-c_v/c_p)} = 0$$

Thus, for adiabatic motion the circulation theorem evaluated on a constant θ surface reduces to the same form as in a barotropic fluid, that is,

$$\frac{d}{dt}(C + 2\Omega A \sin \phi) = 0 \tag{4.10}$$

where C is evaluated for a closed loop encompassing the area A on an adiabatic surface. If we assume that the adiabatic surface is approximately horizontal, and recall that the vertical relative vorticity component is given by

$$\zeta = \lim_{A \to 0} \frac{C}{A}$$

we can write the integral of (4.10) for an infinitesimal parcel of air as

$$A(\zeta + f) = \text{const} \tag{4.11}$$

where $f = 2\Omega \sin \phi$ is the Coriolis parameter. Now suppose that the parcel is confined between potential temperature surfaces θ_0 and $\theta_0 + \delta\theta$ which are separated by a distance δp as shown in Fig. 4.7. The mass of the parcel $M = A\,\delta p/g$ must be conserved following the motion. Therefore

$$A = \frac{Mg}{\delta p} = \left(\frac{\delta\theta}{\delta p}\right)\left(\frac{Mg}{\delta\theta}\right) = \text{const} \times \left(\frac{\delta\theta}{\delta p}\right)$$

since $\delta\theta$ is a constant. Substituting into (4.11) to eliminate A and taking the limit $\delta p \to 0$, we obtain

$$(\zeta + f)\frac{\partial\theta}{\partial p} = \text{const} \tag{4.12}$$

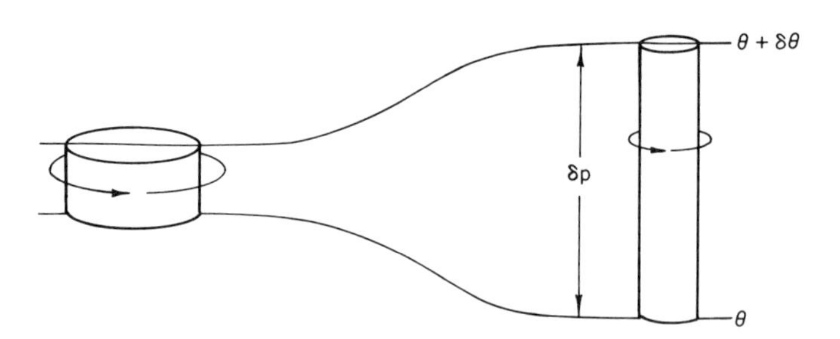

Fig. 4.7 A cylindrical column of air moving adiabatically, conserving potential vorticity.

Equation (4.12) expresses the conservation of *potential vorticity* in adiabatic, frictionless motion. The term potential vorticity is used, as we shall see later, in connection with several other slightly different mathematical expressions. In essence, however, the potential vorticity is always in some sense a measure of the ratio of the absolute vorticity to the effective depth of the vortex. In (4.12), for example, the effective depth is just the distance between potential temperature surfaces measured in pressure units.

In a homogeneous incompressible fluid the theorem of potential vorticity conservation takes a somewhat simple form. In this case since density is a constant we have

$$A = \frac{M}{\rho\,\delta z} = \frac{\text{const}}{\delta z}$$

where δz is the depth of the parcel. Again substituting to eliminate A in (4.11) we get

$$\frac{\zeta + f}{\delta z} = \text{const} \tag{4.13}$$

The conservation of potential vorticity is a powerful constraint on the large-scale motions of the atmosphere. This can be illustrated by considering the flow of air over a large mountain barrier.

However, in order to facilitate understanding it is worthwhile to first consider the simpler case where $\partial\theta/\partial p$ (or δz) is constant so that the absolute vorticity $\eta = \zeta + f$ is conserved following the motion. We consider a uniform flow field in which the motion at a given longitude (denoted by the zero subscript) is purely zonal so that $\zeta_0 = 0$. Then by conservation of absolute vorticity the motion at any point downstream must satisfy $\zeta + f = f_0$. Thus, recalling that f increases toward the north, we see that if the flow trajectories curve northward downstream then $\zeta = f_0 - f < 0$; for trajectories curving towards the south, on the other hand, $\zeta = f_0 - f > 0$. But, as indicated in Fig. 4.8, for westerly flow northward curvature will produce

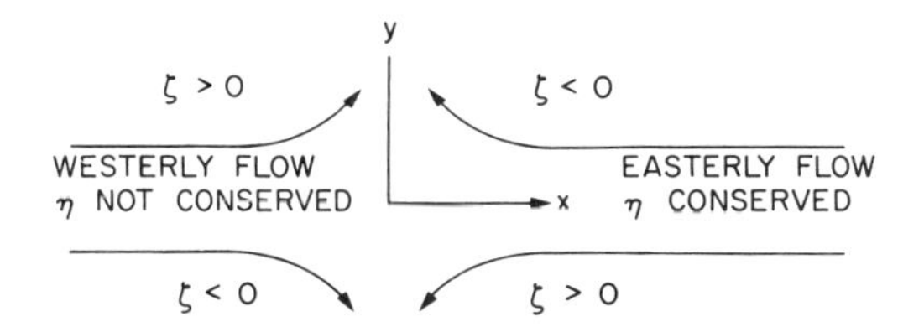

Fig. 4.8 Absolute vorticity conservation for curved flow trajectories.

positive relative vorticity and southward curvature negative vorticity. Thus, westerly zonal flow must remain purely zonal if absolute vorticity is conserved following the motion. The easterly flow case, also shown in Fig. 4.8, is just the opposite. Northward and southward curvatures are associated with negative and positive relative vorticities, respectively. Hence an easterly current can curve either to the north or south and still conserve absolute vorticity.

When $\partial\theta/\partial p$ (or δz) changes following the motion it is potential vorticity which is conserved. But again westerly and easterly flows behave differently. The situation for westerly flow is shown in Fig. 4.9. In Fig. 4.9a a vertical cross section of the flow is shown. We suppose that upstream of the mountain barrier the flow is a uniform zonal flow so that $\zeta = 0$. If the flow is adiabatic, each column of air confined between the potential temperature surfaces θ_0 and $\theta_0 + \delta\theta$ remains between those surfaces as it crosses the mountain. For this reason, a potential temperature surface near the ground must approximately follow the ground contours. However, ordinarily, a potential

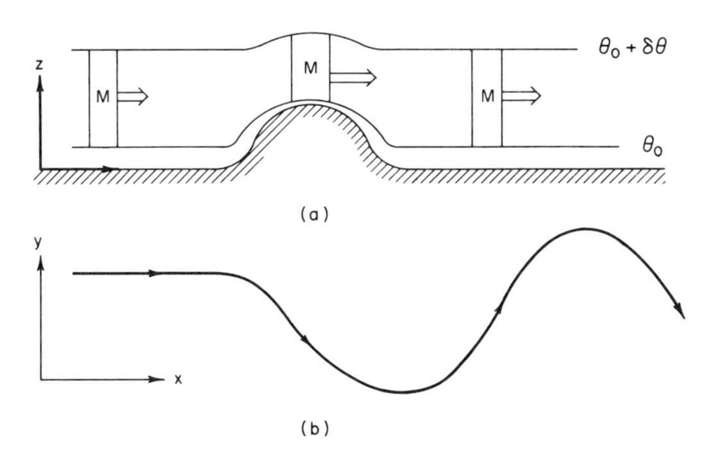

Fig. 4.9 Westerly flow over a topographic barrier: (a) the depth of a column as a function of x; (b) the trajectory of a parcel in the x,y plane.

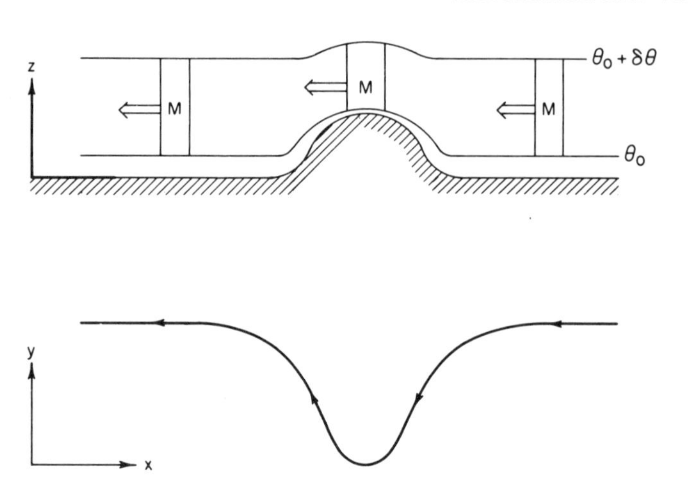

Fig. 4.10 Easterly flow over a topographic barrier.

temperature surface several kilometers above the ground will not be deflected much in the vertical by the ground contours. Thus, as an air column begins to cross the mountain barrier its vertical extent decreases; hence, as is obvious from (4.12), in order to conserve potential vorticity the relative vorticity must become negative. Thus, the air will acquire anticyclonic vorticity and move southward as shown in the x,y plane profile in Fig. 4.9b. Now when the air parcel has passed over the mountain and the column has returned to its original depth it will be south of its original latitude so that f will be smaller and the relative vorticity must now be positive—that is, the trajectory must have cyclonic curvature. When the parcel returns to its original latitude, it will still have a northward velocity component and will continue northward gradually acquiring anticyclonic curvature until its meridional motion direction is reversed. Thus, the parcel will move downstream conserving potential vorticity by following a wavelike trajectory in the horizontal plane. Therefore, steady westerly flow over a mountain barrier will result in a cyclonic flow pattern immediately to the east of the barrier (the lee side trough) followed by an alternating series of ridges and troughs downstream.

The situation for easterly flow impinging on a mountain barrier is quite different. Let us assume that the motion were to remain uniform until the flow reached the mountain. In that case as an air column began to move over the barrier from the east it would have to acquire anticyclonic curvature in order to conserve potential vorticity. Thus, the parcel trajectory would begin to curve northward toward higher latitudes, and the resulting increase in f would force the relative vorticity to become even more strongly negative to conserve potential vorticity. As a result the parcel would eventually curve back toward the east and no steady-state potential vorticity conserving flow pattern would be possible.

For this reason, easterly flow, unlike westerly flow, must feel the influence of the mountain barrier upstream from the barrier. As indicated schematically in Fig. 4.10, a westward-moving air column begins to curve cyclonically before it reaches the mountain. This cyclonic curvature is produced as the flow follows the isobars of the pressure field generated by the mountain forcing. The resulting positive relative vorticity is balanced by a decrease in f so that potential vorticity is conserved. As the column moves to the top of the mountain, it continues to move equatorward so that the decrease in depth is balanced by a decrease in f. Finally, as the parcel moves down the mountain toward the west, the process is simply reversed with the result that some distance downstream from the mountain the air column again is moving westward at its original latitude. Thus, the dependence of the Coriolis parameter on latitude creates a dramatic difference between westerly and easterly flow over a topographic barrier. In the case of a westerly wind the barrier generates a wavelike disturbance in the streamlines which extends far downstream. But in the case of an eastery wind, the disturbance in the streamlines damps out away from the barrier.

Equation (4.13) also indicates that in a barotropic fluid a change in the depth is dynamically analogous to a change in the Coriolis parameter. This effect can easily be demonstrated in a rotating cylindrical vessel filled with an incompressible fluid. For such a fluid in solid-body rotation, the equilibrium shape of the free surface, determined by a balance between the radial pressure gradient and centrifugal forces, is parabolic. Thus, as shown in Fig. 4.11, if a column of fluid moves radially outward it must stretch vertically. According to (4.13) the relative vorticity must then increase to keep the ratio $(\zeta + f)/\delta z$ constant. The same result would apply if a column of fluid on a rotating

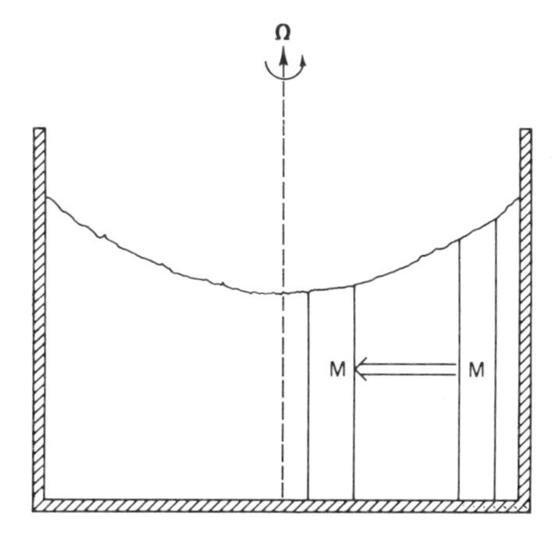

Fig. 4.11 The dependence of depth on radius in a rotating cylindrical vessel.

sphere were moved equatorward without change in depth. In this case ζ would have to increase to offset the decrease of f. Therefore, in a barotropic fluid a decrease of depth with increasing latitude has the same effect on the relative vorticity as the increase of the Coriolis force with latitude.

4.4 The Vorticity Equation

In the previous section we have discussed the time evolution of the vertical component of vorticity for the special case of adiabatic frictionless flow. In the present section we will use the equations of motion to derive an equation for the time rate of change of vorticity without limiting the validity to adiabatic motion.

4.4.1 Cartesian Coordinate Form

For motions of synoptic scale, the vorticity equation can be derived using the approximate horizontal momentum equations (2.24) and (2.25). We differentiate the x component equation with respect to y and the y component equation with respect to x:

$$\frac{\partial}{\partial y}\left\{\frac{\partial u}{\partial t} + u\frac{\partial u}{\partial x} + v\frac{\partial u}{\partial y} + w\frac{\partial u}{\partial z} - fv = -\frac{1}{\rho}\frac{\partial p}{\partial x}\right\} \tag{4.14}$$

$$\frac{\partial}{\partial x}\left\{\frac{\partial v}{\partial t} + u\frac{\partial v}{\partial x} + v\frac{\partial v}{\partial y} + w\frac{\partial v}{\partial z} + fu = -\frac{1}{\rho}\frac{\partial p}{\partial y}\right\} \tag{4.15}$$

Subtracting (4.14) from (4.15) and recalling that $\zeta = \partial v/\partial x - \partial u/\partial y$ we obtain the vorticity equation

$$\frac{\partial \zeta}{\partial t} + u\frac{\partial \zeta}{\partial x} + v\frac{\partial \zeta}{\partial y} + w\frac{\partial \zeta}{\partial z} + (\zeta + f)\left(\frac{\partial u}{\partial x} + \frac{\partial v}{\partial y}\right)$$

$$+ \left(\frac{\partial w}{\partial x}\frac{\partial v}{\partial z} - \frac{\partial w}{\partial y}\frac{\partial u}{\partial z}\right) + v\frac{df}{dy} = \frac{1}{\rho^2}\left[\frac{\partial \rho}{\partial x}\frac{\partial p}{\partial y} - \frac{\partial \rho}{\partial y}\frac{\partial p}{\partial x}\right] \tag{4.16}$$

Using the fact that the Coriolis parameter depends only on y so that $df/dt = v(df/dy)$, (4.16) may be rewritten in the form

$$\frac{d}{dt}(\zeta + f) = -(\zeta + f)\left(\frac{\partial u}{\partial x} + \frac{\partial v}{\partial y}\right) - \left(\frac{\partial w}{\partial x}\frac{\partial v}{\partial z} - \frac{\partial w}{\partial y}\frac{\partial u}{\partial z}\right)$$

$$+ \frac{1}{\rho^2}\left(\frac{\partial \rho}{\partial x}\frac{\partial p}{\partial y} - \frac{\partial \rho}{\partial y}\frac{\partial p}{\partial x}\right) \tag{4.17}$$

Equation (4.17) states that the rate of change of the absolute vorticity following the motion is given by the sum of the three terms on the right, called the divergence term, the tilting or twisting term, and the solenoidal term, respectively.

Generation of vorticity by horizontal divergence, the first term on the right in (4.17), is the fluid analog of the change in angular velocity resulting from a change in the moment of inertia of a solid body when angular momentum is conserved. If there is positive horizontal divergence, the area enclosed by a chain of fluid parcel will increase with time; and if circulation is to be conserved the average absolute vorticity of the enclosed fluid must decrease. This mechanism for changing vorticity is very important in synoptic scale disturbances.

The second term on the right in (4.17) represents vertical vorticity which is generated by the "tilting" of horizontally oriented components of vorticity into the vertical by a nonuniform vertical motion field. This mechanism is illustrated in Fig. 4.12. As shown in the figure we consider a region where the y component of velocity is increasing with height so that there is a component of shear vorticity oriented in the negative x direction. This is indicated by the double arrow in Fig. 4.12. If at the same time there is a vertical motion field in which w decreases with increasing x, advection by the vertical motion will tend to tilt the vorticity vector initially oriented parallel to x so that it has a component in the vertical. Thus, if $\partial v/\partial z > 0$ and $\partial w/\partial x < 0$, there will be a generation of positive vertical vorticity.

Finally, the third term on the right in (4.17) is just the microscopic equivalent of the solenoidal term in the circulation theorem (4.5). To show this

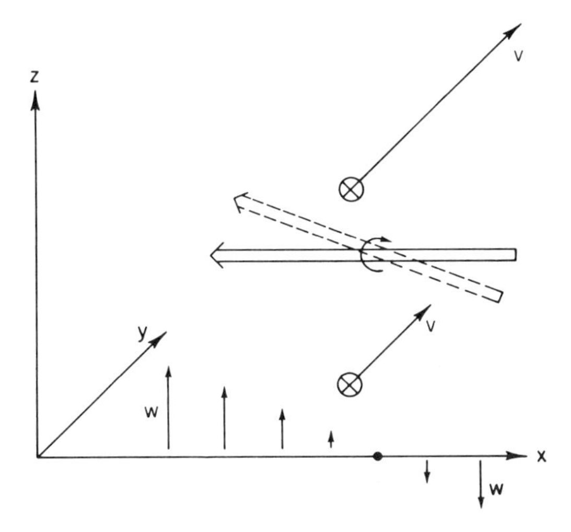

Fig. 4.12 Vorticity generation by the tilting of a horizontal vorticity vector (double arrow).

equivalence we may apply Stokes' theorem to the solenoidal term to get

$$\oint -\alpha\, dp \equiv -\oint \alpha \nabla p \cdot d\mathbf{l} = -\int_A\!\!\int \nabla \times (\alpha\, \nabla p) \cdot \mathbf{k}\, dA$$

where A is the horizontal area bounded by the curve $\mathbf{l}$. Applying the vector identity $\nabla \times (\alpha\, \nabla p) \equiv \nabla\alpha \times \nabla p$ we see that

$$\oint -\alpha\, dp = -\int_A\!\!\int (\nabla\alpha \times \nabla p) \cdot \mathbf{k}\, dA$$

But the solenoidal term in the vorticity equation can be written

$$-\left(\frac{\partial\alpha}{\partial x}\frac{\partial p}{\partial y} - \frac{\partial\alpha}{\partial y}\frac{\partial p}{\partial x}\right) = -\mathbf{k} \cdot (\nabla\alpha \times \nabla p)$$

Comparing the right-hand sides of these two expressions, we see that the solenoidal term in the vorticity equation is just the limit of the solenoidal term in the circulation theorem divided by the area when the area goes to zero.

4.4.2 The Vorticity Equation in Isobaric Coordinates

A somewhat simpler form of the vertical component vorticity equation arises when the motion is referred to the isobaric coordinate system. Starting with the horizontal momentum equation (3.2) we can derive the vorticity equation by operating on (3.2) with the operator $\mathbf{k} \cdot \nabla \times$, where ∇ now indicates the horizontal gradient on a surface of constant pressure. However to facilitate this process it is desirable to first use the vector identity

$$(\mathbf{V} \cdot \nabla)\mathbf{V} = \nabla\left(\frac{\mathbf{V} \cdot \mathbf{V}}{2}\right) + \mathbf{k} \times \mathbf{V}\zeta, \tag{4.18}$$

where $\zeta = \mathbf{k} \cdot (\nabla \times \mathbf{V})$, to rewrite (3.2) as

$$\frac{\partial \mathbf{V}}{\partial t} = -\nabla\left(\frac{\mathbf{V} \cdot \mathbf{V}}{2} + \Phi\right) - \mathbf{k} \times \mathbf{V}\zeta - \omega\frac{\partial \mathbf{V}}{\partial p} + f\mathbf{V} \times \mathbf{k} \tag{4.19}$$

We now apply the operator $\mathbf{k} \cdot \nabla \times$ to (4.19). Using the facts that for any scalar A, $\nabla \times \nabla A = 0$ and for any vectors $\mathbf{a}$, $\mathbf{b}$

$$\nabla \times (\mathbf{a} \times \mathbf{b}) = (\nabla \cdot \mathbf{b})\mathbf{a} - (\mathbf{a} \cdot \nabla)\mathbf{b} - (\nabla \cdot \mathbf{a})\mathbf{b} + (\mathbf{b} \cdot \nabla)\mathbf{a} \tag{4.20}$$

we can eliminate the first term on the right and simplify the second term so that the resulting vorticity equation becomes

$$\frac{\partial \zeta}{\partial t} = -\mathbf{V} \cdot \nabla(\zeta + f) - \omega\frac{\partial \zeta}{\partial p} - (\zeta + f)\,\nabla \cdot \mathbf{V} + \mathbf{k} \cdot \left(\frac{\partial \mathbf{V}}{\partial p} \times \nabla\omega\right) \tag{4.21}$$

Comparing (4.17) and (4.21) we see that in the isobaric system there is no vorticity generation by pressure–density solenoids. This difference arises because in the isobaric system the vertical component of vorticity is $\zeta = (\partial v/\partial x - \partial u/\partial y)_p$, while in height coordinates it is $\zeta = (\partial v/\partial x - \partial u/\partial y)_z$. In practice the difference is generally unimportant because as we shall see in the next section the solenoidal term is usually sufficiently small so that it can be neglected for synoptic scale motions.

4.5 Scale Analysis of the Vorticity Equation

In Section 2.4 we showed that the equations of motion could be simplified for synoptic scale motions by evaluating the order of magnitude of various terms. The same technique can also be applied to the vorticity equation. To scale the vorticity equation we choose characteristic scales for the field variables based on typical observed magnitudes for synoptic scale motions as follows:

$U \sim 10 \quad \text{m s}^{-1}$	horizontal velocity
$W \sim 1 \quad \text{cm s}^{-1}$	vertical velocity
$L \sim 10^6 \quad \text{m}$	length scale
$H \sim 10^4 \quad \text{m}$	depth scale
$\delta p \sim 1 \quad \text{kPa}$	horizontal pressure fluctuations
$\delta\rho/\rho \sim 10^{-2}$	fractional density fluctuation
$L/U \sim 10^5 \quad \text{s}$	time scale
$f_0 \sim 10^{-4} \quad \text{s}^{-1}$	Coriolis parameter
$df/dy \equiv \beta \sim 10^{-11} \quad \text{m}^{-1}\,\text{s}^{-1}$	"beta" parameter
$\rho \sim 1 \quad \text{kg m}^{-3}$	mean density

Again we have chosen an advective time scale because the vorticity pattern, like the pressure pattern, tends to move at a speed comparable to the horizontal wind speed. Using these scales to evaluate the magnitude of the terms in (4.16) we note first that

$$\zeta = \frac{\partial v}{\partial x} - \frac{\partial u}{\partial y} \lesssim \frac{U}{L} \sim 10^{-5} \quad \text{s}^{-1}$$

where $\lesssim$ means less than or equal to in order of magnitude. Comparing with the Coriolis parameter we see that $\zeta/f_0 \lesssim U/(f_0 L) \equiv Ro \sim 10^{-1}$. Thus, for midlatitude synoptic scale systems the relative vorticity is small (order Rossby number) compared to the earth's vorticity. Hence, ζ may be neglected compared to f in the divergence term:

$$(\zeta + f)\left(\frac{\partial u}{\partial x} + \frac{\partial v}{\partial y}\right) \approx f\left(\frac{\partial u}{\partial x} + \frac{\partial v}{\partial y}\right)$$

The magnitudes of the various terms in (4.16) are as follows:

$$\frac{\partial\zeta}{\partial t}, u\frac{\partial\zeta}{\partial x}, v\frac{\partial\zeta}{\partial y} \sim \frac{U^2}{L^2} \sim 10^{-10} \quad \mathrm{s}^{-2}$$

$$w\frac{\partial\zeta}{\partial z} \sim \frac{UW}{LH} \sim 10^{-11} \quad \mathrm{s}^{-2}$$

$$v\frac{df}{dy} \sim U\beta \sim 10^{-10} \quad \mathrm{s}^{-2}$$

$$f\left(\frac{\partial u}{\partial x} + \frac{\partial v}{\partial y}\right) \lesssim \frac{f_0 U}{L} \sim 10^{-9} \quad \mathrm{s}^{-2}$$

$$\left(\frac{\partial w}{\partial x}\frac{\partial v}{\partial z} - \frac{\partial w}{\partial y}\frac{\partial u}{\partial z}\right) \lesssim \frac{WU}{HL} \sim 10^{-11} \quad \mathrm{s}^{-2}$$

$$\frac{1}{\rho^2}\left(\frac{\partial\rho}{\partial x}\frac{\partial p}{\partial y} - \frac{\partial\rho}{\partial y}\frac{\partial p}{\partial x}\right) \lesssim \frac{\delta\rho\,\delta p}{\rho^2 L^2} \sim 10^{-11} \quad \mathrm{s}^{-2}$$

The symbol $\lesssim$ is used for the last three terms because in each case it is possible that the two parts of the expression might partially cancel so that the actual magnitude is less than indicated. In fact, comparing the magnitudes of the various terms we see that this must be true for the divergence term, because if $\partial u/\partial x$ and $\partial v/\partial y$ were not nearly equal and opposite, the divergence term would be an order of magnitude greater than any other term and the equation could not be satisfied. Therefore, scale analysis of the vorticity equation indicates that synoptic scale motions must be quasi-nondivergent. In order that the divergence term be small enough to be balanced by the vorticity advection terms we see that

$$\left(\frac{\partial u}{\partial x} + \frac{\partial v}{\partial y}\right) \lesssim 10^{-6} \quad \mathrm{s}^{-1}$$

so that the horizontal divergence must be small compared to the vorticity in synoptic scale systems. From the above scalings and the definition of the Rossby number we see that

$$\left|\left(\frac{\partial u}{\partial x} + \frac{\partial v}{\partial y}\right)\Big/ f_0\right| \lesssim Ro^2$$

and

$$\left|\left(\frac{\partial u}{\partial x} + \frac{\partial v}{\partial y}\right)\Big/ \zeta\right| \lesssim Ro$$

Thus the ratio of the horizontal divergence to the relative vorticity is the same magnitude as the ratio of relative vorticity to planetary vorticity.

Retaining now only the terms of order $10^{-10}\ \mathrm{s}^{-2}$ in the vorticity equation, we obtain as the first approximation for synoptic scale motions

$$\frac{d_h(\zeta + f)}{dt} = -f\left(\frac{\partial u}{\partial x} + \frac{\partial v}{\partial y}\right) \tag{4.22}$$

where

$$\frac{d_h}{dt} = \frac{\partial}{\partial t} + u\frac{\partial}{\partial x} + v\frac{\partial}{\partial y}$$

Equation (4.22) states that as a first approximation the change of absolute vorticity following the horizontal motion on the synoptic scale is entirely due to the divergence effect. This approximation does not remain valid, however, in the vicinity of atmospheric fronts. The horizontal scale of variation in frontal zones is only ~ 100 km, and for this scale in which $W \sim 10\ \mathrm{cm\ s^{-1}}$ the vertical advection, tilting, and solenoidal terms all may become as large as the divergence term.

For the case of an incompressible barotropic fluid, the potential vorticity theorem can be derived from the approximate vorticity equation quite easily. Now for a homogeneous incompressible fluid the continuity equation takes the form

$$\frac{\partial u}{\partial x} + \frac{\partial v}{\partial y} = -\frac{\partial w}{\partial z}$$

so that the vorticity equation (4.22) may be written

$$\frac{d_h(\zeta + f)}{dt} = (\zeta + f)\frac{\partial w}{\partial z} \tag{4.23}$$

(we have here retained the relative vorticity in the divergence term for reasons which will become apparent). We showed in Section 3.4 that in a barotropic fluid the thermal wind vanishes so that the geostrophic wind is independent of height. Letting the vorticity in (4.23) be approximated by the geostrophic vorticity ζ_g and the wind by the geostrophic wind (u_g, v_g), we can integrate vertically from $z = D_1$ to $z = D_2$ to get

$$(D_2 - D_1)\frac{d_h(\zeta_g + f)}{dt} = (\zeta_g + f)[w(D_2) - w(D_1)]$$

But since $w \equiv dz/dt$ we can write

$$w(D_2) = \frac{dD_2}{dt} \qquad \text{and} \qquad w(D_1) = \frac{dD_1}{dt}$$

so that letting $H = D_2 - D_1$ be the depth of the fluid column we get

$$\frac{1}{(\zeta_g + f)} \frac{d_h(\zeta_g + f)}{dt} = \frac{1}{H} \frac{d_h(H)}{dt}$$

or

$$\frac{d_h}{dt}[\ln(\zeta_g + f)] = \frac{d_h}{dt}[\ln H]$$

which implies that

$$\frac{d_h}{dt}\left(\frac{\zeta_g + f}{H}\right) = 0 \tag{4.24}$$

which is just the potential vorticity theorem for a barotropic fluid.

If the flow is purely horizontal ($w = 0$), then (4.23) reduces to the *barotropic vorticity equation*

$$\frac{d_h(\zeta + f)}{dt} = 0 \tag{4.25}$$

which states that the absolute vorticity is conserved following the motion. (Note that in this case the flow need not be geostrophic.)

Problems

1. What is the circulation about a square of 1000 km on a side for an easterly (that is, westward flowing) current which decreases in magnitude toward the north at a rate of 10 m s^{-1}/500 km? What is the mean relative vorticity in the square?
2. A cylindrical column of air at 30°N with radius 100 km expands to twice its original radius. If the air is initially at rest, what is the mean tangential velocity at the perimeter after expansion?
3. An air parcel at 30°N moves northward conserving absolute vorticity. If its initial relative vorticity is $5 \times 10^{-5}\ s^{-1}$, what is its relative vorticity upon reaching 90°N?
4. An air column at 60°N with $\zeta = 0$ initially stretches from the surface to a fixed tropopause at 10-km height. If the air column moves until it is over a mountain barrier 2.5 km high at 45°N, what is its absolute vorticity and relative vorticity as it passes the mountain top?
5. Find the average vorticity within a cylindrical annulus of inner radius 200 km and outer radius 400 km if the tangential velocity distribution

is given by $V = 10^6/r$ m s^{-1} where r is in meters. What is the average vorticity within the inner circle of radius 200 km?

6. Compute the rate of change of circulation about a square in the x,y plane with sides of 1000-km length if temperature increases eastward at a rate of 1°C/200 km and pressure increases northward at a rate of 1 mb/200 km. The pressure at the origin is 1000 mb.

7. Verify the identity (4.18) by expanding the vectors in cartesian components.

***8.** Derive a formula for the dependence of depth on radius for an incompressible fluid in solid-body rotation in a cylindrical tank. Let H be the depth at the center of the tank, Ω the angular velocity of the tank, and a the radius of the tank.

9. By how much does the relative vorticity change for a column of fluid in a rotating cylinder if the column is moved from the center of the tank to a distance 50 cm from the center? The tank is rotating at the rate of 20 revolutions per minute, the depth of the fluid at the center is 10 cm, and the fluid is initially in solid-body rotation.

10. A cyclonic vortex is in cyclostrophic balance with a tangential velocity profile given by the expression $V = V_0(r/r_0)^n$ where V_0 is the tangential velocity component at the distance r_0 from the vortex center. Compute the circulation about a streamline at radius r, the vorticity at radius r, and the pressure at radius r. (Let p_0 be the pressure at r_0 and assume that density is a constant).

***11.** A westerly zonal flow at 45° is forced to rise adiabatically over a north–south oriented mountain barrier. Before striking the mountain the westerly wind increases linearly toward the south at a rate of 10 m s^{-1}/1000 km. The crest of the mountain range is at the 80-kPa (800-mb) level and the tropopause, located at 30 kPa (300 mb), remains undisturbed by the forced ascent of the air. The surface pressure to the west of the mountain barrier is 100 kPa (1000 mb). What is the initial relative vorticity of the air? What is its relative vorticity when it reaches the crest if it is deflected 5° latitude toward the south during the forced ascent? If the current assumes a uniform speed of 20 m s^{-1} during its ascent to the crest, what is the radius of curvature of the streamlines at the crest?

12. A cylindrical vessel of radius a and constant depth H rotating at an angular velocity $\mathbf{\Omega}$ about its vertical axis of symmetry is filled with a homogeneous, incompressible fluid which is initially at rest with respect to the vessel. A volume of fluid V is then withdrawn through a point sink at the center of the cylinder, thus creating a vortex. Neglecting friction, derive an expression for the resulting relative azimuthal velocity as a function of radius i.e., the velocity in a coordinate system rotating with the tank. (Assume that the motion is independent of depth and that $V \ll \pi a^2 H$). Also compute the relative vorticity and the relative circulation.

13. (a) How far must a zonal ring of air initially at rest with respect to the earth's surface at 60° latitude and 100-km height be displaced latitudinally in order to acquire an easterly (east to west) component of 10 m s^{-1} with respect to the earth's surface?
(b) To what height must it be displaced vertically in order to acquire the same velocity? Assume a frictionless atmosphere.

***14.** The horizontal motion within a cylindrical annulus with permeable walls of inner radius 10 cm, outer radius 20 cm, and 10-cm depth is independent of height and azimuth and is represented by the expressions

$$u = 7 - 0.2r, \qquad v = 40 + 2r$$

where u and v are the radial and tangential velocity components in cm s^{-1}, positive outward and counterclockwise, respectively, and r is distance from the center of the annulus in cm. Assuming an incompressible fluid, find
(a) the circulation about the annular ring
(b) the average vorticity within the annular ring
(c) the average divergence within the annular ring, and
(d) the average vertical velocity at the top of the annulus if it is zero at the base.

Suggested References

Batchelor, *An Introduction to Fluid Dynamics*, has an extensive treatment of the general applications of the concepts of vorticity and circulation in fluid dynamics. Of special interest is the section on motion in a thin layer on a rotating sphere.

Haltiner, *Numerical Weather Prediction*, presents a more complete scale analysis of the vorticity equation.

Chapter

5 The Planetary Boundary Layer

In our discussions of large-scale flow fields in Chapters 3 and 4 we have in all cases neglected any effects of momentum or heat diffusion. The neglect of frictional forces due to molecular viscosity and heating due to molecular diffusion can be readily justified on the basis of scale analysis. However, near the ground strong wind shears and surface heating continually lead to the development of turbulent eddies. These eddies are very effective mixing agents which serve to transfer heat and water vapor away from the earth's surface, and momentum towards the earth's surface, at a rate many orders of magnitude faster than the mixing rate for molecular diffusion. This turbulent transport has an appreciable influence on the motion throughout a layer, referred to as the *planetary boundary layer*, whose depth may range from about 30 m in conditions of large static stability to more than 3 km in highly convective conditions. For average midlatitude conditions the planetary boundary layer extends through the lowest kilometer of the atmosphere, and thus contains about 10% of the mass of the atmosphere.

For a statically stable atmosphere, the turbulent mixing in the boundary layer is generated primarily by dynamical instability due to the strong vertical shear of the wind near the ground. Thus, the turbulent mixing is mechanically driven—not thermally driven. In this case it is useful to divide the planetary boundary layer into two sublayers, the *surface layer* and the

Ekman layer. The surface layer, which is confined to the lowest few meters of the atmosphere, is a layer in which the velocity profile is adjusted so that the horizontal frictional stress is nearly independent of height. The Ekman layer, which extends from the top of the surface layer to a height of about 1 km, is a layer in which there is a three-way balance between the Coriolis force, the pressure gradient force, and the viscous stress. Even though the ideal Ekman layer wind profile is seldom, if ever, observed, the Ekman layer does provide a qualitatively valid description of those dynamical properties of the boundary layer flow which are important for the analysis of midlatitude synoptic scale motions. It should be mentioned, however, that in neutral or unstably stratified conditions the structure of the boundary layer is generally determined by convectively driven turbulent mixing. Such conditions are typical throughout most of the tropics. However, in the present discussion we will restrict our attention to the midlatitude mechanically driven Ekman layer.

Because the planetary boundary layer is a turbulent layer, a rigorous mathematical theory for the structure of the velocity field in the layer is not possible. In order to satisfactorily study the structure of the planetary boundary layer we would require detailed knowledge of the structure and amplitude of the turbulent eddies which are responsible for the vertical momentum transport. Even a cursory examination of this subject is outside the scope of this text. In the present chapter it will be assumed that the frictional force in turbulent flow may be represented in the same manner as done in laminar flow by introducing an eddy viscosity coefficient.

5.1 The Mixing Length Theory

The concept of an eddy viscosity coefficient was mentioned briefly in Section 1.4.3. In this section we will discuss in more detail the heuristic argument of the famous fluid dynamicist L. Prandtl which provides some theoretical basis for estimating the magnitude of the eddy viscosity. Prandtl's basic idea was that the momentum transport of the small-scale eddy motions may be parameterized in terms of the large-scale mean flow.

In order to understand the basis of this parameterization we must first consider the derivation of mean flow equations for a turbulent fluid. (In fact we have been dealing with such equations throughout this book; however, we have not previously explicitly considered the partitioning of the flow between turbulent eddies and mean fields.) In a turbulent fluid the velocity measured at a point generally fluctuates rapidly in time as eddies of various scales pass the point. In order that our velocity measurements be truly representative of the large-scale flow it is thus necessary to average the flow over an interval of time long enough to average out the eddy fluctuations but still short enough to preserve the trends in the large-scale flow field. To do this we use angle brackets to define the mean velocity $\langle \mathbf{U} \rangle$ as the average in time at a given point. The instantaneous velocity is then

$$\mathbf{U} = \langle \mathbf{U} \rangle + \mathbf{U}'$$

where $\mathbf{U}'$ designates the deviation from the mean at any time. $\mathbf{U}'$ is thus associated with the turbulent eddies.

We now apply this averaging scheme to the horizontal equations of motion. We have previously written these equations in the approximate form:

$$\frac{\partial u}{\partial t} + u\frac{\partial u}{\partial x} + v\frac{\partial u}{\partial y} + w\frac{\partial u}{\partial z} - fv = -\frac{1}{\rho}\frac{\partial p}{\partial x} \tag{5.1}$$

$$\frac{\partial v}{\partial t} + u\frac{\partial v}{\partial x} + v\frac{\partial v}{\partial y} + w\frac{\partial v}{\partial z} + fu = -\frac{1}{\rho}\frac{\partial p}{\partial y} \tag{5.2}$$

Using the continuity equation,

$$\frac{\partial \rho}{\partial t} + \frac{\partial}{\partial x}(\rho u) + \frac{\partial}{\partial y}(\rho v) + \frac{\partial}{\partial z}(\rho w) = 0 \tag{5.3}$$

Eqs. (5.1) and (5.2) can be transformed to a more convenient form. If we multiply (5.1) by ρ and (5.3) by u, then add the resulting equations, we obtain the *flux* form of the x-momentum equation:

$$\frac{\partial}{\partial t}(\rho u) + \frac{\partial}{\partial x}(\rho u^2) + \frac{\partial}{\partial y}(\rho uv) + \frac{\partial}{\partial z}(\rho uw) - f\rho v = -\frac{\partial p}{\partial x} \tag{5.4}$$

An analogous operation gives the flux form of the y-momentum equation:

$$\frac{\partial}{\partial t}(\rho v) + \frac{\partial}{\partial x}(\rho uv) + \frac{\partial}{\partial y}(\rho v^2) + \frac{\partial}{\partial z}(\rho vw) + f\rho u = -\frac{\partial p}{\partial y} \tag{5.5}$$

We next substitute in place of u, v, and w in (5.4) and (5.5), letting

$$u = \langle u \rangle + u', \qquad v = \langle v \rangle + v', \qquad w = \langle w \rangle + w' \tag{5.6}$$

If we neglect the small fluctuations in density associated with the turbulence,[1] the resulting equations can be averaged in time to get a simple partition of the flow between the mean flow and turbulent fields. Thus, for example, the term ρuw becomes

$$\langle \rho uw \rangle = \rho\langle(\langle u \rangle + u')(\langle w \rangle + w')\rangle = \rho\langle u \rangle\langle w \rangle + \rho\langle u'w' \rangle$$

[1] More precisely we assume that $|u'/\langle u \rangle|, |v'/\langle v \rangle|, |w'/\langle w \rangle| \gg |\rho'/\langle \rho \rangle|$. This assumption is quite valid in the planetary boundary layer.

where terms like $\langle\langle u\rangle w'\rangle$ and $\langle u'\langle w\rangle\rangle$ vanish because $\langle u\rangle$ and $\langle w\rangle$ are constant over the averaging interval so that, for example, $\langle\langle u\rangle w'\rangle = \langle u\rangle\langle w'\rangle = 0$ since $\langle w'\rangle = 0$. Carrying out this averaging process for all terms in (5.4) and (5.5) we obtain with the aid of the averaged continuity equation

$$\frac{\partial\langle u\rangle}{\partial t} + \langle u\rangle\frac{\partial\langle u\rangle}{\partial x} + \langle v\rangle\frac{\partial\langle v\rangle}{\partial y} + \langle w\rangle\frac{\partial\langle u\rangle}{\partial z} - f\langle v\rangle$$

$$= -\frac{1}{\rho}\frac{\partial\langle p\rangle}{\partial x} - \frac{1}{\rho}\left[\frac{\partial}{\partial x}(\rho\langle u'u'\rangle) + \frac{\partial}{\partial y}(\rho\langle u'v'\rangle) + \frac{\partial}{\partial z}(\rho\langle u'w'\rangle)\right] \quad (5.7)$$

$$\frac{\partial\langle v\rangle}{\partial t} + \langle u\rangle\frac{\partial\langle v\rangle}{\partial x} + \langle v\rangle\frac{\partial\langle v\rangle}{\partial y} + \langle w\rangle\frac{\partial\langle v\rangle}{\partial z} + f\langle u\rangle$$

$$= -\frac{1}{\rho}\frac{\partial\langle p\rangle}{\partial y} - \frac{1}{\rho}\left[\frac{\partial}{\partial x}(\rho\langle u'v'\rangle) + \frac{\partial}{\partial y}(\rho\langle v'v'\rangle) + \frac{\partial}{\partial z}(\rho\langle v'w'\rangle)\right] \quad (5.8)$$

The terms in square brackets on the right-hand side in (5.7) and (5.8) which depend on the turbulent fluctuations are called the *eddy stress terms.*

In the mixing length theory these eddy stress terms are parameterized in terms of the mean field variables by assuming that the eddy stresses are proportional to the gradient of the mean wind. Since we are here primarily concerned with the planetary boundary layer where vertical gradients are much larger than the horizontal gradients, the discussion will be limited to the vertical eddy stress terms.

According to the mixing length hypothesis a parcel of fluid which is displaced vertically will carry the mean horizontal velocity of its original level for a characteristic distance l' analogous to the mean free path in molecular viscosity. This displacement will create a turbulent fluctuation whose magnitude will depend upon l' and the shear of the mean velocity. Thus, for example,

$$u' = -l'\frac{\partial\langle u\rangle}{\partial z}$$

where it must be understood that $l' > 0$ for *upward* parcel displacement and $l' < 0$ for *downward* parcel displacement. The vertical eddy stress $-\rho\langle u'w'\rangle$ can then be written as

$$-\rho\langle u'w'\rangle = \rho\langle w'l'\rangle\frac{\partial\langle u\rangle}{\partial z} \quad (5.9)$$

In order to estimate w' in terms of the mean fields we assume that the vertical stability of the atmosphere is nearly neutral so that buoyancy effects are small (see Section 2.7.3). The horizontal scale of the eddies should then be comparable to the vertical scale so that $|w'| \sim |\mathbf{V}'|$ and we can set

$$w' = l' \left| \frac{\partial \langle \mathbf{V} \rangle}{\partial z} \right|$$

where $\mathbf{V}'$ and $\langle \mathbf{V} \rangle$ designate the turbulent and mean parts of the horizontal velocity field, respectively. Here the absolute value of the mean velocity gradient is needed because if $l' > 0$ we must have $w' > 0$ (that is, upward parcel displacement by the eddy fluctuations). Thus the eddy stress can be written

$$-\rho \langle u'w' \rangle = \rho \langle l'^2 \rangle \left| \frac{\partial \langle \mathbf{V} \rangle}{\partial z} \right| \frac{\partial \langle u \rangle}{\partial z} = A_z \frac{\partial \langle u \rangle}{\partial z} \tag{5.10}$$

where $A_z \equiv \rho \langle l'^2 \rangle |\partial \langle \mathbf{V} \rangle / \partial z|$ is called the *eddy exchange coefficient*. Similarly, we may show that the vertical eddy stress due to motion in the y direction may be written

$$-\rho \langle v'w' \rangle = A_z \frac{\partial \langle v \rangle}{\partial z}$$

In the planetary boundary layer it is usually assumed that A_z depends only on the distance from the surface.

5.2 Planetary Boundary Layer Equations

If the horizontal eddy stress terms in (5.7) and (5.8) are neglected and the vertical stresses are rewritten by letting

$$\tau_x = -\rho \langle u'w' \rangle, \qquad \tau_y = -\rho \langle v'w' \rangle$$

the mean flow momentum equations become

$$\frac{\partial u}{\partial t} + u \frac{\partial u}{\partial x} + v \frac{\partial u}{\partial y} + w \frac{\partial u}{\partial z} - fv = -\frac{1}{\rho} \frac{\partial p}{\partial x} + \frac{1}{\rho} \frac{\partial \tau_x}{\partial z} \tag{5.11}$$

$$\frac{\partial v}{\partial t} + u \frac{\partial v}{\partial x} + v \frac{\partial v}{\partial y} + w \frac{\partial v}{\partial z} + fu = -\frac{1}{\rho} \frac{\partial p}{\partial y} + \frac{1}{\rho} \frac{\partial \tau_y}{\partial z} \tag{5.12}$$

where we have omitted the angle brackets since all dependent variables are here mean flow variables. For midlatitude synoptic scale motions, we showed in Section 2.4 that the acceleration terms du/dt and dv/dt in (5.11) and (5.12) were small compared to the Coriolis force and pressure gradient force terms. Outside the boundary layer the first approximation was then simply geostrophic balance. In the boundary layer the acceleration terms are still small compared to the Coriolis and pressure gradient terms. Thus to a first approximation the planetary boundary layer equations express a three-way balance between the Coriolis force, the pressure gradient force, and the viscosity force:

$$-fv = -\frac{1}{\rho}\frac{\partial p}{\partial x} + \frac{\partial}{\partial z}\left(\frac{\tau_x}{\rho}\right) \tag{5.13}$$

$$+fu = -\frac{1}{\rho}\frac{\partial p}{\partial y} + \frac{\partial}{\partial z}\left(\frac{\tau_y}{\rho}\right) \tag{5.14}$$

where we have here assumed that the vertical variation of ρ is negligible in the boundary layer.

5.2.1 The Surface Layer

For simplicity we assume that the flow close to the ground is directed parallel to the x axis. The surface stress divided by density can then be written in terms of the friction velocity u_* which is defined by the identity $u_*^2 \equiv (\tau_x/\rho)_s$, where the subscript s indicates the value at the surface. Measurements indicate that the surface stress in the atmosphere has a typical magnitude of $\tau_x \sim 0.1\ \mathrm{N\,m^{-2}}$. Hence, $(\tau_x/\rho)_s \sim 0.1\ \mathrm{m^2\,s^{-2}}$ and $u_* \sim 0.3\ \mathrm{m\,s^{-1}}$.

According to the scale analysis in Section 2.4 the Coriolis and pressure gradient force terms in (5.13) have magnitudes of about $10^{-3}\ \mathrm{m\,s^{-2}}$ in midlatitudes. Thus if these terms are to balance the eddy stress it is necessary that

$$\frac{\Delta(\tau_x/\rho)}{\Delta z} \lesssim 10^{-3} \quad \mathrm{m\,s^{-2}}$$

Thus, for $\Delta z = 10$ m, $\Delta(\tau_x/\rho) \lesssim 10^{-2}\ \mathrm{m^2\,s^{-2}}$. Hence, the change in the vertical eddy stress in the lowest 10 m of the atmosphere is less than 10% of the surface stress. To a first approximation it is then permissible to assume that in the lowest several meters of the atmosphere the stress remains constant at its surface value:

$$\tau_x/\rho = \frac{A_z}{\rho}\frac{\partial u}{\partial z} = u_*^2 \tag{5.15}$$

where we have parameterized the surface stress in terms of the eddy exchange coefficient A_z defined in (5.10). In deriving A_z we assumed that the horizontal and vertical scales of the eddies were approximately equal. Near the surface the vertical eddy scale is limited by the distance to the surface. Thus, a logical choice for the mixing length is $l' = kz$ where k is a constant. Hence we have

$A_z = \rho(kz)^2 |\partial u/\partial z|$. Substituting this expression for A_z into (5.15) and taking the square root of the result we find

$$\frac{\partial u}{\partial z} = \frac{u_*}{kz}$$

Integrating with respect to z yields the *logarithmic wind profile*

$$u = \frac{u_*}{k} \ln\left(\frac{z}{z_0}\right) \tag{5.16}$$

where z_0, the *roughness length*, is a constant of integration chosen so that $u = 0$ at $z = z_0$. The constant k in (5.16) is a universal constant called *von Karman's constant* which has an experimentally determined value of $k \simeq 0.4$. The roughness length z_0 varies widely depending on the physical characteristics of the surface. For grassy fields typical values are in the range of 1–4 cm. Although a number of assumptions are required in the derivation of (5.16) many experimental programs have shown that the logarithmic profile provides a generally satisfactory fit to observed wind profiles in the surface layer.

5.2.2 The Ekman Layer

Above the surface layer the structure of the boundary layer is determined by (5.13) and (5.14). Observations indicate that in this region the characteristic scale of the turbulent eddies is not simply proportional to the distance from the ground, but tends rather to be nearly constant with height. Thus, in order to simplify the discussion we now introduce an *eddy* viscosity coefficient $K \equiv A_z/\rho$ which we assume to be independent of height. In this case the turbulent Ekman layer equations becomes exactly analogous to the laminar flow equations but with the eddy viscosity K replacing the molecular kinematic viscosity coefficient.

Introducing the eddy viscosity parameterization (5.15) into (5.13) and an analogous expression into (5.14) we obtain as the approximate equations for the planetary boundary layer

$$K \frac{\partial^2 u}{\partial z^2} + f(v - v_g) = 0 \tag{5.17}$$

$$K \frac{\partial^2 v}{\partial z^2} - f(u - u_g) = 0 \tag{5.18}$$

where we have used the definitions

$$u_g \equiv -\frac{1}{\rho f}\frac{\partial p}{\partial y}, \qquad v_g \equiv \frac{1}{\rho f}\frac{\partial p}{\partial x}.$$

The Ekman layer equations (5.17) and (5.18) can be solved to determine the departure of the wind field from geostrophic balance in the boundary layer. In order to keep the analysis as simple as possible we will for the present assume that (5.17) and (5.18) with constant K apply throughout the depth of the boundary layer (i.e., we will neglect the surface layer). The boundary conditions on u and v then require that both horizontal velocity components vanish at the ground and approach their geostrophic values far from the ground:

$$\begin{aligned} u &= 0, \qquad v = 0 \qquad \text{at } z = 0 \\ u &\to u_g, \qquad v \to v_g \qquad \text{as } z \to \infty \end{aligned} \tag{5.19}$$

To solve (5.17)–(5.18) it is convenient to first multiply (5.18) by $i \equiv \sqrt{-1}$ and add the result to (5.17) to obtain a second-order equation in the *complex velocity* $u + iv$:

$$K \frac{\partial^2}{\partial z^2}(u + iv) - if(u + iv) = -if(u_g + iv_g) \tag{5.20}$$

For simplicity, we assume that the geostrophic wind is independent of height and that the flow is oriented so that the geostrophic wind is entirely in the zonal direction ($v_g = 0$). Then the general solution of (5.20) may be written

$$u + iv = A \exp[(if/K)^{1/2}z] + B \exp[-(if/K)^{1/2}z] + u_g$$

It can be shown that $\sqrt{i} = (1 + i)/\sqrt{2}$. Using this relationship and applying the boundary conditions (5.19), we find that for the Northern Hemisphere ($f > 0$) $A = 0$ and $B = -u_g$. Thus

$$u + iv = -u_g e^{-\gamma(1+i)z} + u_g$$

where

$$\gamma = (f/2K)^{1/2}$$

Applying the Euler formula $e^{-i\theta} = \cos\theta - i \sin\theta$ and separating the real from the imaginary part we obtain for the Northern Hemisphere

$$\begin{aligned} u &= u_g(1 - e^{-\gamma z} \cos \gamma z) \\ v &= u_g e^{-\gamma z} \sin \gamma z \end{aligned} \tag{5.21}$$

This solution is the famous *Ekman spiral* named for the Swedish oceanographer V. W. Ekman who first derived an analogous solution for the surface wind drift current in the ocean. The structure of the solution (5.21) is best illustrated by a hodograph as shown in Fig. 5.1. In this figure the components of the wind velocity are plotted as a function of height. Thus the points on the curve in Fig. 5.1 correspond to u and v in (5.21) for values of γz increasing as

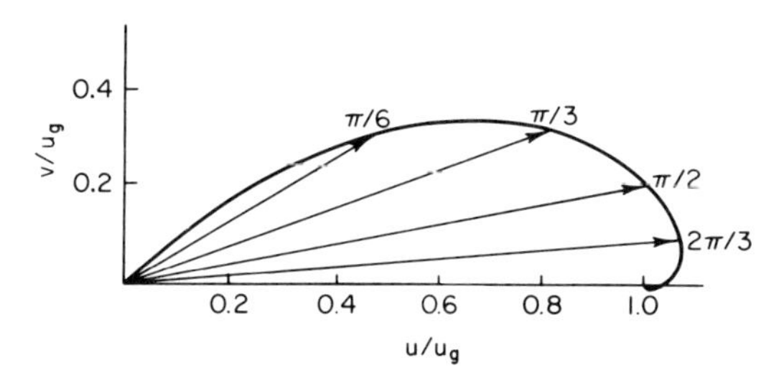

Fig. 5.1 Hodograph of the Ekman spiral solution. Points marked on the curve are values of γz, which is a nondimensional measure of height.

one moves away from the origin along the spiral. It can be seen from Fig. 5.1 that when $z = \pi/\gamma$, the wind is parallel to the geostrophic wind although slightly greater than geostrophic in magnitude. It is conventional to designate this level as the top of the planetary boundary layer.[2] Thus the depth of the *Ekman layer* is

$$De \equiv \pi/\gamma \tag{5.22}$$

Observations indicate that the wind approaches its geostrophic value at about one kilometer above the ground. Substituting $De = 1$ km and $f = 10^{-4}$ into (5.22), we can solve for the eddy viscosity K. The result is that $K \simeq 5\ \text{m}^2\ \text{s}^{-1}$. Referring back to (5.10) we see that $K \simeq \langle l'^2 \rangle |\partial \langle \mathbf{V} \rangle / \partial z|$. Thus if the mean wind shear is of the order of $5\ \text{m s}^{-1}\ \text{km}^{-1}$, the mixing length l' must be about 30 m in order that $K \simeq 5\ \text{m}^2\ \text{s}^{-1}$. This mixing length is small compared to the depth of the boundary layer, as it should be if the mixing length concept is to be useful.

Qualitatively the most striking feature of the Ekman layer solution is the fact that the wind in the boundary layer has a component directed toward lower pressure. This is a direct result of the three-way balance between the pressure gradient force, the Coriolis force, and the viscous force. This balance is illustrated in Fig. 5.2 for a level well within the boundary layer. Since the Coriolis force is always normal to the velocity and the frictional force is mainly a retarding force, their sum can only exactly balance the pressure gradient force if the wind is directed toward lower pressure (that is, to the left of the geostrophic wind in the Northern Hemisphere). Furthermore, it is easy to see that as the frictional force becomes increasingly important the cross isobar angle must increase.

The ideal Ekman layer discussed here is rarely, if ever, observed in the atmospheric boundary layer partly because, as mentioned above, the eddy mixing coefficient must vary rapidly with height near the ground. In other

[2] It is often referred to as the *gradient wind level* by synoptic meteorologists.

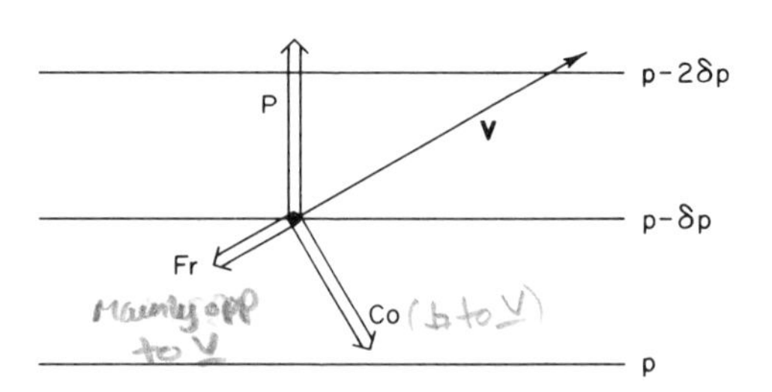

Fig. 5.2 Balance of forces in the Ekman layer. *P* designates the pressure gradient force, *Co* the Coriolis force, *Fr* the frictional force.

words, the Ekman layer solution is applicable only above the surface layer. Thus, a more satisfactory representation for the planetary boundary layer can be obtained by combining the logarithmic surface layer profile with the Ekman spiral. We again treat the eddy viscosity coefficient as a constant but now apply (5.20) only to the region above the surface layer. Thus, in place of the lower boundary condition $u + iv = 0$ we must let

$$u + iv = C_0 e^{i\alpha} \tag{5.23}$$

where C_0 is the magnitude of the wind velocity at the top of the surface layer (usually taken to be the conventional anemometer level) and α is the angle between the wind and the isobars in the surface layer. To determine the constant C_0 a second condition is needed at the bottom of the spiral layer. From (5.16) we see that in the surface layer $\partial u/\partial z = u/[z \ln(z/z_0)]$, with a similar expression for the v component. Thus, matching of the Ekman layer solution to the surface solution requires that we must have at the top of the surface layer (bottom of the spiral layer)

$$u + iv = C \frac{\partial}{\partial z}(u + iv) \tag{5.24}$$

where C is a real constant.

For convenience we let $z = 0$ designate the bottom of the spiral layer. Then using (5.23) plus the condition $u + iv \to u_g$ for $z \to \infty$, the solution of (5.20) may be written as

$$u + iv = (C_0 e^{i\alpha} - u_g)e^{-(1+i)\gamma z} + u_g \tag{5.25}$$

Substituting (5.25) into the boundary condition (5.24) we find by equating the real and imaginary parts

$$C_0 \cos\alpha = \gamma C[C_0(\sin\alpha - \cos\alpha) + u_g]$$

$$C_0 \sin\alpha = \gamma C[-C_0(\sin\alpha + \cos\alpha) + u_g]$$

Eliminating C then yields

$$C_0 = u_g(\cos\alpha - \sin\alpha) \tag{5.26}$$

Thus (5.25) becomes

$$u + iv = u_g + \sqrt{2}u_g \sin\alpha\, e^{-(1+i)\gamma z + i[\alpha + (3\pi/4)]}$$

so that the spiral components in the Northern Hemisphere are

$$\begin{aligned} u &= u_g[1 - \sqrt{2}\sin\alpha\, e^{-\gamma z}\cos(\gamma z - \alpha + \pi/4)] \\ v &= u_g\sqrt{2}\sin\alpha\, e^{-\gamma z}\sin(\gamma z - \alpha + \pi/4) \end{aligned} \tag{5.27}$$

For $\alpha = \pi/4$ the modified spiral (5.27) reduces to the classical Ekman spiral (5.21). The surface wind angle α is a parameter which like the eddy viscosity coefficient should be chosen to give the best fit to observations. For typical wind profiles, such as that of Fig. 5.3, $\alpha \simeq \pi/8$.

Although the modified Ekman spiral provides a somewhat better fit to observations than the classical Ekman spiral, observed boundary layer winds generally deviate from the spiral pattern. Both transience and baroclinic effects (i.e., vertical shear of the geostrophic wind in the boundary layer) may cause deviations from the Ekman solution. But even in steady state barotropic situations the Ekman spiral is seldom observed.

It turns out that the Ekman layer wind profile is generally unstable for a neutrally buoyant atmosphere. The circulations which develop as a result of this instability have horizontal and vertical scales comparable to the depth of the boundary layer. Thus, it is not possible to parameterize them by a simple mixing length theory. However, these circulations do in general transport considerable momentum vertically. The net result is usually to decrease the angle between the boundary layer wind and the geostrophic wind. A typical observed wind hodograph is shown in Fig. 5.3. Although the

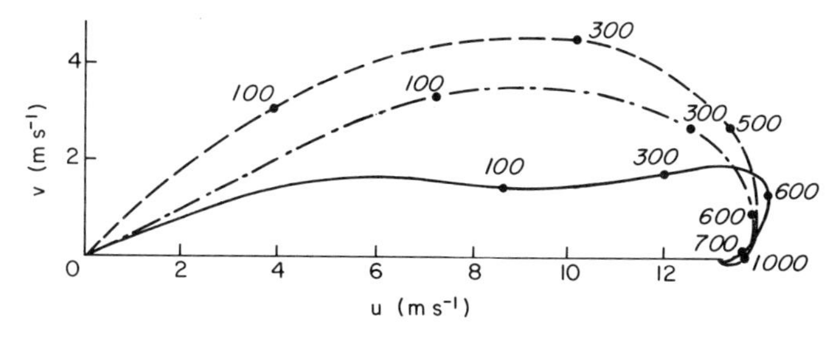

Fig. 5.3 Mean wind hodograph for Jacksonville, Florida ($\simeq 30°$N), April 4, 1968 (solid line) compared with the Ekman spiral (dashed line) and the modified Ekman spiral (dash-dot line) computed with $De \simeq 1200$ m. Heights shown in meters. (Adapted from Brown, 1970. Reproduced with permission of the American Meteorological Society.)

detailed structure is rather different from the Ekman spiral, the vertically integrated horizontal mass transport in the boundary layer is still directed toward lower pressure. And as we shall see in the next section it is this fact which is of primary importance for synoptic scale systems.

5.3 Secondary Circulations and Spin-Down

In the idealized Ekman spiral solution of (5.21) the v component of the wind multiplied by ρ gives the cross isobaric mass transport per unit area at any level in the boundary layer. Thus the net mass transport toward lower pressure in the Ekman layer for a column of unit width extending vertically through the entire layer is

$$M = \int_0^{De} \rho v \, dz = \int_0^{De} \rho u_g e^{-\pi z/De} \sin \frac{\pi z}{De} \, dz \tag{5.28}$$

Neglecting local variations of density in the boundary layer the continuity equation may be written as

$$\frac{\partial}{\partial z}(\rho w) = -\frac{\partial}{\partial x}(\rho u) - \frac{\partial}{\partial y}(\rho v) \tag{5.29}$$

Integrating (5.29) through the depth of the boundary layer we find that

$$(w\rho)_{De} = -\int_0^{De} \left[\frac{\partial}{\partial x}(\rho u) + \frac{\partial}{\partial y}(\rho v) \right] dz$$

Here we have assumed that the ground is level so that $w = 0$ at $z = 0$. Substituting from (5.21) we can rewrite this expression for the vertical mass flux at the top of the Ekman layer, where we have again assumed that $v_g = 0$ so that u_g is independent of x,

$$(\rho w)_{De} = -\frac{\partial}{\partial y} \int_0^{De} \rho u_g e^{-\pi z/De} \sin \frac{\pi z}{De} \, dz \tag{5.30}$$

Comparing (5.30) with (5.28) we see that the vertical flux at the top of the boundary layer is equal to the horizontal convergence of mass in the boundary layer, which is simply $-\partial M/\partial y$ in the above example. Noting that $-\partial u_g/\partial y \equiv \zeta_g$ is just the geostrophic vorticity in this case, we have after integrating (5.30)[3]

$$w_{De} = \zeta_g |K/2f|^{1/2} \tag{5.31}$$

where we have neglected the variation of density with height in the boundary layer and have assumed that $1 + e^{-\pi} \simeq 1$. Hence, we obtain the important

[3] The absolute value is used in (5.31) so that the formula will be valid in both hemispheres.

result that the vertical velocity at the top of the boundary layer is proportional to the geostrophic vorticity. In this way the effect of friction in the boundary layer is communicated directly to the free atmosphere through a forced *secondary circulation* rather than indirectly by the slow process of viscous diffusion. For a typical synoptic scale system with $\zeta_g \sim 10^{-5}\ \text{s}^{-1}$, $f \sim 10^{-4}\ \text{s}^{-1}$, and $De \sim 1$ km, the vertical velocity given by (5.31) is of the order of a few tenths of a centimeter per second.

An analogous secondary circulation is responsible for the decay of the circulation created when a cup of tea is stirred. Away from the boundary of the cup there is an approximate balance between the radial pressure gradient and the centrifugal force of the spinning fluid. However, near the bottom of the cup viscosity slows the motion and the centrifugal force is not sufficient to balance the radial pressure gradient. (Note that the radial pressure gradient is independent of depth since tea is an incompressible fluid.) Therefore, radial inflow takes place near the bottom of the cup. Because of this inflow the tea leaves always are observed to cluster near the center at the bottom of the cup if the tea has been stirred. By continuity the radial inflow in the bottom boundary layer requires upward motion and a slow compensating outward radial flow throughout the remaining portion of the cup. This slow outward flow approximately conserves angular momentum, and by replacing high angular momentum fluid by low angular momentum fluid serves to *spin-down* the vorticity in the cup far more rapidly than could mere molecular diffusion.

This spin-down effect is also important in the atmosphere. It is most easily illustrated in the case of a barotropic atmosphere. We showed previously in Section 4.5 that for synoptic scale motions the vorticity equation could be written approximately as

$$\frac{d}{dt}(\zeta + f) = -f\left(\frac{\partial u}{\partial x} + \frac{\partial v}{\partial y}\right) = f\frac{\partial w}{\partial z} \tag{5.32}$$

where we have neglected ζ compared to f in the divergence term. Neglecting the latitudinal variation of f we next evaluate the integral of (5.32) from the top of the boundary layer $z = De$ to the tropopause $z = H$:

$$\int_{De}^{H} \frac{d\zeta}{dt}\,dz = f \int_{w(De)}^{w(H)} dw \tag{5.33}$$

Assuming that $w = 0$ at $z = H$ and that the vorticity may be approximated by its geostrophic value (which in the barotropic case is independent of height) we obtain from (5.33)

$$\frac{d\zeta_g}{dt} = \frac{-f}{(H - De)}\,w(De)$$

Substituting from (5.31) and noting that $H \gg De$ we obtain a differential equation for the time dependence of ζ_g:

$$\frac{d\zeta_g}{dt} = -\left|\frac{fK}{2H^2}\right|^{1/2} \zeta_g \tag{5.34}$$

Equation (5.34) may be immediately integrated to give

$$\zeta_g = \zeta_g(0) \exp\{-|fK/2H^2|^{1/2} t\} \tag{5.35}$$

where $\zeta_g(0)$ is the value of the geostrophic vorticity at time $t = 0$. From (5.35) we see that $\tau_e \equiv H|2/fK|^{1/2}$ is the time which it takes a barotropic vortex of height H to spin-down to e^{-1} of its original value (this "e-folding" time scale is what is meant by the *spin-down time*). Taking typical values of the parameters as follows: $H = 10$ km, $f = 10^{-4}\ \mathrm{s}^{-1}$, and $K = 10\ \mathrm{m}^2\ \mathrm{s}^{-1}$, we find that $\tau_e \approx 4$ d. Thus, for midlatitude synoptic scale disturbances in a barotropic atmosphere the characteristic spin-down time is a few days. This decay time scale should be compared to the time scale for ordinary viscous diffusion. It can be shown that the time for eddy diffusion to penetrate a depth H is of the order $\tau_d \sim H^2/K$. For the above values of H and K, the diffusion time scale is thus $\tau_d \sim 100$ d. Hence, the spin-down process is a far more effective mechanism for destroying vorticity in a rotating atmosphere than is eddy diffusion.

Physically the spin-down process in the atmospheric case is similar to that described for the teacup, except that in synoptic scale systems it is primarily the Coriolis force which balances the pressure gradient force away from the boundary, not the centrifugal force. Again the role of the secondary circulation driven by forces resulting from eddy viscosity in the boundary layer is to provide a slow radial flow in the interior which is superposed on the azimuthal circulation of the vortex. This secondary circulation is directed outward in a cyclone so that the horizontal area enclosed by any chain of fluid particles gradually increases. Since the circulation is conserved, the azimuthal velocity at any distance from the vortex center must decrease in time. Or, from another point of view, the Coriolis force for the outward-flowing fluid is directed clockwise, and this force thus exerts a torque opposite to the direction of the circulation of the vortex. In Fig. 5.4 a qualitative sketch of the streamlines of this secondary flow is shown.

It should now be obvious exactly what is meant by the term *secondary circulation*. It is simply a circulation superposed on the primary circulation (in this case the azimuthal circulation of the vortex) by the physical constraints of the system. In the case of the boundary layer it is viscosity which is responsible for the presence of the secondary circulation. However, other processes such as temperature advection and diabatic heating may also lead to secondary circulations as we shall see later.

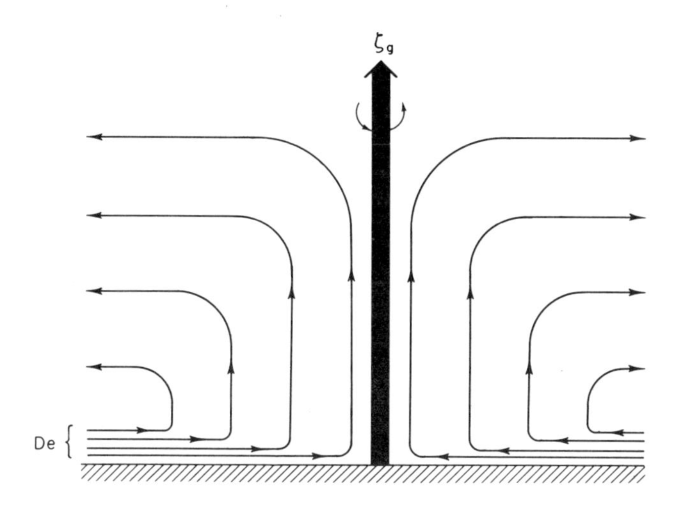

Fig. 5.4 Streamlines of the secondary circulation forced by frictional convergence in the planetary boundary layer for a cyclonic vortex in a barotropic atmosphere.

The above discussion has concerned only the neutrally stratified barotropic atmosphere. An analysis for the more realistic case of a stably stratified baroclinic atmosphere would be much more complicated. However, qualitatively the effects of stratification may be easily understood. The buoyancy force (see Section 2.7.3) will act to suppress vertical motion since air lifted vertically in a stable environment will be denser than the environmental air. As a result the interior secondary circulation will be restricted in vertical extent as shown in Fig. 5.5. Most of the return flow will take place just above the boundary layer. This secondary flow will rather quickly spin-down the vorticity at the top of the Ekman layer without appreciably affecting the higher levels. When the geostrophic vorticity at the top of the boundary layer is reduced to zero, the "pumping" action of the Ekman layer is eliminated. The result is a baroclinic vortex with a vertical shear of the azimuthal velocity which is just strong enough to bring ζ_g to zero at the top of the boundary layer. This vertical shear of the geostrophic wind requires a radial temperature gradient to satisfy the thermal wind relationship. This radial temperature gradient is in fact produced during the spin-down phase by the adiabatic cooling of the air forced out of the Ekman layer. Thus, the secondary circulation in the baroclinic atmosphere serves two purposes: (1) it changes the azimuthal velocity field of the vortex through the action of the Coriolis force; and (2) it changes the temperature distribution so that a thermal wind balance is always maintained between the vertical shear of the azimuthal velocity and the radial temperature gradient.

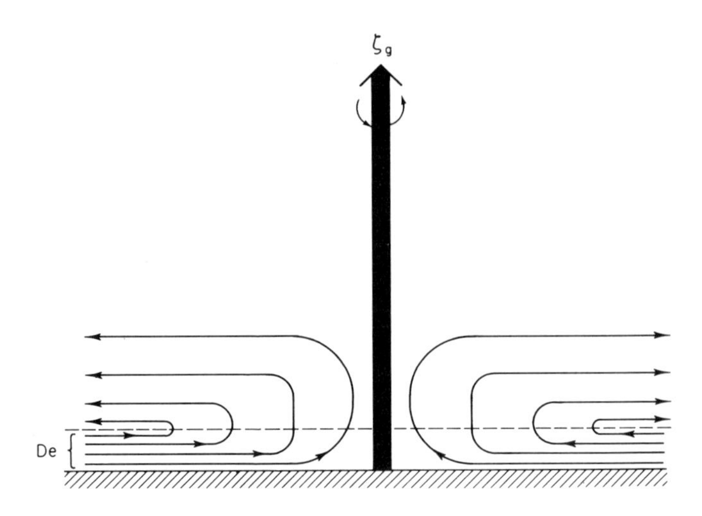

Fig. 5.5 Streamlines of the secondary circulation forced by frictional convergence in the planetary boundary layer for a cyclonic vortex in a stably stratified baroclinic atmosphere.

Problems

1. Verify by direct substitution that the Ekman spiral expression (5.21) is indeed a solution of the boundary layer equations (5.17) and (5.18) for the case $v_g = 0$.

2. Derive the Ekman spiral solution for the more general case where the geostrophic wind has both x and y components (u_g and v_g, respectively) which are independent of height.

3. Letting the Coriolis parameter and density be constants, show that Eq. (5.31) is correct for the more general Ekman spiral solution obtained in Problem 2.

4. For laminar flow in a rotating cylindrical vessel filled with water (molecular kinematic viscosity $\nu = 0.01\ \text{cm}^2\ \text{s}^{-1}$), compute the depth of the Ekman layer and the spin-down time if the depth of the fluid is 30 cm and the rotation rate of the tank is ten revolutions per minute.

5. For the situation of Problem 4, how small would the radius of the tank have to be in order that the time scale for viscous diffusion from the side walls be comparable to the spin-down time?

6. For the modified Ekman spiral (5.27) the shear stress at the bottom of the spiral layer,

$$\tau_{zx} + i\tau_{zy} = \rho K \frac{\partial}{\partial z}(u + iv)\bigg|_{z=0}$$

must be equal to the surface stress. Show that the magnitude of the surface stress in this case can be expressed in terms of the geostrophic wind as $\tau_0 = \rho u_g \sin\alpha\,(2fK)^{1/2}$.

7. Derive an expression for the Ekman spiral which is valid in the Southern Hemisphere.

8. Suppose that in a certain region the geostrophic wind is westerly at $15\ \mathrm{m\ s^{-1}}$. Compute the net cross isobaric transport in the Ekman layer using both the ideal solution (5.21) and modified solution (5.27). Let $K = 5\ \mathrm{m^2\ s^{-1}}$, $\alpha = \pi/8$ rad, $\rho = 1\ \mathrm{kg\ m^{-3}}$, and $f = 10^{-4}\ \mathrm{s^{-1}}$.

***9.** Derive an expression for the wind driven surface Ekman layer in the ocean. Assume that the wind stress τ_w is constant and directed along the x axis. Continuity of stress at the air–sea interface ($z = 0$) requires that the wind stress equal the water stress so that the boundary condition at the surface becomes

$$\rho K \frac{\partial u}{\partial z} = \tau_w, \qquad K \frac{\partial v}{\partial z} = 0 \qquad \text{at } z = 0$$

where K is the eddy viscosity in the ocean (assumed constant). As a lower boundary condition assume that $u, v \to 0$ as $z \to -\infty$. If $K = 10^{-3}\ \mathrm{m^2\ s^{-1}}$ what is the depth of the surface Ekman layer at 45°N latitude?

***10.** Show that the vertically integrated mass transport in the wind driven oceanic surface Ekman layer is directed 90° to the right of the surface wind stress in the Northern Hemisphere. Explain this result physically.

***11.** A homogeneous barotropic ocean of depth $H = 3$ km has a zonally symmetric geostrophic jet whose profile is given by the expression

$$\bar{u}_g = U \exp[-(y/L)^2]$$

where $U = 1\ \mathrm{m\ s^{-1}}$ and $L = 200$ km are constants. Compute the vertical velocity produced by convergence in the Ekman layer at the ocean bottom and show that the meridional profile of the secondary cross stream motion forced in the interior is the same as the meridional profile of $\bar{u}_g$. What are the maximum values of $\bar{v}$ and $\bar{w}$ if $K = 10^{-3}\ \mathrm{m^2\ s^{-1}}$ and $f = 10^{-4}\ \mathrm{s^{-1}}$? (Assume that $\bar{w}$ and the viscous stress vanish at the surface.)

***12.** Using the approximate zonally averaged momentum equation

$$\frac{\partial \bar{u}}{\partial t} \simeq f\bar{v}$$

compute the "spin-down" time for the zonal jet in Problem 11.

13. Derive a formula for the vertical velocity at the top of the boundary layer equivalent to (5.31) using the modified Ekman spiral solution (5.27). What is the vertical velocity at the top of the boundary layer for a synoptic scale disturbance with $\zeta_g = 10^{-5}\ \mathrm{s}^{-1}$, $K = 10\ \mathrm{m}^2\ \mathrm{s}^{-1}$, $f = 10^{-4}\ \mathrm{s}^{-1}$, and $\alpha = \pi/8$ rad?

Suggested References

Currie, *Fundamental Mechanics of Fluids*, provides an excellent elementary introduction to laminar viscous flows including various types of boundary layers.

Batchelor, *An Introduction to Fluid Dynamics*, has a useful discussion of the molecular basis of viscosity as well as a succinct derivation of the complete stress tensor. This book also has an excellent treatment of viscous boundary layers including the Ekman layer.

Tennekes and Lumley, *A First Course in Turbulence*, is by far the best introduction to the physics of turbulent flows. This text contains an interesting discussion of the planetary boundary layer including thermal effects.

Cole, *Perturbation Methods in Applied Mathematics*, provides an excellent graduate level introduction to the mathematical foundations of boundary layer theory.

Greenspan, *The Theory of Rotating Fluids*, gives a unified account of the role of viscosity in rotating fluids at an advanced level. This is the only book available which focuses solely on phenomena which occur only in rotating fluids.

Lumley and Panofsky, *The Structure of Atmospheric Turbulence*, covers both observational and theoretical aspects of the subject including the effects of thermal stratification in the boundary layer.

Chapter

6 The Dynamics of Synoptic Scale Motions in Middle Latitudes

A primary goal of dynamic meteorology is to interpret the observed structure of large-scale atmospheric motion systems in terms of the physical laws governing the motions. In Chapter 2 we derived the basic conservation laws for momentum, mass, and energy which atmospheric motions must obey. Taken together these conservation laws completely define the relationship between the mass and velocity fields in atmospheric circulation systems.

In this chapter we will show from scaling considerations that the laws of motion, mass continuity, and energy conservation constrain synoptic scale disturbances so that to a good approximation *the three-dimensional velocity field is uniquely determined by the geopotential field.*

We first discuss the observed structure of midlatitude synoptic systems and the mean circulations in which they are embedded. We then develop two diagnostic equations, the geopotential tendency equation and the "omega" equation. Together, these equations enable us to diagnose the three-dimensional structure of a synoptic system. Finally, we shall use these diagnostic relationships as aids in obtaining an idealized model for a typical developing synoptic disturbance.

6.1 The Observed Structure of Midlatitude Synoptic Systems

In the real atmosphere circulation systems plotted on a synoptic map rarely resemble the simple circular vortices discussed in Chapter 3. Rather, they are generally highly asymmetric in form with the highest winds and largest temperature gradients concentrated along narrow bands called *fronts*. Also, such systems generally are highly baroclinic with both the amplitudes and phases of the geopotential and velocity perturbations changing substantially with height. Part of this complexity is due to the fact that these synoptic systems are not superposed on a uniform mean flow, but are embedded in a planetary scale flow which is itself highly baroclinic. Furthermore, this planetary scale flow is influenced by *orography* (that is, by large-scale terrain variations) and continent–ocean heating contrasts so that it is highly longitude dependent. Therefore, it is not accurate to view synoptic systems as disturbances superposed on a zonal flow which varies only with latitude and height. However, as we shall see in Chapter 9, such a point of view is very useful as a first approximation in theoretical analysis of synoptic wave disturbances.

Indeed, zonally averaged cross sections do provide some useful information on the gross structure of the planetary scale circulation. In Fig. 6.1 we show mean meridional cross sections for the longitudinally averaged flow in the Northern Hemisphere during the months of January and July for the 0–30-km altitude region. Meteorologists divide this region of the atmosphere into two layers based on the vertical gradient of temperature. In the lower layer, the *troposphere*, temperature generally decreases with height, while in the upper layer, the *stratosphere*, the temperature generally gradually increases with height. The troposphere and the stratosphere are separated by the *tropopause*, a level of temperature minimum which varies in height from about 16 km near the equator to 9 km near the poles. In the present chapter we will be concerned only with the structure of the wind and temperature fields in the troposphere. The stratosphere will be discussed in Chapter 11.

As would be expected the pole to equator temperature gradient in the troposphere is much larger in the winter than in the summer. Since the zonal wind and temperature fields satisfy the thermal wind relationship (3.30) to a high degree of accuracy, the maximum zonal wind speed is much larger in the winter than in the summer. Furthermore, in both seasons the core of maximum zonal wind speed (called the jet stream axis) is located just below the tropopause at the height where the meridional temperature gradient vanishes and at the latitude where the average meridional temperature gradient in the troposphere divided by the Coriolis parameter is a maximum.

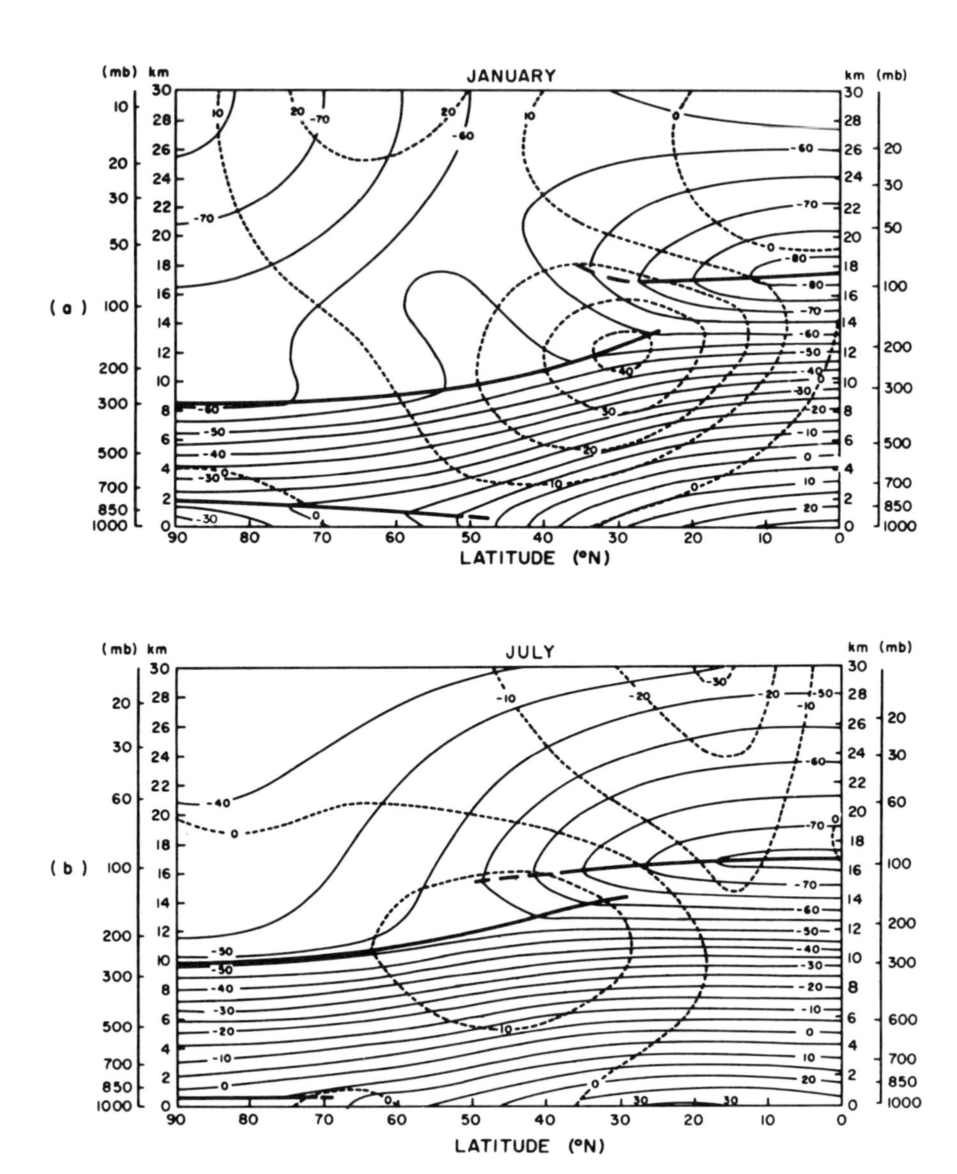

Fig. 6.1 Mean meridional cross sections of wind and temperature for (a) January and (b) July. Thin solid temperature lines in degrees centigrade. Dashed wind lines in meters per second. Heavy solid lines represent tropopause and inversion discontinuities. (After *Arctic Forecast Guide*, Navy Weather Research Facility, April 1962.)

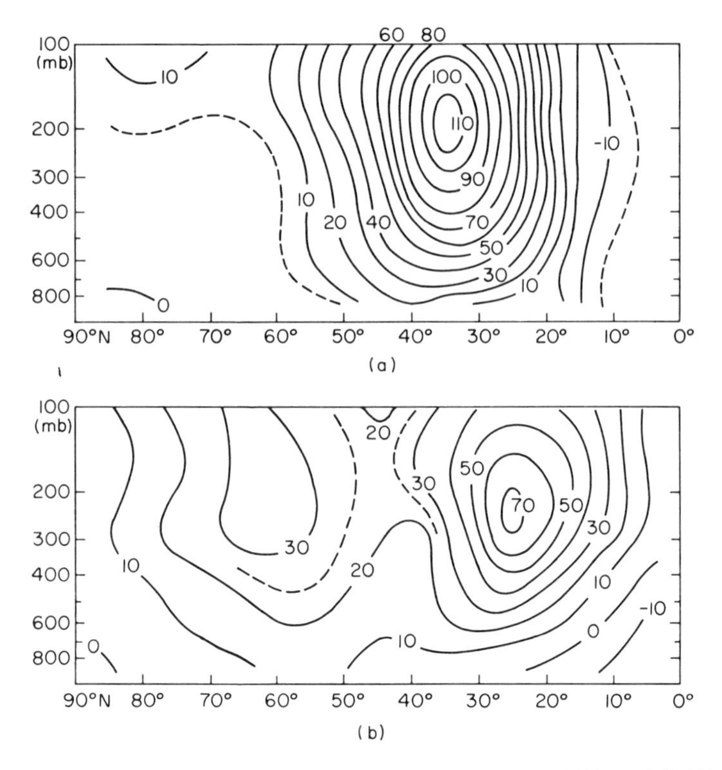

Fig. 6.2 Winter mean zonal winds in the Northern Hemisphere at (a) 140°E and (b) 0° longitude. Speeds shown are in knots (2 knots approximately equals 1 m s^{-1}). (After Palmén and Newton, 1969.)

That the zonally averaged meridional cross sections of Fig. 6.1 are not representative of the mean wind structure at all longitudes can be seen in Fig. 6.2, which shows meridional cross sections of the zonal wind averaged over the months December–February at two widely separated longitudes in the Northern Hemisphere. The explanation for the large differences in the jet stream structure at these two longitudes is readily seen from examination of Fig. 6.3, which shows the mean 500-mb geopotential contours for January in the Northern Hemisphere. Even after averaging the height field for a month, very striking departures from zonal symmetry remain. These are clearly linked to the distributions of continents and oceans. The most prominent asymmetries are the troughs to the east of the American and Asian continents. Referring back to Fig. 6.2 we see that the intense jet at 35°N in the 140°E cross section is a result of the semipermanent trough at that longitude. Thus, it is apparent that the mean flow in which the synoptic systems are embedded should really be regarded as longitude dependent.

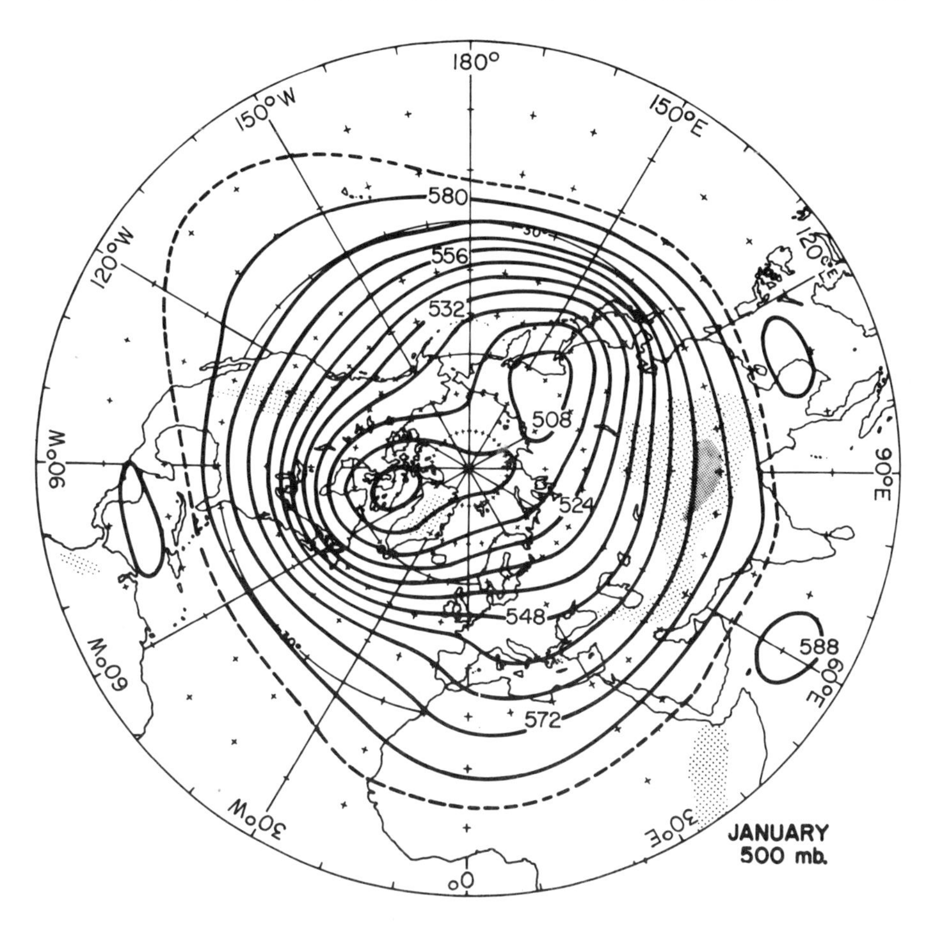

Fig. 6.3 Mean 500-mb contours in January, Northern Hemisphere. Heights shown in tens of meters. (After Palmén and Newton, 1969.)

In addition to its longitudinal dependence the planetary scale flow also varies somewhat from day to day due to its interactions with transient synoptic scale disturbances. As a result, monthly mean charts tend to smooth out the actual structure of the jet stream since the position and intensity of the jet vary. Thus, at any time the planetary scale flow in the region of the jet stream has much greater baroclinicity than indicated on time-averaged charts. This point is illustrated schematically in Fig. 6.4, which shows an idealized vertical cross section through the jet stream. At any instant the axis of the jet stream tends to coincide with a narrow zone of strong temperature gradients called the *polar front*. This is the zone which in general separates the cold air of polar origin from warm tropical air. The occurrence

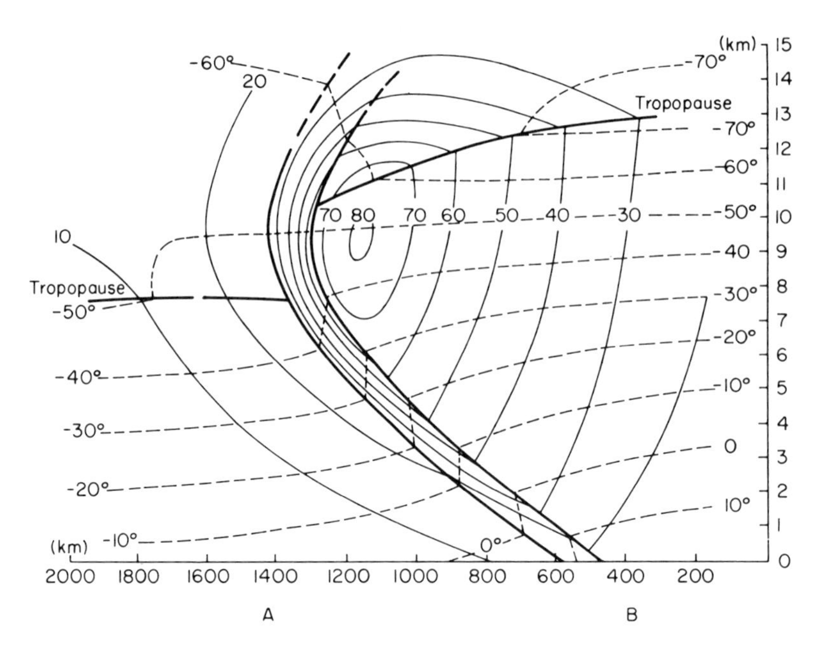

Fig. 6.4 Schematic isotherms (dashed lines, degrees centigrade) and isotachs (thin solid lines, meters per second) in the vicinity of the polar front. Heavy lines indicate boundaries of the frontal zone and tropopauses. (After Palmén and Newton, 1969.)

of an intense jet core above the frontal zone is, of course, not mere coincidence, but rather a consequence of the thermal wind balance.

It is a common observation in fluid dynamics that jet flows in which strong velocity shears occur may be unstable with respect to small perturbations. By this is meant that any small disturbance introduced into the jet flow will tend to amplify, drawing energy from the jet as it grows. Most synoptic scale systems in midlatitudes appear to develop as the result of an instability of the jet stream flow. This instability, called *baroclinic instability*, depends primarily on vertical shear in the jet stream, and hence tends to occur primarily in the region of the frontal zone. Baroclinic instability is not, however, identical to frontal instability since most baroclinic instability models describe only geostrophically scaled motions while disturbances in the vicinity of strong frontal zones must be highly nongeostrophic. As we shall see in Chapter 10, baroclinic disturbances may themselves act to intensify preexisting temperature gradients and hence generate frontal zones.

The stages in the development of a typical baroclinic cyclone which develops as a result of baroclinic instability along the polar front are shown schematically in Fig. 6.5. In the stage of rapid development strong cold advection is seen to occur west of the trough at the surface, with weaker

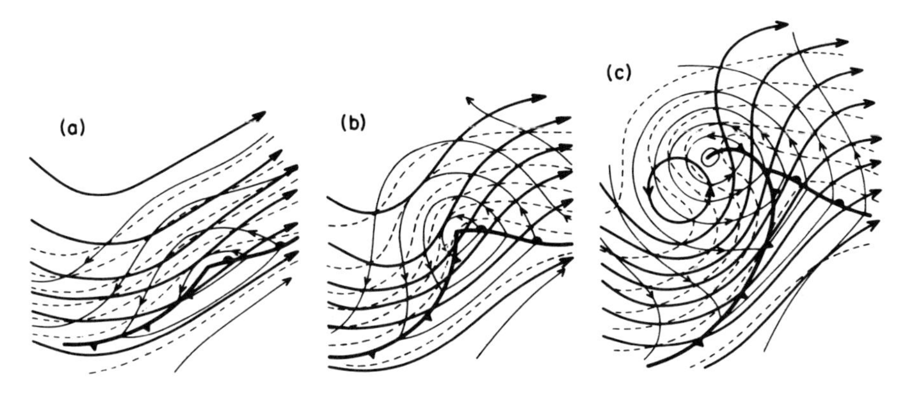

Fig. 6.5 Schematic 500-mb contours (heavy solid lines), 1000-mb contours (thin lines), and 1000–500-mb thickness (dashed) for a developing baroclinic wave at three stages of development. (After Palmén and Newton, 1969.)

warm advection to the east. This pattern of thermal advection is a direct consequence of the fact that the trough at 500 mb lags (lies to the west of) the surface trough so that the mean geostrophic wind in the 500–1000-mb layer is directed across the 1000–500-mb thickness lines toward larger thickness west of the surface trough, and toward small thickness east of the surface trough. This dependence of the phase of the disturbance on height is better illustrated by Fig. 6.6, which shows a schematic downstream (or west–east) cross section through a developing baroclinic system. Throughout the

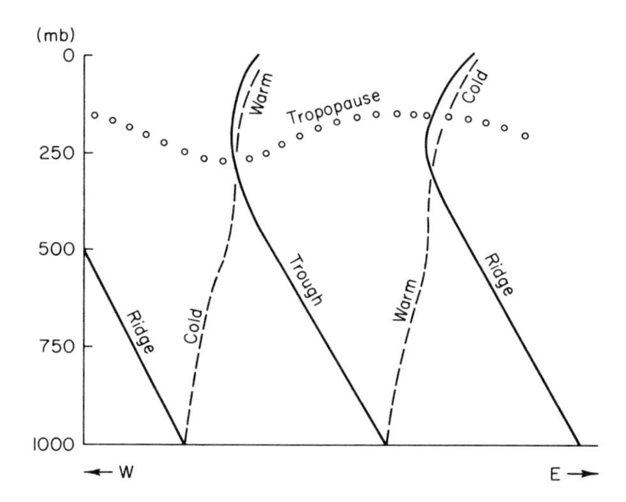

Fig. 6.6 West–east cross section through developing baroclinic wave. Solid lines are trough and ridge axes; dashed lines are axes of temperature extrema; the chain of open circles denotes the tropopause.

troposphere the trough and ridge axes tilt westward (or upstream) with height, while the axes of warmest and coldest air are observed to have the opposite tilt. As we shall see later the westward tilt of the troughs and ridges is necessary in order that the mean flow give up potential energy to the developing wave. In the mature stage (not shown in Fig. 6.5) the troughs at 500 and 1000 mb are nearly in phase. As a consequence, the thermal advection and energy conversion are quite weak.

6.2 Development of the Quasi-Geostrophic System

In the remainder of this chapter we wish to show from scaling considerations that the observed structure of midlatitude synoptic systems may be understood to be a consequence of the constraints imposed on the motions by Newton's second law, mass continuity, and energy conservation. For this purpose, it is convenient to develop the governing equations which are the mathematical expressions for these laws using the isobaric coordinate system. Meteorological measurements are generally referred to constant-pressure surfaces. In addition, the continuity equation takes a simple form in isobaric coordinates. Thus, use of the isobaric coordinate system will simplify the development of diagnostic equations.

6.2.1 Scale Analysis in Isobaric Coordinates

The basic equations in isobaric coordinates were developed in Section 3.1, and for reference will be repeated here. The approximate horizontal equations of motion are

$$\frac{du}{dt} - fv = -\frac{\partial \Phi}{\partial x} \tag{6.1}$$

$$\frac{dv}{dt} + fu = -\frac{\partial \Phi}{\partial y} \tag{6.2}$$

Here the total derivative is defined by

$$\frac{d}{dt} \equiv \left(\frac{\partial}{\partial t}\right)_p + u\left(\frac{\partial}{\partial x}\right)_p + v\left(\frac{\partial}{\partial y}\right)_p + \omega\frac{\partial}{\partial p} \tag{6.3}$$

where $\omega \equiv dp/dt$ is the individual pressure change. The hydrostatic equation may be written as

$$\frac{\partial \Phi}{\partial p} = -\alpha = -\frac{RT}{p} \tag{6.4}$$

The continuity equation is

$$\frac{\partial u}{\partial x} + \frac{\partial v}{\partial y} + \frac{\partial \omega}{\partial p} = 0 \tag{6.5}$$

And, finally, the thermodynamic energy equation is

$$\left(\frac{\partial T}{\partial t} + u\frac{\partial T}{\partial x} + v\frac{\partial T}{\partial y}\right) - S_p\omega = \dot{q}/c_p \tag{6.6}$$

where $S_p \equiv -T\,\partial \ln \theta/\partial p$.

This set of equations, although already considerably simplified, is still very awkward to use for acquiring an understanding of the structure of synoptic scale systems. We now show how scale analysis may be used to derive a simplified set suitable for diagnostic analysis of synoptic systems. In fact, we will show that for midlatitude synoptic scale systems the fields of geopotential tendency and vertical motion (that is $\partial\Phi/\partial t$ and ω) are uniquely determined for any given distribution of geopotential Φ.

To obtain this set of diagnostic equations, we first eliminate T by rewriting the thermodynamic energy equation (6.6) in terms of $\partial\Phi/\partial p$ using (6.4), the hydrostatic relationship.

With the aid of the definition of entropy (2.43) the resulting equation can be written as follows:

$$\frac{\partial}{\partial t}\left(-\frac{\partial\Phi}{\partial p}\right) + u\frac{\partial}{\partial x}\left(-\frac{\partial\Phi}{\partial p}\right) + v\frac{\partial}{\partial y}\left(-\frac{\partial\Phi}{\partial p}\right) - \sigma\omega = \frac{\alpha}{c_p}\frac{ds}{dt} \tag{6.7}$$

where σ, the *static stability parameter*, is defined by

$$\sigma \equiv -\frac{\alpha}{\theta}\frac{\partial\theta}{\partial p} = \frac{RS_p}{p}$$

For a statically stable atmosphere $\partial\theta/\partial p < 0$ so that $\sigma > 0$.

Equation (6.7) may be further simplified if we note that for synoptic scale systems the horizontal velocity is approximately equal to the geostrophic velocity

$$\mathbf{V} = \mathbf{i}u + \mathbf{j}v \simeq \mathbf{V}_g \equiv \frac{\mathbf{k}\times\nabla\Phi}{f}$$

Thus, to a first approximation, the horizontal velocity components in (6.7) may be replaced by their geostrophic values so that in vectorial notation

$$u\frac{\partial}{\partial x}\left(-\frac{\partial\Phi}{\partial p}\right) + v\frac{\partial}{\partial y}\left(-\frac{\partial\Phi}{\partial p}\right) \simeq \mathbf{V}_g\cdot\nabla\left(-\frac{\partial\Phi}{\partial p}\right)$$

Further, following the scaling arguments of Section 2.7.4 we assume that the diabatic heating ds/dt is small compared to the terms on the left in (6.7). The approximate thermodynamic energy equation is thus

$$\frac{\partial}{\partial t}\left(-\frac{\partial \Phi}{\partial p}\right) = -\mathbf{V}_g \cdot \nabla\left(-\frac{\partial \Phi}{\partial p}\right) + \sigma\omega \tag{6.8}$$

It will be shown in Problem 1 that σ can be expressed in terms of Φ so that (6.8) contains only the two dependent variables Φ and ω.

In giving a physical interpretation to the terms in (6.8), we will refer to $-\partial\Phi/\partial p$ as the "temperature" since it is proportional to temperature on an isobaric surface in a hydrostatic atmosphere. Alternatively, we could call $-\partial\Phi/\partial p$ the "thickness" since it is equal to the thickness $\delta\Phi$ divided by the pressure interval δp in the limit $\delta p \to 0$. Thus, the term on the left-hand side in (6.8) is proportional to the local rate of change of temperature on an isobaric surface. Similarly, the first term on the right-hand side is proportional to the advection of temperature by the geostrophic wind on an isobaric surface. The second term on the right in (6.8) is usually called the adiabatic cooling (heating) term. This term expresses the adiabatic temperature changes which result from rising and expansion (sinking and compression) of air parcels in a stable environment. The diabatic heating rate is usually small compared to the horizontal temperature advection and adiabatic cooling terms for midlatitude synoptic scale motions. For simplicity, we shall generally neglect diabatic heating in subsequent discussions although for time scales of more than a couple of days it should be included.

To further simplify our system of equations, it is convenient to replace the horizontal equations of motion by the vorticity equation. In Section 4.4.2 we previously discussed the vorticity equation in the isobaric coordinate system. Referring to (4.21), we see that the vorticity equation in the isobaric system can be written as follows:

$$\frac{\partial \zeta}{\partial t} = -\mathbf{V} \cdot \nabla(\zeta + f) - \omega\frac{\partial \zeta}{\partial p} - (\zeta + f)\,\nabla \cdot \mathbf{V} + \left\{\frac{\partial u}{\partial p}\frac{\partial \omega}{\partial y} - \frac{\partial v}{\partial p}\frac{\partial \omega}{\partial x}\right\} \tag{6.9}$$

where $\zeta = \mathbf{k} \cdot (\nabla \times \mathbf{V})$ and all horizontal derivatives are evaluated at constant pressure. The terms in (6.9) in order reading from left to right are as follows:

1. the local rate of change of relative vorticity;
2. the horizontal advection of absolute vorticity;
3. the vertical advection of relative vorticity;
4. the divergence term;
5. the twisting or tilting term.

As shown by the scale analysis of Chapter 4, we may simplify the vorticity equation for midlatitude synoptic scale motions by

1. neglecting the vertical advection and twisting terms,
2. neglecting ζ compared to f in the divergence term,
3. approximating the horizontal velocity by the geostrophic wind in the *advection term*, and
4. replacing the relative vorticity by its geostrophic value.

As a further simplification, we may expand the Coriolis parameter in a Taylor series about the latitude ϕ_0 as

$$f = f_0 + \beta y + \text{(higher-order terms)}$$

where $\beta \equiv (df/dy)_{\phi_0}$, and $y = 0$ at ϕ_0. If we let L designate the latitudinal scale of the motions, then the ratio of the first two terms in the expansion of f has order of magnitude

$$\frac{\beta L}{f_0} \sim \frac{\cos \phi_0}{\sin \phi_0} \frac{L}{a}$$

Thus, when the latitudinal scale of the motions is small compared to the radius of the earth ($L/a \ll 1$) we can let the Coriolis parameter have a constant value f_0 except where it appears differentiated in the advection term, in which case $df/dy \equiv \beta$ is assumed to be constant. This approximation is usually referred to as the *beta-plane approximation.*

Applying all the above approximations, we obtain the *quasi-geostrophic vorticity equation*

$$\frac{\partial \zeta_g}{\partial t} = -\mathbf{V}_g \cdot \nabla(\zeta_g + f) - f_0 \nabla \cdot \mathbf{V} \tag{6.10}$$

where $\zeta_g = \nabla^2\Phi/f_0$ and $\mathbf{V}_g = \mathbf{k} \times \nabla\Phi/f_0$ are both evaluated using the constant Coriolis parameter f_0.

It is very important to note that the horizontal wind is *not* replaced by its geostrophic value in the divergence term. In fact, when the geostrophic wind is computed using a constant Coriolis parameter, it is just the small departures of the horizontal wind from geostrophy which account for the divergence. As we shall see later, this divergence and its corresponding vertical motion field are dynamically necessary to keep the temperature changes hydrostatic and vorticity changes geostrophic in synoptic scale systems.

The horizontal divergence in (6.10) can easily be eliminated using the continuity equation

$$\nabla \cdot \mathbf{V} = -\frac{\partial \omega}{\partial p}$$

to obtain an alternative form of the quasi-geostrophic vorticity equation,

$$\frac{\partial \zeta_g}{\partial t} = -\mathbf{V}_g \cdot \nabla(\zeta_g + f) + f_0 \frac{\partial \omega}{\partial p} \tag{6.11}$$

Since ζ_g and $\mathbf{V}_g$ are both defined in terms of Φ, (6.11) can be used to diagnose the ω field provided that the fields of both Φ and $\partial\Phi/\partial t$ are known. Since both the local change of the geostrophic vorticity and the advection of vorticity by the geostrophic wind can be estimated with reasonable accuracy estimates of ω based on (6.11) usually are better than the estimates based on the continuity equation discussed in Chapter 3.

Because ζ_g and $\mathbf{V}_g$ are both functions of Φ the quasi-geostrophic vorticity equation (6.11) and the hydrostatic thermodynamic energy equation (6.8) each contain only the two dependent variables Φ and ω. Therefore, (6.8) and (6.11) form a closed set of prediction equations in Φ and ω. It is thus possible to eliminate ω between these two equations and obtain an equation relating Φ to $\partial\Phi/\partial t$. This equation is called the *geopotential tendency equation.* It is actually just a form of the potential vorticity equation (see Section 6.2.3). Alternatively, we can eliminate the time derivative terms between (6.8) and (6.11) and obtain an equation which relates the ω field at any instant to the Φ field at that instant. This equation is called the vertical motion or *omega equation.*

Thus, by suitable manipulation of (6.8) and (6.11) we can obtain expressions which allow us to compute the geopotential tendency $\partial\Phi/\partial t$ and the vertical motion ω from observations of the instantaneous field of Φ alone. Hence, to a first approximation, the evolution of midlatitude synoptic scale flow can be predicted without direct measurement of the velocity field. This set of equations, which really constitutes the core of modern dynamic meteorology, is called the *quasi-geostrophic system.*

6.2.2 The Tendency Equation

If we define the geopotential tendency as $\chi \equiv \partial\Phi/\partial t$, then (6.8) and (6.11) may be written as

$$\frac{\partial \chi}{\partial p} = -\mathbf{V}_g \cdot \nabla\left(\frac{\partial \Phi}{\partial p}\right) - \sigma\omega \tag{6.12}$$

$$\nabla^2 \chi = -f_0 \mathbf{V}_g \cdot \nabla\left(\frac{1}{f_0}\nabla^2\Phi + f\right) + f_0^2 \frac{\partial \omega}{\partial p} \tag{6.13}$$

where we have used the relationship $\zeta_g = (1/f_0)\,\nabla^2\Phi$ so that after changing the order of partial differentiation,

$$\frac{\partial \zeta_g}{\partial t} = \frac{1}{f_0}\nabla^2\chi$$

If we multiply (6.12) by f_0^2/σ, then differentiate with respect to pressure and add the result to (6.13), we obtain

$$\underbrace{\left(\nabla^2 + \frac{f_0^2}{\sigma}\frac{\partial^2}{\partial p^2}\right)\chi}_{\text{A}} = \underbrace{-f_0\mathbf{V}_g\cdot\nabla\left(\frac{1}{f_0}\nabla^2\Phi + f\right)}_{\text{B}} + \underbrace{\frac{f_0^2}{\sigma}\frac{\partial}{\partial p}\left(-\mathbf{V}_g\cdot\nabla\frac{\partial\Phi}{\partial p}\right)}_{\text{C}} \tag{6.14}$$

where we have assumed σ to be a constant.[1] Equation (6.14) is the geopotential tendency equation.

We shall now discuss in turn the physical interpretation of each term in (6.14). Term A involves only second derivatives in space of the χ field. For wavelike disturbances, this term can be shown to be proportional to $-\chi$. To demonstrate this fact, we assume that the fields of Φ and χ vary sinusoidally in x and y.[2] Thus we may write

$$\chi = X(p, t)\sin kx \sin ly \tag{6.15}$$

where the *wave numbers* k and l are defined as

$$k = \frac{2\pi}{L_x}, \qquad l = \frac{2\pi}{L_y}$$

with L_x and L_y the wavelengths in the x and y directions, respectively. The horizontal Laplacian of χ is then simply

$$\nabla^2\chi = -(k^2 + l^2)X(p, t)\sin kx \sin ly = -(k^2 + l^2)\chi$$

The vertical dependence $X(p)$ can be approximated in a similar fashion. Referring back to Fig. 6.6 we see that for developing synoptic scale disturbances trough and ridge lines slope with height. Thus the longitude of the surface trough corresponds approximately to the longitude of the 200-mb

[1] Actually, σ varies substantially with pressure even in the troposphere. However, the qualitative discussion in this section would not be changed if we were to include this additional complication.

[2] This assumption is not as restrictive as it may appear because any bounded continuous function can be expanded in a double Fourier series in x and y. Thus our discussion would apply to any single component of the Fourier expansion for any distribution of Φ.

ridge. This phase shift of nearly 180° from the surface to the tropopause can be approximated by assuming that

$$X(p, t) = X_0(t)\cos(\pi p/p_0)$$

where $p_0 = 100$ kPa (1000 mb). Thus $\partial^2 X/\partial p^2 \simeq -(\pi/p_0)^2 X$. Term A can now be written approximately

$$\left(\nabla^2 + \frac{f_0^2}{\sigma}\frac{\partial^2}{\partial p^2}\right)\chi \simeq -\left[k^2 + l^2 + \frac{1}{\sigma}\left(\frac{f_0 \pi}{p_0}\right)^2\right]\chi$$

so that the left side in (6.14) is just proportional to the *negative* of the geopotential tendency.

We next consider the first term on the right-hand side in (6.14). This term is proportional to the advection of absolute vorticity by the geostrophic wind. For physical interpretation, it is convenient to divide term B into two parts by writing

$$\mathbf{V}_g \cdot \nabla\left(\frac{1}{f_0}\nabla^2\Phi + f\right) = \mathbf{V}_g \cdot \nabla\left(\frac{1}{f_0}\nabla^2\Phi\right) + v_g\frac{df}{dy}$$

These two parts represent the geostrophic advections of relative vorticity and planetary vorticity, respectively. For disturbances in the westerlies, these two effects tend to oppose each other as illustrated by the 500-mb wave disturbance shown schematically in Fig. 6.7.

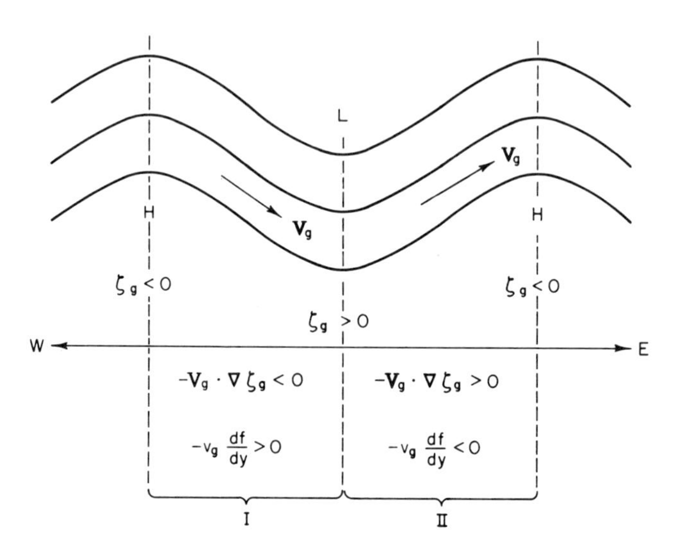

Fig. 6.7 Schematic 500-mb geopotential field showing regions of positive and negative advections of relative and planetary vorticity.

In region I upstream from the 500-mb trough, the geostrophic wind is directed from the negative vorticity maximum at the ridge toward the positive vorticity maximum at the trough so that

$$-\mathbf{V}_g \cdot \nabla\left(\frac{1}{f_0}\nabla^2\Phi\right) < 0$$

But at the same time, since $v_g < 0$ in region I, the geostrophic wind has its y component directed down the gradient of planetary vorticity so that $-v_g df/dy > 0$. Hence in region I the advection of relative vorticity tends to decrease the vorticity, whereas advection of planetary vorticity tends to increase the vorticity. Similar arguments (but with reversed signs) apply to region II. Therefore, advection of relative vorticity tends to move the vorticity pattern and hence the troughs and ridges eastward (downstream). But advection of planetary vorticity tends to move the troughs and ridges westward against the advecting wind field. This type of motion is called *retrograde* motion or *retrogression.*

The actual displacement of the pattern will obviously depend upon which type of vorticity advection dominates (as well as on the divergence term). In order to estimate the relative magnitudes of the relative vorticity and planetary vorticity advections, we assume that Φ has a sinusoidal dependence like that given for χ in (6.15). We can then write

$$\zeta_g = \frac{1}{f_0}\nabla^2\Phi \simeq -\frac{(k^2 + l^2)}{f_0}\Phi$$

Thus, for a disturbance of given amplitude the absolute value of the relative vorticity increases for increasing wave number or *decreasing* wavelength. As a consequence, the advection of relative vorticity tends to dominate for short waves ($L_x \lesssim 3000$ km) while for long waves ($L_x \gtrsim 10{,}000$ km) the planetary vorticity advection tends to dominate. Therefore, as a general rule short-wavelength synoptic scale systems should move rapidly eastward with the advecting zonal flow while long planetary waves should tend to retrograde.[3] Waves of intermediate wavelength may be quasi-stationary, or move eastward much slower than the mean geostrophic wind speed.

Returning to (6.14) we see that for short waves term B is negative in region I. Thus, as a result of vorticity advection χ will tend to be positive (recall that term A is proportional to $-\chi$). Hence, the geopotential height tendency will

[3] The observed long waves in the atmosphere actually appear to be fixed in position rather than retrograde. This is believed to be a result of the forcing due to topographic influences and land continent heating contrasts as was previously mentioned in Section 6.1. However, detailed studies indicate that there do exist long-wave components of the motion which retrograde rapidly. In general these contain much less energy then the stationary component.

be positive and a ridge will tend to develop in this region. This ridging is, of course, necessary for the development of a negative geostrophic vorticity. Similar arguments, but with the signs reversed, apply to region II downstream from the trough where falling geopotential heights are associated with a positive relative vorticity advection. It is also important to note that the vorticity advection term is zero along both the trough and ridge axes since both $\nabla\zeta_g$ and v_g are zero at the axes. Thus, vorticity advection cannot change the strength of this type of disturbance, but only acts to propagate the disturbance horizontally.

The mechanism for amplification or decay of midlatitude synoptic systems is contained in term C of (6.14). This term, called the *differential thickness advection*, tends to be a maximum at the trough and ridge lines in a developing baroclinic wave. Now since

$$\mathbf{V}_g \cdot \nabla\left(\frac{\partial\Phi}{\partial p}\right)$$

is the thickness advection, which is proportional to the hydrostatic temperature advection, it is clear that

$$-\frac{\partial}{\partial p}\left[\mathbf{V}_g \cdot \nabla\left(\frac{\partial\Phi}{\partial p}\right)\right]$$

is proportional to the rate of change of temperature advection with height or the *differential temperature advection.*

To examine the influence of differential temperature advection on the geopotential tendency we consider the idealized developing wave shown in Fig. 6.5. Below the 500-mb ridge there is strong warm advection associated with the warm front, while below the 500-mb trough there is strong cold advection associated with the cold front. Above the 500-mb level the thickness pattern and geopotential pattern become nearly parallel so that thermal advection tends to be small in the upper troposphere. Now in the region of warm advection

$$\mathbf{V}_g \cdot \nabla\left(\frac{\partial\Phi}{\partial p}\right) > 0$$

since $\mathbf{V}_g$ has a component down the temperature gradient. But as explained above the warm advection decreases with height (increases pressure) so that

$$\frac{\partial}{\partial p}\left[\mathbf{V}_g \cdot \nabla\left(\frac{\partial\Phi}{\partial p}\right)\right] > 0$$

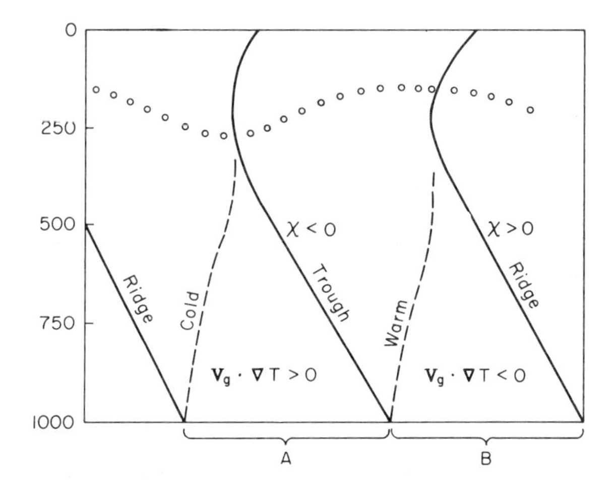

Fig. 6.8 East–west section through a developing synoptic disturbance showing the relationship of temperature advection to the upper level height tendencies, A and B designate, respectively, regions of cold advection and warm advection in the lower troposphere.

Conversely, beneath the 500-mb trough where there is cold advection decreasing with height the opposite signs obtain. Therefore, along the 500-mb trough and ridge axes where the vorticity advection is zero the tendency equation states that for a developing wave

$$\chi \sim \frac{\partial}{\partial p}\left[\mathbf{V}_g \cdot \nabla\left(\frac{\partial \Phi}{\partial p}\right)\right] \begin{matrix} > 0 & \text{at the ridge} \\ < 0 & \text{at the trough} \end{matrix}$$

Therefore, as indicated in Fig. 6.8, the effect of cold advection below the 500-mb trough is to *deepen* the trough at 500 mb, and the effect of warm advection below the 500-mb ridge is to *build* the ridge at 500 mb. Hence, it is the differential temperature or thickness advection which intensifies the upper-level troughs and ridges in a developing system.

Qualitatively the effects of differential temperature advection may be easily understood since the advection of cold air into the air column below the 500-mb trough will reduce the thickness of that column, and hence will lower the height of the 500-mb surface unless there is a compensating rise in the surface pressure. Obviously warm advection into the air column below the ridge will have the opposite effect.

In summary, we have shown that in the absence of diabatic heating the horizontal temperature advection must be nonzero in order that a midlatitude synoptic system intensify through baroclinic processes. As we shall see later in Chapter 9 the temperature advection pattern described above indirectly implies conversion of potential energy to kinetic energy.

6.2.3 The Quasi-Geostrophic Potential Vorticity Equation

The geopotential tendency equation which we discussed in the previous subsection may be regarded as a diagnostic equation which relates $\chi \equiv \partial\Phi/\partial t$ to the distribution of Φ. The form of this equation given in (6.14) is ideal for discussing the physical processes which generate the geopotential tendency field. However, since the tendency is just the local time rate of change of geopotential (6.14) can also be regarded as a *prognostic* equation for the time evolution of the Φ field. To use (6.14) prognostically it is convenient to simplify the right-hand side by using the chain rule of differentiation to write term C as

$$-\frac{f_0^2}{\sigma}\mathbf{V}_g\cdot\nabla\frac{\partial^2\Phi}{\partial p^2}-\frac{f_0^2}{\sigma}\frac{\partial\mathbf{V}_g}{\partial p}\cdot\nabla\frac{\partial\Phi}{\partial p} \tag{6.16}$$

But $f_0\partial\mathbf{V}_g/\partial p = \mathbf{k}\times\nabla(\partial\Phi/\partial p)$, which is perpendicular to $\nabla(\partial\Phi/\partial p)$. Thus, the second part of (6.16) vanishes and the first part can be combined with term B in (6.14) to yield

$$\left(\frac{\partial}{\partial t}+\mathbf{V}_g\cdot\nabla\right)\left[\frac{1}{f_0}\nabla^2\Phi+f+\frac{f_0}{\sigma}\frac{\partial^2\Phi}{\partial p^2}\right]=0 \tag{6.17}$$

This equation can be written in a more compact form as

$$\frac{Dq}{Dt}=0$$

$$q\equiv\frac{1}{f_0}\nabla^2\Phi+f+\frac{f_0}{\sigma}\frac{\partial^2\Phi}{\partial p^2},\qquad \frac{D}{Dt}\equiv\frac{\partial}{\partial t}+\mathbf{V}_g\cdot\nabla \tag{6.18}$$

Thus, the scalar quantity q is conserved following the geostrophic wind in isobaric coordinates. The scalar q, often called the *quasi-geostrophic potential vorticity*, is a linearized form of the potential vorticity $(\zeta + f)\,\partial\theta/\partial p$ discussed in Section 4.3. Given the distribution of Φ, (6.18) can be integrated in time to provide a forecast of the evolution of the Φ field. However, because V_g depends on the distribution of Φ the equation is highly nonlinear and numerical methods must be used for obtaining solutions.

6.2.4 The Omega Equation

A diagnostic equation for the vertical motion field may be obtained by eliminating χ between (6.12) and (6.13). To do this, we take the horizontal Laplacian of (6.12) to obtain

$$\nabla^2\frac{\partial\chi}{\partial p}=-\nabla^2\left[\mathbf{V}_g\cdot\nabla\left(\frac{\partial\Phi}{\partial p}\right)\right]-\sigma\nabla^2\omega \tag{6.19}$$

where we have again assumed that σ is a constant. We next differentiate (6.13) with respect to pressure yielding

$$\frac{\partial}{\partial p}(\nabla^2\chi) = -f_0 \frac{\partial}{\partial p}\left[\mathbf{V}_g \cdot \nabla\left(\frac{1}{f_0}\nabla^2\Phi + f\right)\right] + f_0^2 \frac{\partial^2 \omega}{\partial p^2} \tag{6.20}$$

Since the order of the operators on the left-hand side in (6.19) and (6.20) may be reversed, the result of subtracting (6.19) from (6.20) is to eliminate χ. After some rearrangement of terms, we obtain the *omega equation*

$$\underbrace{\left(\nabla^2 + \frac{f_0^2}{\sigma}\frac{\partial^2}{\partial p^2}\right)\omega}_{\text{A}} = \underbrace{\frac{f_0}{\sigma}\frac{\partial}{\partial p}\left[\mathbf{V}_g \cdot \nabla\left(\frac{1}{f_0}\nabla^2\Phi + f\right)\right]}_{\text{B}} + \underbrace{\frac{1}{\sigma}\nabla^2\left[\mathbf{V}_g \cdot \nabla\left(-\frac{\partial \Phi}{\partial p}\right)\right]}_{\text{C}} \tag{6.21}$$

Equation (6.21) involves only derivatives in space. It is, therefore, a diagnostic equation for the field of ω in terms of the instantaneous Φ field. The omega equation, unlike the continuity equation, gives a measure of ω which does not depend on accurate observations of the horizontal wind. In fact, direct wind observations are not required at all. This method is also superior to the vorticity equation method since no knowledge of the vorticity tendency is required. In fact, only observations of Φ at a single time are required to determine the ω field using (6.21).

As was the case for the geopotential tendency equation, the terms in (6.21) are all subject to straightforward physical interpretation. The differential operator in A is identical to the operator in term A of the tendency equation (6.14). Assuming that ω has a horizontal spatial dependence similar to that of χ and a vertical dependence similar to that of $\partial\chi/\partial p$, i.e.,

$$\omega = \sin(\pi p/p_0)\sin kx \sin ly$$

we can write

$$\left(\nabla^2 + \frac{f_0^2}{\sigma}\frac{\partial^2}{\partial p^2}\right)\omega \simeq \left[-(k^2 + l^2) - \frac{1}{\sigma}\left(\frac{f_0 \pi}{p_0}\right)^2\right]\omega$$

from which we can see that term A is proportional to $-\omega$.

Term B is called the *differential vorticity advection*. Clearly this term is proportional to the rate of increase with height of the advection of absolute vorticity. To understand the role of differential vorticity advection we again consider an idealized developing baroclinic system. Figure 6.9 indicates schematically the geopotential contours at 500 and 1000 mb for such a system. At the centers of the surface high and surface low, designated H and L, respectively, the vorticity advection at 1000 mb must be very small. However, at 500 mb the positive relative vorticity advection is a *maximum* above

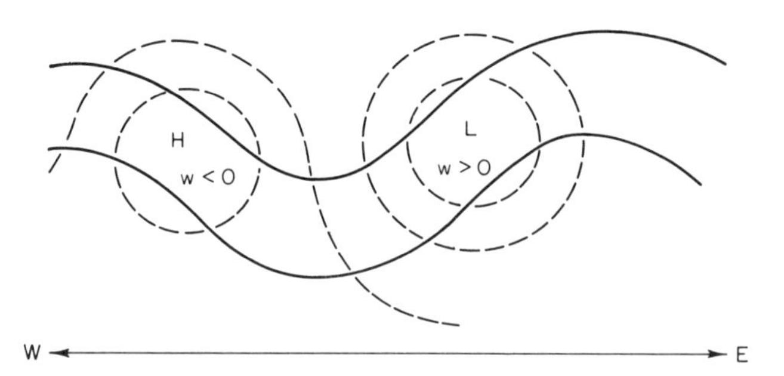

Fig. 6.9 Schematic 500-mb contours (solid lines) and 1000-mb contours (dashed lines) indicating regions of strong vertical motion due to differential vorticity advection.

the surface low, while negative relative vorticity advection is strongest above the surface high. Thus, for a short-wave system where relative vorticity advection is larger than the planetary vorticity advection

$$\frac{\partial}{\partial p}[\mathbf{V}_g \cdot \nabla(\zeta_g + f)] \begin{matrix} < 0 & \text{above point H} \\ > 0 & \text{above point L} \end{matrix}$$

Recalling that $\omega \simeq -w\rho g$ so that $\omega < 0$ implies upward vertical motion, we see from the omega equation (6.21) that differential vorticity advection is associated with rising motion above the surface low and subsidence above the surface high. This pattern of vertical motion is in fact just what is required to produce the thickness tendencies in the 1000–500-mb layer above the surface highs and lows. For example, above the surface low the positive vorticity advection creates a positive vorticity tendency. Since the geostrophic vorticity is proportional to the Laplacian of geopotential, increasing vorticity implies a falling geopotential. Thus, above the surface low $\chi < 0$. Hence the 500–1000-mb thickness is decreasing in that region. Since horizontal temperature advection is small above the center of the surface low, the only way to cool the atmosphere as required by the thickness tendency is by adiabatic cooling through the vertical motion field. Thus, the vertical motion maintains a hydrostatic temperature field (that is, a field in which temperature and thickness are proportional) in the presence of differential vorticity advection. Without this compensating vertical motion either the vorticity changes at 500 mb could not remain geostrophic or the temperature changes in the 1000–500-mb layer could not remain hydrostatic.

Term C of (6.21), which is merely the negative of the horizontal Laplacian of the thickness advection, is proportional to the thickness advection

$$\nabla^2\left[\mathbf{V}_g \cdot \nabla\left(-\frac{\partial\Phi}{\partial p}\right)\right] \propto -\mathbf{V}_g \cdot \nabla\left(-\frac{\partial\Phi}{\partial p}\right)$$

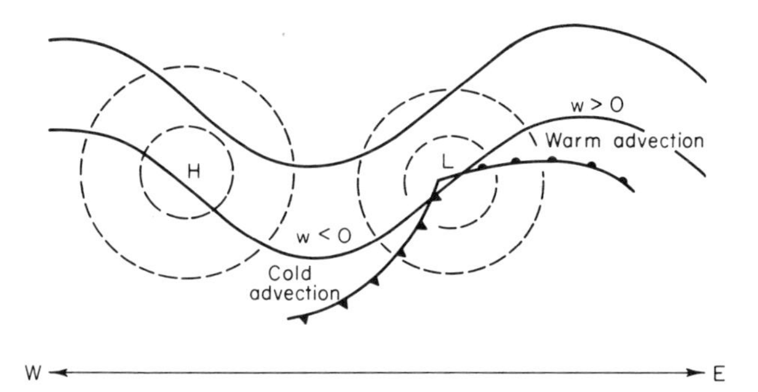

Fig. 6.10 Schematic 500-mb contours (thin solid lines), 1000 mb contours (dashed lines), and surface fronts (heavy lines) indicating regions of strong vertical motion due to temperature advection.

If there is warm (cold) advection, term C will be positive (negative) so that in the absence of differential vorticity advection ω would be negative (positive). Thus, as indicated in Fig. 6.10 rising motion will occur to the east of the surface low in the warm front zone, and sinking motion will occur west of the surface low behind the cold front.

Physically, this vertical motion pattern is required to keep the upper-level vorticity field geostrophic in the presence of the height changes caused by the thermal advection. For example, warm advection increases the 500–1000-mb thickness in the region of the 500-mb ridge. Thus, the geopotential height rises at the ridge and the anticyclonic vorticity must increase if geostrophic balance is to be maintained. Since vorticity advection cannot produce additional anticyclonic vorticity at the ridge, horizontal divergence is required to account for the negative vorticity tendency. Continuity of mass then requires that there be upward motion to replace the diverging air at the upper levels. By analogous arguments it can be shown that subsidence is required in the cold advection region beneath the 500-mb trough.

To summarize, we have shown as a result of scaling arguments that for synoptic scale motions where vorticity is constrained to be geostrophic and temperature is constrained to be hydrostatic, the vertical motion field is determined uniquely by the geopotential field. Further, we have shown that this vertical motion field is just that required to ensure that changes in vorticity will be geostrophic and changes in temperature will be hydrostatic. These constraints, whose importance can hardly be overemphasized, will be elaborated in Section 9.2.1.

Finally, it is worth mentioning that the two terms on the right-hand side of the omega equation (6.21), although they can be clearly associated with separate physical processes, often tend in the real atmosphere to have a large degree of cancellation between them. It is thus sometimes more convenient to rewrite the right hand side to combine terms B and C. This subject is discussed further in Chapter 9.

6.3 Idealized Model of a Developing Baroclinic System

In the previous section, we have shown that for synoptic scale systems the fields of vertical motion and geopotential tendency are determined to a first approximation by the three-dimensional distribution of geopotential. The results of our diagnostic analyses using the geopotential tendency and omega equations can now be combined to illustrate the essential structural characteristics of a developing baroclinic wave. For reference, we restate here the qualitative content of the tendency and omega equations:

Geopotential Tendency Equation

$$\text{Geopotential}\begin{pmatrix}\text{fall}\\ \text{rise}\end{pmatrix} \propto (\pm)\ \text{Vorticity advection}$$

$$+\ \text{Rate of decrease with height of}\begin{pmatrix}\text{cold}\\ \text{warm}\end{pmatrix}\text{advection}$$

Omega Equation

$$\begin{pmatrix}\text{Rising}\\ \text{Sinking}\end{pmatrix}\text{motion} \propto \text{Rate of increase with height of}\ (\pm)\ \text{vorticity advection}$$

$$+\begin{pmatrix}\text{warm}\\ \text{cold}\end{pmatrix}\text{advection}$$

In Fig. 6.11 the vertical motion and horizontal divergence–convergence fields and their relationship to the 500- and 1000-mb geopotential fields are illustrated schematically for a developing baroclinic wave. Also indicated are the physical processes which give rise to the vertical circulation in various regions.

Additional structural features, including those which can be diagnosed with the tendency equation, are summarized in Table 6.1. In this table the signs of various physical parameters are indicated for vertical columns located at the position of (A) the 500-mb trough, (B) the surface low, and (C) the 500-mb ridge. It can be seen from this table that in all cases the vertical motion and divergence fields act to keep the temperature changes hydrostatic and vorticity changes geostrophic. Following the nomenclature of Chapter 5, we may regard the vertical and divergent motions as a *secondary circulation* imposed by the simultaneous constraints of geostrophic and hydrostatic balance. Thus, as was stated in Chapter 5, secondary circulations can be driven by processes other than friction. The secondary circulation described in this chapter is completely independent of the circulation driven

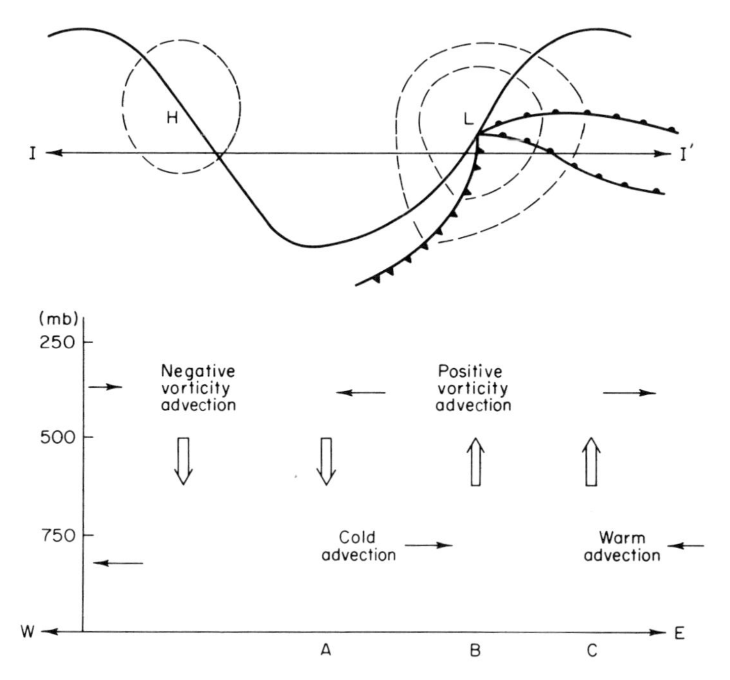

Fig. 6.11 Secondary circulation associated with a developing baroclinic wave: (top) schematic 500-mb contour (solid line), 1000-mb contours (dashed lines), and surface fronts; (bottom) vertical profile through the line II′ indicating the divergent and vertical motion fields.

by Ekman layer pumping. In fact, it is observed that in midlatitude synoptic scale systems, the vertical velocity forced by frictional convergence in the Ekman layer is generally much smaller than the vertical velocity due to differential vorticity advection. For this reason we have neglected Ekman layer friction in developing the equations of the quasi-geostrophic system.

It is also of interest to note that the secondary circulation in a developing baroclinic system always acts to oppose the horizontal advection fields. Thus, the divergent motions tend partly to cancel the vorticity advection and the adiabatic temperature changes due to vertical motion tend to cancel partly the thermal advection. This tendency of the secondary flow to cancel partly the advective changes has important implications for the evolution of the systems which will be discussed in Chapter 9.

It should now be clear that a secondary divergent circulation is necessary to satisfy the twin constraints of geostrophic and hydrostatic balance. We have, however, not yet discussed the driving force which produces this secondary circulation. Referring again to Fig. 6.11, we see that in the region of the 500-mb trough (column A) cold advection causes the geopotential

Table 6.1 *Characteristics of a Developing Baroclinic Disturbance*

Physical parameter	A 500-mb trough	B Surface low	C 500-mb ridge
$\dfrac{\partial\,\delta\Phi}{\partial t}$ (500–1000 mb)	Negative (thickness advection partly cancelled by adiabatic warming)	Negative (adiabatic cooling)	Positive (thickness advection partly cancelled by adiabatic cooling)
w (500 mb)	Negative	Positive	Positive
$\dfrac{\partial\Phi}{\partial t}$ (500 mb)	Negative (differential thickness advection)	Negative (vorticity advection)	Positive (differential thickness advection)
$\dfrac{\partial\zeta_g}{\partial t}$ (1000 mb)	Negative (divergence)	Positive (convergence)	Positive (convergence)
$\dfrac{\partial\zeta_g}{\partial t}$ (500 mb)	Positive (convergence)	Positive (advection partly canceled by divergence)	Negative (divergence)

height to fall and thus intensifies the horizontal pressure gradient. The wind therefore becomes slightly subgeostrophic and experiences an acceleration across the isobars toward lower pressure. It is this cross isobaric ageostrophic wind component which is responsible for the convergence which spins up the vorticity at 500 mb so that it adjusts geostrophically to the new geopotential distribution. In terms of the momentum balance, the cross isobaric flow is accelerated by the pressure gradient force so that the wind speed adjusts back toward geostrophic balance. As the system evolves there is a continuous interplay between the changing geopotential and wind distributions with the secondary circulation always acting to restore geostrophic balance. In the region of the 500-mb ridge analogous arguments apply, but in this case the subgeostrophic flow leads to a divergent secondary circulation. In both cases, as we will see in Chapter 10, the ageostrophic flow towards lower pressure is associated with conversion of energy from potential energy to kinetic energy.

Finally, it should be kept in mind that the actual trajectories of air parcels moving through a synoptic system are influenced strongly by the secondary circulation. Thus, trajectory computations based on an assumption of geostrophic motion on an isobaric surface would lead to large errors in the vicinity of the warm and cold fronts as shown in Fig. 6.12. This figure further illustrates the profound difference between trajectories and streamlines for transient motion systems.

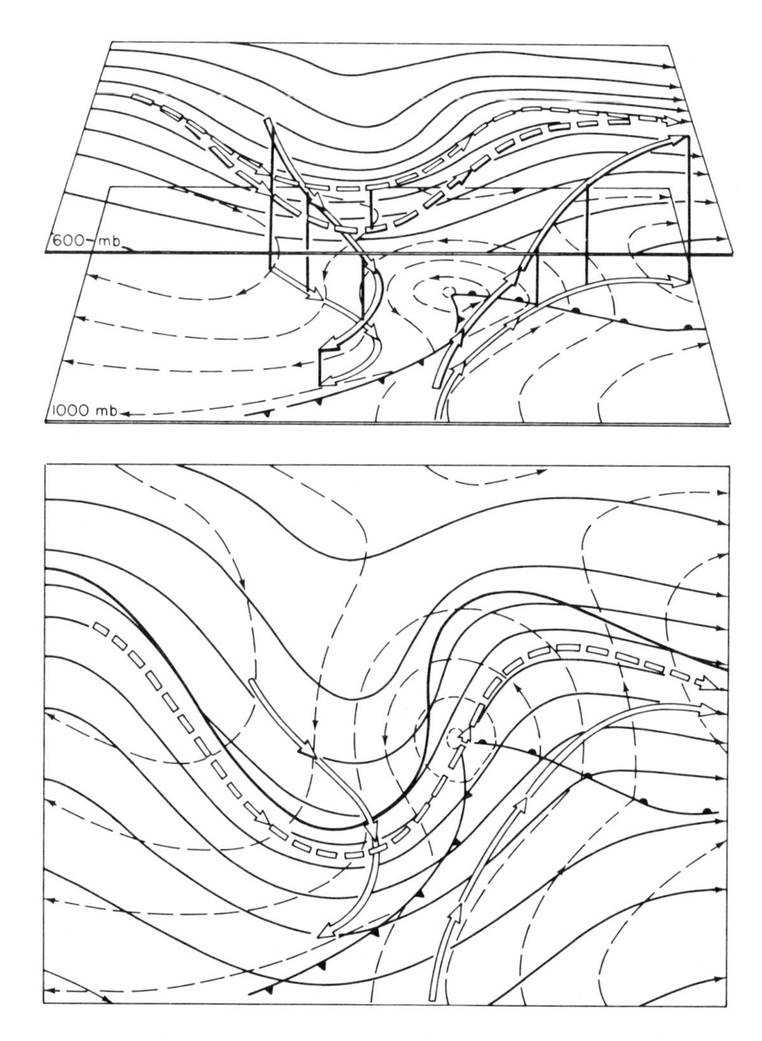

Fig. 6.12 Schematic 600-mb contours (solid) and 1000-mb contours (dashed) for a developing baroclinic wave: upper part gives persepective view indicating selected three-dimensional trajectories (arrows) and their projections on the 1000- or 600-mb surfaces. Dashed trajectory indicates that parcels centered along the 600-mb jet upstream of the developing wave pass through the disturbance approximately along the 600-mb streamlines. However, as shown by the other trajectories, air parcels originating either poleward or equatorward of the jet core are strongly influenced by the vertical motion field associated with the fronts. (After Palmén and Newton, 1969.)

Problems

1. Show that the static stability parameter

$$\sigma = -\frac{\alpha}{\theta}\frac{\partial\theta}{\partial p}$$

may be written in terms of Φ as

$$\sigma = \frac{\partial^2\Phi}{\partial p^2} - \frac{1}{p}\frac{\partial\Phi}{\partial p}\left(\frac{R}{c_p} - 1\right) = \frac{1}{p^2}\left(\frac{\partial}{\partial \ln p} - \frac{R}{c_p}\right)\frac{\partial\Phi}{\partial \ln p}$$

2. Show that for an isothermal atmosphere σ as defined in Problem 1 varies inversely as the square of the pressure.

3. Suppose that on the 50-kPa (500-mb) surface the relative vorticity at a certain location at 45°N latitude is increasing at a rate of 3×10^{-6} s^{-1}/3 h. The wind is from the southwest at 20 m s^{-1} and the relative vorticity decreases towards the northeast at a rate of 4×10^{-6} s^{-1}/100 km. Use the quasi-geostrophic vorticity equation to estimate the horizontal divergence at this location on a β plane.

4. Given the following expression for the geopotential field

$$\Phi = \Phi_0(p) + cf_0\{y[\cos(\pi p/p_0) - 1] + k^{-1}\sin k(x - ct)\}$$

where Φ_0 is a function of p alone, c is a constant speed, k a zonal wave number, and $p_0 = 100$ kPa (1000 mb):

(a) Use the quasi-geostrophic vorticity equation to obtain the horizontal divergence field consistent with this Φ field. (Assume that $df/dy = 0$.)

(b) Assuming that $\omega(p_0) = 0$ obtain an expression for $\omega(x, y, p, t)$ by integrating the continuity equation with respect to pressure.

(c) Sketch the geopotential fields at 75 kPa (750 mb) and 25 kPa (250 mb). Indicate regions of maximum divergence and convergence and positive and negative vorticity advection.

5. For the geopotential distribution of Problem 4 obtain an alternative expression for ω by using the adiabatic thermodynamic energy equation (6.8). Assume that σ is a constant. For what value of k does this expression for ω agree with that obtained in Problem 4?

6. As an additional check on the results of Problems 4 and 5 use the omega equation (6.21) to obtain an expression for ω. Note that the three expressions for ω agree only for one value of k. Thus, the geopotential field $\Phi(x, y, p, t)$ of Problem 4 is consistent with quasi-geostrophic dynamics only for one value of the zonal wave number.

7. Suppose that the geopotential distribution at a certain time has the form

$$\Phi(x, y, p) = \Phi_0(p) - f_0 U_0 y \cos(\pi p/p_0) + f_0 c k^{-1} \sin kx$$

Where U_0 is a constant zonal speed and all other constants are as in Problem 4. Assuming that f and σ are constants, show by evaluating the terms in the right-hand side of the tendency equation (6.14) that $\chi = 0$ provided that $k^2 = \sigma^{-1}(f_0 \pi/p_0)^2$. Make qualitative sketches of the geopotential fields at 750 mb and 250 mb for this case. Indicate regions of maximum positive and negative vorticity advection at each level. (Note: the wavelength corresponding to this value of k is called the *radius of deformation.*)

8. For the geopotential field of Problem 7 use the omega equation (6.21) to find an expression for ω for the conditions when $\chi = 0$. *Hint*: let $\omega = W_0 \cos kx \sin(\pi p/p_0)$ where W_0 is a constant to be determined. Sketch a cross section in the x, p plane indicating trough and ridge lines, vorticity maxima and minima, vertical motion and divergence patterns, and locations of maximum cold and warm temperature advection.

9. Given the following expression for the geopotential field,

$$\Phi = \Phi_0(p) + f_0[-Uy + k^{-1}V \cos(\pi p/p_0) \sin k(x - ct)]$$

where U, V, and c are constant speeds, use the quasi-geostrophic vorticity equation (6.10) to obtain an estimate of ω. Assume that $df/dy = \beta$ is a constant (*not* zero), and that ω vanishes for $p = p_0$.

10. For the conditions given in Problem 9 use the adiabatic thermodynamic energy equation to obtain an alternative estimate for ω. Determine the value of c for which this estimate of ω agrees with that found in Problem 9.

11. For the conditions given in Problem 9 use the omega equation (6.21) to obtain an expression for ω. Verify that this result agrees with the results of Problems 9 and 10. Sketch the phase relationship between Φ and ω at 250 mb and 750 mb. What is the amplitude of ω if $\beta = 2 \times 10^{-11}$ $m^{-1} s^{-1}$, $U = 25$ m s^{-1}, $V = 8$ m s^{-1}, $k = 2\pi/(10^4$ km$)$, $f_0 = 10^{-4}$ s^{-1}, $\sigma = 10^{-4}$ Pa^{-2} m^2 s^{-2}, and $p_0 = 10^2$ kPa?

Suggested Reference

Palmén and Newton, *Atmospheric Circulation Systems*, contains excellent descriptions of the observed structure of the mean zonal winds, planetary waves, and synoptic scale disturbances.

Chapter

7 Atmospheric Oscillations: Linear Perturbation Theory

In Chapter 8 we will discuss numerical techniques for solving the equations governing large-scale atmospheric motions. If the objective is to produce an accurate forecast of the circulation at some future time, a detailed numerical model based on the primitive equations and including processes such as latent heating, radiative transfer, and frictional dissipation should produce the best results. However, the inherent complexity of such a model generally precludes any simple interpretation of the physical processes which produce the predicted circulation. If we wish to gain physical insight into the fundamental nature of atmospheric motions, it is necessary to employ simplified models in which certain processes are omitted, and compare the results with those of more complete models. This is, of course, just what has been done in deriving the filtered quasi-geostrophic model. However, the quasi-geostrophic system still requires numerical solution of a complicated nonlinear system of equations. It is difficult to gain an appreciation for the processes essential to the development of baroclinic wave disturbances by examining the results of such numerical integrations.

In this chapter we will discuss a simple technique, the *perturbation method*, which is ideally suited for *qualitative* analysis of the nature of atmospheric motions. We then use perturbation theory to examine several types of pure waves in the atmosphere. In Chapter 9 perturbation theory will be applied to the problem of the development of synoptic wave disturbances.

7.1 The Perturbation Method

In the perturbation method all field variables are divided into two parts, a *basic state* portion which is usually assumed to be independent of time and longitude and a *perturbation* portion which is the local deviation of the field from the basic state. Thus, for example, if $\bar{u}$ designates a time and longitude averaged zonal velocity and u' is the deviation from that average, then the complete zonal velocity field is $u(x, t) = \bar{u} + u'(x, t)$. In that case, for example, the inertial acceleration $u\,\partial u/\partial x$ can be written

$$u\frac{\partial u}{\partial x} = (\bar{u} + u')\frac{\partial}{\partial x}(\bar{u} + u') = \bar{u}\frac{\partial u'}{\partial x} + u'\frac{\partial u'}{\partial x}$$

The basic assumptions of perturbation theory are that the basic state variables must themselves satisfy the governing equations when the perturbations are set to zero, and the perturbation fields must be small enough so that all terms in the governing equations which involve products of the perturbation variables can be neglected. The latter requirement would be met in the above example if $|u'/\bar{u}| \ll 1$ so that

$$\left|\bar{u}\frac{\partial u'}{\partial x}\right| \gg \left|u'\frac{\partial u'}{\partial x}\right|$$

By neglecting all terms which are nonlinear in the perturbations, the nonlinear governing equations are reduced to linear differential equations in the perturbation variables in which the basic state variables are specified coefficients. These equations can then be solved by standard methods to determine the character and structure of the perturbations in terms of the known basic state. Usually the perturbations are assumed to be sinusoidal waves and solution of the perturbation equations determines such characteristics as the propagation speed, vertical structure, and conditions for growth or decay of the waves. The perturbation technique is especially useful in studying the stability of a given basic state flow with respect to small superposed perturbations. This application will be the subject of Chapter 9.

7.2 Properties of Waves

Wave motions are oscillations which propagate in space. In this chapter we are concerned with linear sinusoidal wave motions. Many of the mechanical properties of such waves are also features of a simpler system—the simple harmonic oscillator. An important property of the harmonic oscillator is that the period, or time required to execute a single oscillation, is independent of the amplitude of the oscillation. For most natural vibratory systems this

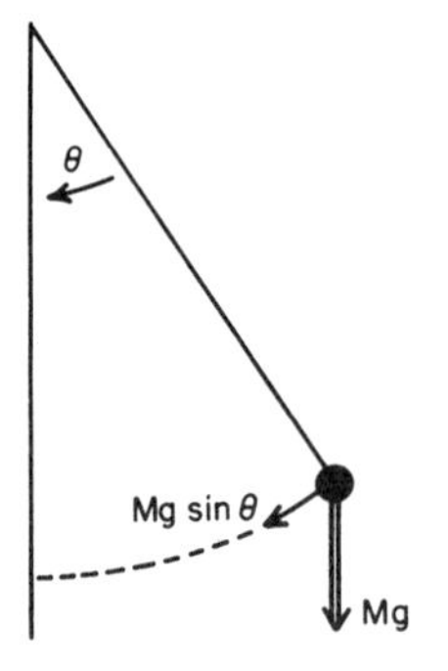

Fig. 7.1 A simple pendulum.

condition holds only for oscillations of sufficiently small amplitude. The classical example of such a system is the simple pendulum (Fig. 7.1) consisting of a mass M suspended by a massless string of length l, free to perform small oscillations about the equilibrium position $\theta = 0$. The component of the gravity force parallel to the direction of motion is $-Mg \sin \theta$. Thus, the equation of motion is

$$Ml \frac{d^2\theta}{dt^2} = -Mg \sin \theta$$

Now for small displacements $\sin \theta \simeq \theta$ so that the governing equation becomes

$$\frac{d^2\theta}{dt^2} + \nu^2\theta = 0 \tag{7.1}$$

where $\nu^2 \equiv g/l$. The harmonic oscillator equation (7.1) has the general solution

$$\theta = \theta_1 \cos \nu t + \theta_2 \sin \nu t = \theta_0 \cos(\nu t - \alpha)$$

where θ_1, θ_2, θ_0, and α are constants determined by the initial conditions (see Problem 1, page 168), and ν is the frequency of oscillation. The complete solution can thus be expressed in terms of an amplitude θ_0 and a *phase* $\phi \equiv \nu t - \alpha$. The phase varies linearly in time by a factor of 2π radians per wave period.

Propagating waves may also be characterized by their amplitudes and phases. However, in propagating waves the phase depends not only on time but on one or more space variables as well. Thus, for a one-dimensional wave propagating in the x direction, $\phi = kx - \nu t + \alpha$. Here, the *wave number* k is defined as 2π divided by the wavelength. For propagating waves the phase is constant for an observer moving at the *phase speed* $c = \nu/k$. This may be

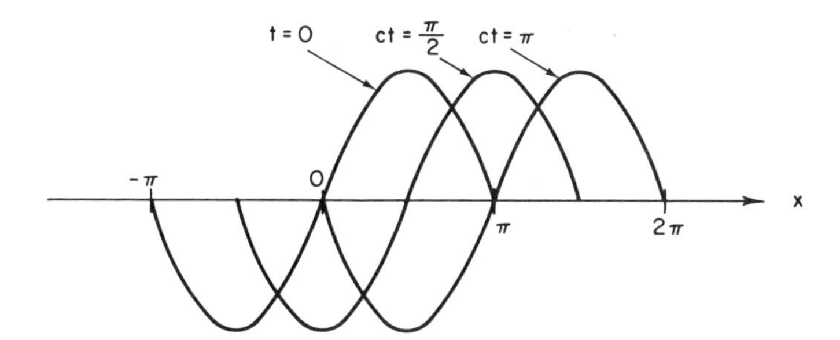

Fig. 7.2 A sinusoidal wave traveling in the positive x direction at speed c. (Wave number is assumed to be unity.)

verified by observing that if phase is to remain constant following the motion,

$$\frac{d\phi}{dt} = \frac{d}{dt}(kx - vt - \alpha) = k\frac{dx}{dt} - v = 0.$$

Thus, $dx/dt = v/k$ for phase to be constant. For $v > 0$ and $k > 0$ we have $c > 0$ so that phase propagates in the positive direction as illustrated for a sinusoidal wave in Fig. 7.2.

7.2.1 Fourier Series

The representation of a perturbation as a simple sinusoidal wave might seem an oversimplification since disturbances in the atmosphere are never purely sinusoidal. It turns out, however, that any reasonably well-behaved function of longitude can be represented in terms of a zonal mean plus a *Fourier* series of sinusoidal components:

$$f(x) = \sum_{m=1}^{\infty} (A_m \sin k_m x + B_m \cos k_m x) \tag{7.2}$$

where $k_m = 2\pi m/L$ is the zonal wave number, L is the distance around a latitude circle, and m is an integer designating the number of waves around a circle of latitude. The coefficients A_m are calculated by multiplying both sides of (7.2) by $\sin(2\pi nx/L)$ and integrating around a latitude circle. Applying the orthogonality relationships

$$\int_0^l \sin\frac{2\pi mx}{L}\sin\frac{2\pi nx}{L}\,dx = \begin{Bmatrix} 0, & m \neq n \\ L/2, & m = n \end{Bmatrix}$$

we obtain

$$A_m = \frac{2}{L}\int_0^L f(x)\sin\frac{2\pi mx}{L}\,dx$$

In a similar fashion, multiplying both sides in (7.2) by $\cos(2\pi nx/L)$ and integrating gives

$$B_m = \frac{2}{L}\int_0^L f(x)\cos\frac{2\pi mx}{L}\,dx$$

A_m and B_m are called the *Fourier coefficients*, whereas

$$f_m(x) = A_m \sin k_m x + B_m \cos k_m x \tag{7.3}$$

is called the *mth Fourier component* or *mth harmonic* of the function $f(x)$. If the Fourier coefficients are computed for a quantity such as the longitudinal dependence of the observed geopotential perturbation, it turns out that the largest amplitude Fourier components will be those for which m is close to the observed number of troughs or ridges around a latitude circle. When only qualitative information is desired, it is usually sufficient to limit the analysis to a single typical Fourier component, and assume that the behavior of the actual field will be similar to that of the component. The expression for a Fourier component may be written more compactly by using complex exponential notation. According to the Euler formula

$$e^{i\phi} = \cos\phi + i\sin\phi$$

where $i \equiv \sqrt{-1}$ is the imaginary unit. Thus, we can write

$$\begin{aligned} f_m(x) &= \mathrm{Re}\{C_m \exp(ik_m x)\} \\ &= \mathrm{Re}\{C_m \cos k_m x + iC_m \sin k_m x\} \end{aligned} \tag{7.4}$$

where Re{ } denotes "real part of," and C_m is a complex coefficient. Comparing (7.3) and (7.4) we see that the two representations of $f_m(x)$ are identical provided that

$$B_m = \mathrm{Re}\{C_m\} \qquad \text{and} \qquad A_m = -\mathrm{Im}\{C_m\}$$

where Im{ } stands for "imaginary part of." This exponential notation will generally be used for applications of the perturbation theory below as well as in Chapter 9.

7.2.2 Dispersion and Group Velocity

A fundamental property of linear oscillators is that the frequency of oscillation ν depends only on the physical characteristics of the oscillator, not on the motion itself. For propagating waves, however, ν generally depends on the wave number of the perturbation as well as the physical properties of the medium. Thus, since $c = \nu/k$, we see that the phase speed also depends on the wave number except in the special case where $\nu \propto k$. Waves in which the phase speed varies with k are referred to as *dispersive*, and the formula which

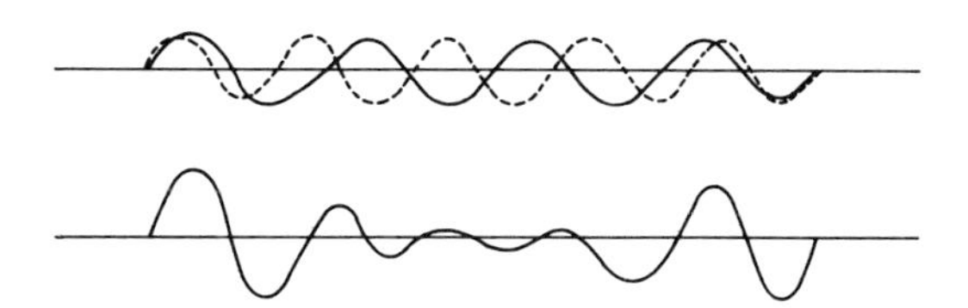

Fig. 7.3 Wave groups formed from two sinusoidal components of slightly different wavelengths. For nondispersive waves the pattern in the lower part of the diagram propagates without change of shape. For dispersive waves the shape of the pattern changes in time.

relates ν and k is called a *dispersion relationship*. Some types of waves, such as acoustic waves, have phase speeds which are independent of the wave number. In these so-called *nondispersive* waves a transient disturbance consisting of a number of Fourier wave components (a *wave group*) will preserve its shape as it propagates in space at the phase speed of the wave.

For dispersive waves, however, the shape of a wave group will not remain constant as the waves propagate. Since the individual Fourier components of a wave group alternately reinforce and cancel each other depending on the relative phases of the components, the energy of the group will be concentrated in limited regions as illustrated in Fig. 7.3.

When waves are dispersive, the speed of the wave group is generally different from the average phase speed of the individual Fourier components. Hence, as shown in Fig. 7.4, individual wave components may move through the group as the group propagates along. Furthermore, the group generally broadens in the course of time, that is, the energy is *dispersed*.

An expression for the *group velocity*, which is the velocity at which the observable disturbance, and hence the energy, propagates can be derived as

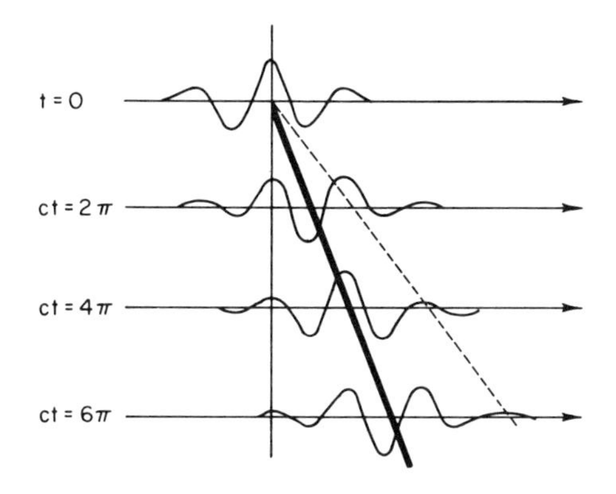

Fig. 7.4 Propagation of a wave group. Heavy line shows group speed, dashed line shows phase speed.

follows: We consider two horizontally propagating waves of equal amplitude but slightly different wavelengths with wave numbers and frequencies differing by $2\,\Delta k$ and $2\,\Delta\nu$, respectively. The amplitude of the total disturbance is thus

$$\psi(x, t) = e^{i[(k+\Delta k)x-(\nu+\Delta\nu)t]} + e^{i[(k-\Delta k)x-(\nu-\Delta\nu)t]}$$

Rearranging terms we get

$$\psi = [e^{i(\Delta kx-\Delta\nu t)} + e^{-i(\Delta kx-\Delta\nu t)}]e^{i(kx-\nu t)}$$

or

$$\psi = 2\cos(\Delta k\, x - \Delta\nu\, t)e^{i(kx-\nu t)} \tag{7.5}$$

The disturbance (7.5) is the product of a high-frequency *carrier wave* of wavelength $2\pi/k$ whose phase velocity is the average for the two Fourier components, and a low frequency *envelope* of wavelength $2\pi/\Delta k$ which travels at the speed $\Delta\nu/\Delta k$. Thus, in the limit as $\Delta k \to 0$, the horizontal velocity of the envelope, or *group velocity*, is just

$$U_g = \frac{\partial\nu}{\partial k}$$

Thus, the wave energy indeed propagates at the group velocity. This result applies generally to arbitrary wave envelopes provided that the wavelength of the wave group, $2\pi/\Delta k$, is large compared to the wavelength of the dominant component, $2\pi/k$.

7.3 Simple Wave Types

Waves in fluids result from the action of restoring forces on fluid parcels which have been displaced from their equilibrium positions. The restoring forces may be due to compressibility, gravity, rotation, or electromagnetic effects. In the present section we consider the two simplest examples of linear waves in fluids, acoustic waves and shallow water gravity waves.

7.3.1 Acoustic or Sound Waves

Sound waves, or acoustic waves, are longitudinal waves—that is, waves in which the particle oscillations are parallel to the direction of propagation. Sound is propagated by the alternating adiabatic compression and expansion of the medium. As an example, in Fig. 7.5 we show a schematic section along a tube which has a diaphragm at its left end. If the diaphragm is deflected to the right by striking it as shown in Fig. 7.5a, the air between points 1 and 2 will be compressed. Thus, there will be a pressure gradient which will accelerate

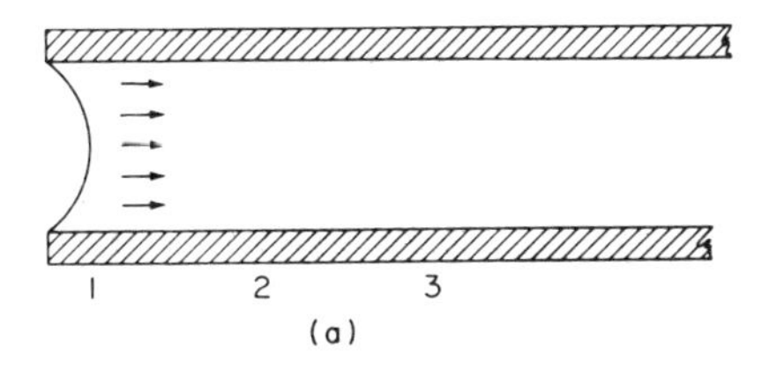

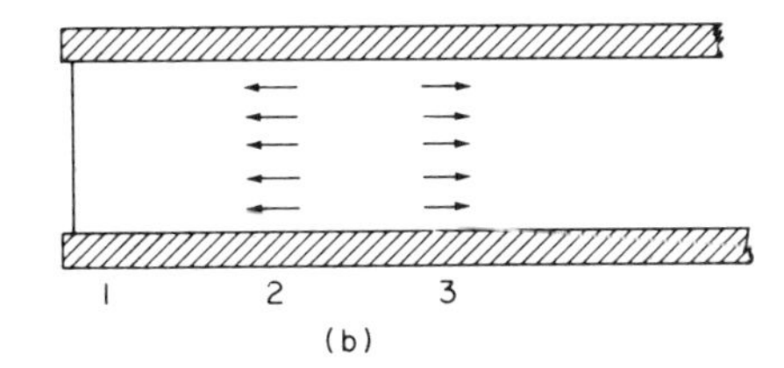

Fig. 7.5 Schematic diagram illustrating the propagation of a sound wave in a tube with a flexible diaphragm at the left end.

the air into the region between points 2 and 3. As a result the air will be compressed between points 2 and 3, and as shown in Fig. 7.5b the pressure gradient force will cause the air to accelerate out of that region, thus creating a new region of compression to the right of point 3. By this continual process of adiabatic increase and decrease of the pressure through alternating compression and rarefaction, the disturbance excited by striking the membrane will move rightward down the pipe. Individual air parcels do not, however, have a net rightward motion; they only oscillate back and forth while the pressure pattern moves rightward at the speed of sound.

To introduce the perturbation method we consider one-dimensional sound waves propagating in a straight pipe parallel to the x axis. To exclude the possibility of *transverse* oscillations (that is, oscillations in which the particle motion is at right angles to the direction of phase propagation), we assume at the outset that $v = w = 0$. In addition we eliminate all dependence on y and z by assuming that $u = u(x, t)$. With these restrictions the momentum equation, continuity equation, and thermodynamic energy equation for adiabatic motion are, respectively,

$$\frac{du}{dt} + \frac{1}{\rho}\frac{\partial p}{\partial x} = 0 \tag{7.6}$$

$$\frac{d\rho}{dt} + \rho\frac{\partial u}{\partial x} = 0 \tag{7.7}$$

$$\frac{d \ln \theta}{dt} = 0 \tag{7.8}$$

where for this case

$$\frac{d}{dt} = \frac{\partial}{\partial t} + u\frac{\partial}{\partial x}$$

Recalling that

$$\theta = \frac{p}{\rho R}\left(\frac{p_0}{p}\right)^{R/c_p} \qquad \text{where } p_0 = 1000 \quad \text{mb}$$

we may eliminate θ in (7.8) to give

$$\frac{1}{\gamma}\frac{d\ln p}{dt} - \frac{d\ln \rho}{dt} = 0 \tag{7.9}$$

where $\gamma = c_p/c_v$. Eliminating ρ between (7.7) and (7.9) gives

$$\frac{1}{\gamma}\frac{d\ln p}{dt} + \frac{\partial u}{\partial x} = 0 \tag{7.10}$$

The dependent variables are now divided into constant basic state portions (denoted by overbars) and perturbation portions (denoted by primes):

$$\begin{aligned} u(x, t) &= \bar{u} + u'(x, t) \\ p(x, t) &= \bar{p} + p'(x, t) \\ \rho(x, t) &= \bar{\rho} + \rho'(x, t) \end{aligned} \tag{7.11}$$

Substituting (7.11) into (7.6) and (7.10) we obtain

$$\frac{\partial}{\partial t}(\bar{u} + u') + (\bar{u} + u')\frac{\partial}{\partial x}(\bar{u} + u') + \frac{1}{\bar{\rho} + \rho'}\frac{\partial}{\partial x}(\bar{p} + p') = 0$$

$$\frac{\partial}{\partial t}(\bar{p} + p') + (\bar{u} + u')\frac{\partial}{\partial x}(\bar{p} + p') + \gamma(\bar{p} + p')\frac{\partial}{\partial x}(\bar{u} + u') = 0$$

We next observe that provided $|\rho'/\bar{\rho}| \ll 1$ we can use the binomial expansion to approximate the density term as

$$\frac{1}{\bar{\rho} + \rho'} = \frac{1}{\bar{\rho}}\left(1 + \frac{\rho'}{\bar{\rho}}\right)^{-1} \simeq \frac{1}{\bar{\rho}}\left(1 - \frac{\rho'}{\bar{\rho}}\right)$$

Neglecting products of the perturbation quantities and noting that the basic state fields are constants, we obtain the linear perturbation equations[1]

$$\left(\frac{\partial}{\partial t} + \bar{u}\frac{\partial}{\partial x}\right)u' + \frac{1}{\bar{\rho}}\frac{\partial p'}{\partial x} = 0 \tag{7.12}$$

$$\left(\frac{\partial}{\partial t} + \bar{u}\frac{\partial}{\partial x}\right)p' + \gamma\bar{p}\frac{\partial u'}{\partial x} = 0 \tag{7.13}$$

[1] It is not necessary that $|u'/\bar{u}| \ll 1$ for this linearization to be valid. In fact $\bar{u}$ may be zero, but (7.12) and (7.13) will still be approximately valid, provided u' and p' are sufficiently small so that the terms $u'\,\partial u'/\partial x$ and $u'\,\partial p'/\partial x$ are small compared to the terms retained in (7.12) and (7.13), respectively.

Eliminating u' by operating on (7.13) with $(\partial/\partial t + \bar{u}\,\partial/\partial x)$ and substituting from (7.12) we get

$$\left(\frac{\partial}{\partial t} + \bar{u}\frac{\partial}{\partial x}\right)^2 p' - \frac{\gamma\bar{p}}{\bar{\rho}}\frac{\partial^2 p'}{\partial x^2} = 0 \tag{7.14}$$

Following the discussion of the previous section, we assume a solution of the form

$$p' = Ae^{ik(x-ct)} \tag{7.15}$$

where for brevity we omit the Re{ } notation; but it is to be understood that only the real part of (7.15) has physical significance. Substituting the assumed solution (7.15) into (7.14) we find that the phase speed c must satisfy

$$(-ikc + ik\bar{u})^2 - \frac{\gamma\bar{p}}{\bar{\rho}}(ik)^2 = 0$$

where we have canceled out the factor $Ae^{ik(x-ct)}$ which is common to both terms. Solving for c gives

$$c = \bar{u} \pm \left(\frac{\gamma\bar{p}}{\bar{\rho}}\right)^{1/2} = \bar{u} \pm (\gamma R\bar{T})^{1/2} \tag{7.16}$$

Therefore (7.15) is a solution of (7.14) provided that the phase speed satisfies (7.16). According to (7.16) the speed of wave propagation relative to the zonal current is $(\gamma R\bar{T})^{1/2}$. This quantity is called the *adiabatic speed of sound*. The mean zonal velocity here plays only a role of *doppler shifting* the sound wave so that the frequency

$$\nu = kc = k\bar{u} \pm k(\gamma R\bar{T})^{1/2}$$

corresponding to a given wave number k appears higher to an observer downstream from the source than to an upstream observer.

7.3.2 Shallow Water Gravity Waves

As a second example of pure wave motion we consider the horizontally propagating oscillations known as shallow water waves. Shallow water gravity waves can exist only if the fluid has a free surface or an internal density discontinuity. As shown in the previous subsection, in acoustic waves the restoring force is parallel to the direction of propagation of the wave. In shallow water gravity waves, however, the restoring force is in the vertical, so that it is transverse to the direction of propagation.

The mechanism for propagation of gravity waves is most easily understood by considering a disturbance on the free surface of an incompressible fluid such as water. Suppose, as shown in Fig. 7.6, there is a depression in the free

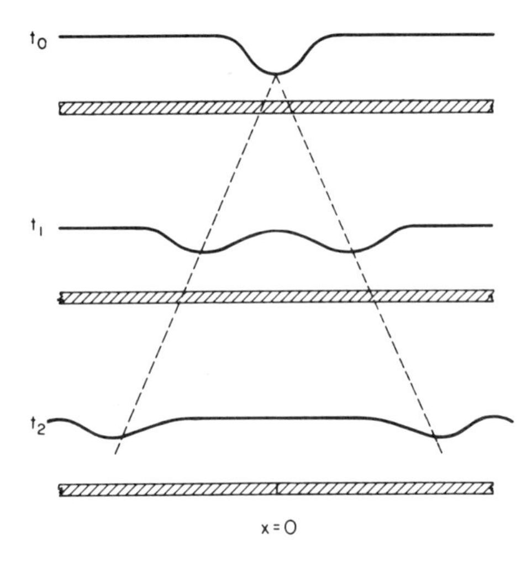

Fig. 7.6 The propagation of a surface gravity wave initiated by a depression at time t_0.

surface centered at the origin $x = 0$ at time $t = t_0$. As a result of this disturbance there will be a horizontal pressure gradient which will cause an acceleration toward the origin. Thus, fluid will converge horizontally into the column of water centered at $x = 0$. However, since the fluid is incompressible, this convergence must be compensated by horizontal divergence on both sides. Thus at time $t = t_1$ there will be depressions on both sides of the original disturbance. Again, the unbalanced horizontal pressure gradients will cause accelerations into the depressions and the result will be a continual outward propagation of the disturbance due to alternating horizontal convergence and divergence in individual columns of fluid.

As a specific example we consider waves propagating along the interface between two homogeneous incompressible fluids of differing density. The assumption of incompressibility is sufficient to exclude sound waves from the system, and we can thus isolate the gravity waves.

We consider a two-layer fluid system as shown in Fig. 7.7. If the density of the lower layer ρ_1 is greater than the density of the upper layer ρ_2, the system is stably stratified. Since both ρ_1 and ρ_2 are constants, the horizontal pressure gradient in each layer is independent of height if the pressure is hydrostatic. This may be verified by differentiating the hydrostatic approximation with respect to x:

$$\frac{\partial}{\partial z}\left(\frac{\partial p}{\partial x}\right) = -\frac{\partial \rho}{\partial x} g = 0$$

For simplicity, we assume that there is no horizontal pressure gradient in the upper layer. The pressure gradient in the lower layer can be obtained by

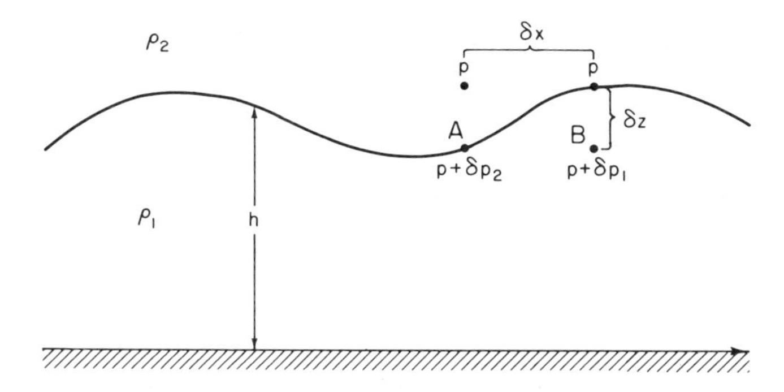

Fig. 7.7 A two-layer fluid system.

vertical integration of the hydrostatic equation. For the points A and B shown in Fig. 7.7, we find, respectively,

$$p + \delta p_1 = p + \rho_1 g \, \delta z = p + \rho_1 g \frac{\partial h}{\partial x} \delta x$$

$$p + \delta p_2 = p + \rho_2 g \, \delta z = p + \rho_2 g \frac{\partial h}{\partial x} \delta x$$

where $\partial h/\partial x$ is the slope of the interface. Taking the limit $\delta x \to 0$, we obtain the pressure gradient in the lower layer

$$\lim_{\delta x \to 0} \left[\frac{(p + \delta p_1) - (p + \delta p_2)}{\delta x} \right] = g \, \Delta\rho \frac{\partial h}{\partial x}$$

where $\Delta\rho = \rho_1 - \rho_2$.

We assume that the motion is two dimensional in the x,z plane. The x-momentum equation for the lower layer is then

$$\frac{\partial u}{\partial t} + u \frac{\partial u}{\partial x} + w \frac{\partial u}{\partial z} = - \frac{g \, \Delta\rho}{\rho_1} \frac{\partial h}{\partial x} \tag{7.17}$$

while the continuity equation is

$$\frac{\partial u}{\partial x} + \frac{\partial w}{\partial z} = 0 \tag{7.18}$$

Now since the pressure gradient in (7.17) is independent of z, u will also be independent of z provided that $u \neq u(z)$ initially. Thus, (7.18) can be integrated vertically from the lower boundary $z = 0$ to the interface $z = h$ to yield

$$w(h) - w(0) = - h \frac{\partial u}{\partial x}$$

But $w(h)$ is just the rate at which the interface height is changing,

$$w(h) = \frac{dh}{dt} = \frac{\partial h}{\partial t} + u\frac{\partial h}{\partial x}$$

and $w(0) = 0$ for a flat lower boundary. Hence the vertically integrated continuity equation can be written

$$\frac{\partial h}{\partial t} + \frac{\partial}{\partial x}(hu) = 0 \tag{7.19}$$

Equations (7.17) and (7.19) are a closed set in the variables u and h. We now apply the perturbation technique by letting

$$u = \bar{u} + u', \qquad h = H + h'$$

where $\bar{u}$ as before is a constant basic state zonal velocity and H is the mean depth of the lower layer. The perturbation forms of (7.17) and (7.19) are then

$$\frac{\partial u'}{\partial t} + \bar{u}\frac{\partial u'}{\partial x} + \frac{g\,\Delta\rho}{\rho_1}\frac{\partial h'}{\partial x} = 0 \tag{7.20}$$

$$\frac{\partial h'}{\partial t} + \bar{u}\frac{\partial h'}{\partial x} + H\frac{\partial u'}{\partial x} = 0 \tag{7.21}$$

where we assume that $H \gg |h'|$ so that products of the perturbation variables can be neglected.

Eliminating u' between (7.20) and (7.21) we get

$$\left(\frac{\partial}{\partial t} + \bar{u}\frac{\partial}{\partial x}\right)^2 h' - \frac{gH\,\Delta\rho}{\rho_1}\frac{\partial^2 h'}{\partial x^2} = 0 \tag{7.22}$$

which is a wave equation similar in form to (7.14). We again assume a wave-type solution by letting

$$h' = Ae^{ik(x-ct)}$$

Substituting into (7.22) we find that the assumed solution satisfies the equation provided that

$$c = \bar{u} \pm \left(\frac{gH\,\Delta\rho}{\rho_1}\right)^{1/2} \tag{7.23}$$

If the upper and lower layers are air and water, respectively, then $\Delta\rho \simeq \rho_1$ and the phase speed formula simplifies to

$$c = \bar{u} \pm \sqrt{gH}$$

The quantity $\sqrt{gH}$ is called the *shallow water* wave speed. It is a valid approximation only for waves whose wavelengths are much greater than the depth of the fluid. This restriction is necessary in order that the vertical velocities be small enough so that the hydrostatic approximation is valid. If the depth of the ocean is taken as 4 km, the shallow water gravity wave speed is $\simeq 200$ m s^{-1}. Thus, long waves on the ocean surface travel very rapidly. It should be emphasized again that this theory applies only to waves of wavelength much greater than H. Such long waves are not ordinarily excited by the wind stresses, but may be produced by very large-scale disturbances such as earthquakes.[2]

Gravity waves may also occur at interfaces within the ocean where there is a very sharp density gradient (diffusion will always prevent formation of a true density discontinuity). In particular, the surface water is separated from the deep water by a narrow region of sharp density contrast called the *thermocline*. If the thermocline is regarded as an interface across which the density changes by an amouunt $\Delta\rho/\rho_1 \simeq 0.01$, then from (7.23) it is clear that the wave speed for waves traveling along the thermocline will be only one-tenth of the surface wave speed for a system of the same depth.

7.4 Internal Gravity (Buoyancy) Waves

We now consider the nature of gravity wave propagation in the atmosphere. Atmospheric gravity waves can only exist when the atmosphere is stably stratified so that a fluid parcel displaced vertically will undergo buoyancy oscillations (see Section 2.7.3). Since the buoyancy force is the restoring force responsible for gravity waves, the term *buoyancy wave* is actually more appropriate as a name for these waves. However, in this text we shall generally use the time honored name *gravity wave*.

In a fluid, such as the ocean, which is bounded both above and below, gravity waves propagate primarily in the horizontal plane since vertically traveling waves are reflected from the boundaries to form standing waves. However, in a fluid which has no upper boundary, such as the atmosphere, gravity waves may propagate *vertically* as well as horizontally. In vertically propagating waves the phase is a function of height—such waves are referred to as *internal waves*. Although vertically propagating internal gravity waves are not important for short-range synoptic scale forecasting (and indeed are nonexistent in the filtered quasi-geostrophic models), they are responsible for the occurrence of mountain *lee waves*. They also are believed to be an important mechanism for transporting energy and momentum to high levels, and are often associated with the formation of clear air turbulence (CAT).

[2] Long waves excited by underwater earthquakes or volcanic eruptions are called *tsunamis*.

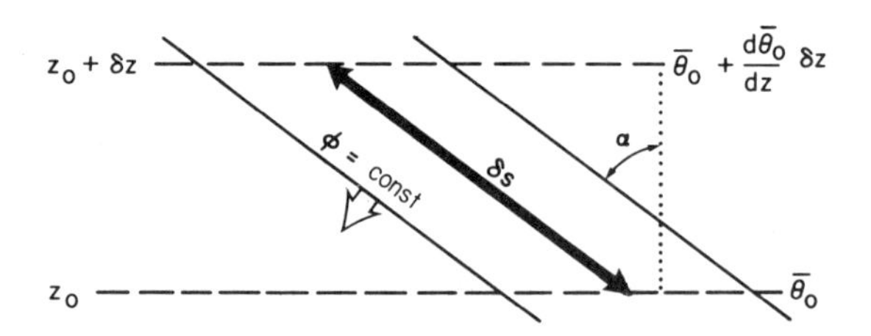

Fig. 7.8 Parcel oscillation path (heavy arrow) for waves with phase lines tilted at an angle α to the vertical.

For simplicity we limit our discussion to two-dimensional internal gravity waves propagating in the x,z plane. An expression for the frequency of such waves can be obtained by modifying the parcel theory developed in Section 2.7.3. Internal gravity waves are *transverse* waves in which the paths of parcel oscillations are parallel to the phase lines in the x,z plane as indicated in Fig. 7.8.

For parcels displaced vertically a distance $\delta z = \delta s \cos \alpha$ the *vertical* buoyancy force per unit mass is just $-N^2\, \delta z$, as was shown in (2.52). Thus, the component of the buoyancy force parallel to the tilted path along which the parcels oscillate is just

$$-N^2(\delta s \cos \alpha) \cos \alpha = -(N \cos \alpha)^2\, \delta s$$

The momentum equation for the parcel oscillation is then

$$\frac{d^2(\delta s)}{dt^2} = -(N \cos \alpha)^2\, \delta s \tag{7.24}$$

which has the general solution

$$\delta s = e^{\pm i(N \cos \alpha)t}$$

Thus the parcels execute a simple harmonic oscillation at the frequency $\nu = N \cos \alpha$. This frequency depends only on the static stability (measured by the buoyancy frequency N) and the angle of the phase lines to the vertical.

The above heuristic derivation can be verified by considering the linearized equations for two-dimensional internal gravity waves. For simplicity we employ the *Boussinesq approximation* in which density is treated as a constant except where it is coupled to gravity in the buoyancy term of the vertical momentum equation. Thus, in this approximation the atmosphere is considered to be incompressible and local density variations are assumed to be small perturbations of the constant basic state density field. Because the vertical variation of the basic state density is neglected except where coupled with gravity, the Boussinesq approximation is only valid for motions in which the vertical scale is less than the atmospheric scale height H ($\simeq 8$ km).

Neglecting effects of rotation, the basic equations for two-dimensional motion of an incompressible atmosphere may be written as follows:

$$\frac{\partial u}{\partial t} + u\frac{\partial u}{\partial x} + w\frac{\partial u}{\partial z} + \frac{1}{\rho}\frac{\partial p}{\partial x} = 0 \tag{7.25}$$

$$\frac{\partial w}{\partial t} + u\frac{\partial w}{\partial x} + w\frac{\partial w}{\partial z} + \frac{1}{\rho}\frac{\partial p}{\partial z} + g = 0 \tag{7.26}$$

$$\frac{\partial u}{\partial x} + \frac{\partial w}{\partial z} = 0 \tag{7.27}$$

$$\frac{\partial \theta}{\partial t} + u\frac{\partial \theta}{\partial x} + w\frac{\partial \theta}{\partial z} = 0 \tag{7.28}$$

where the potential temperature θ is related to pressure and density by

$$\theta = \frac{p}{\rho R}\left(\frac{p_s}{p}\right)^{\kappa} \tag{7.29}$$

We now linearize (7.25)–(7.29) by letting

$$\begin{aligned} \rho &= \rho_0 + \rho', & u &= u' \\ p &= \bar{p}(z) + p', & w &= w' \\ \theta &= \bar{\theta}(z) = \theta' \end{aligned} \tag{7.30}$$

Thus, the basic state is assumed to be motionless with constant density ρ_0. The basic state pressure field must satisfy the hydrostatic equation

$$\frac{d\bar{p}}{dz} = -\rho_0 g \tag{7.31}$$

while the basic state potential temperature must satisfy (7.29), so that

$$\ln \bar{\theta} = \gamma^{-1} \ln \bar{p} - \ln \rho_0 + \text{const} \tag{7.32}$$

The linearized equations are obtained by substituting from (7.30) into (7.25)–(7.29) and neglecting all terms which are products of the perturbation variables. Thus, for example, the last two terms in (7.26) are approximated as follows:

$$\begin{aligned} \frac{1}{\rho}\frac{\partial p}{\partial z} + g &= \frac{1}{(\rho_0 + \rho')}\left(\frac{d\bar{p}}{dz} + \frac{\partial p'}{\partial z}\right) + g \\ &\simeq \frac{1}{\rho_0}\frac{d\bar{p}}{dz}\left(1 - \frac{\rho'}{\rho_0}\right) + \frac{1}{\rho_0}\frac{\partial p'}{\partial z} + g \\ &= \frac{1}{\rho_0}\frac{\partial p'}{\partial z} + \frac{\rho'}{\rho_0} g \end{aligned} \tag{7.33}$$

where (7.31) has been used to eliminate $\bar{p}$. The perturbation form of (7.29) is obtained by noting that

$$\ln\left[\bar{\theta}\left(1 + \frac{\theta'}{\bar{\theta}}\right)\right] = \gamma^{-1}\ln\left[\bar{p}\left(1 + \frac{p'}{\bar{p}}\right)\right] - \ln\left[\rho_0\left(1 + \frac{\rho'}{\rho_0}\right)\right] + \text{const.} \quad (7.34)$$

Now, recalling that for any $\varepsilon \ll 1$ we can write $\ln(1 + \varepsilon) \simeq \varepsilon$, we find with the aid of (7.32) that (7.34) may be approximated by

$$\frac{\theta'}{\bar{\theta}} \simeq \frac{1}{\gamma}\frac{p'}{\bar{p}} - \frac{\rho'}{\rho_0}$$

Solving for ρ' yields

$$\rho' \simeq -\rho_0\frac{\theta'}{\bar{\theta}} + \frac{p'}{c_s^2} \quad (7.35)$$

where $c_s^2 \equiv \bar{p}\gamma/\rho_0$ is the square of the speed of sound. For buoyancy wave motions $|\rho_0\theta'/\theta_0| \gg |p'/c_s^2|$; i.e., density fluctuations due to pressure changes are small compared with those due to temperature changes. Therefore, to a first approximation,

$$\theta'/\bar{\theta} = -\rho'/\rho_0 \quad (7.36)$$

Using (7.33) and (7.36) the linearized version of the set (7.25)–(7.28) can be written as

$$\frac{\partial u'}{\partial t} + \frac{1}{\rho_0}\frac{\partial p'}{\partial x} = 0 \quad (7.37)$$

$$\frac{\partial w'}{\partial t} + \frac{1}{\rho_0}\frac{\partial p'}{\partial z} - \frac{\theta'}{\bar{\theta}}g = 0 \quad (7.38)$$

$$\frac{\partial u'}{\partial x} + \frac{\partial w'}{\partial z} = 0 \quad (7.39)$$

$$\frac{\partial \theta'}{\partial t} + w'\frac{d\bar{\theta}}{dz} = 0 \quad (7.40)$$

Subtracting $\partial(7.37)/\partial z$ from $\partial(7.38)/\partial x$ we can eliminate p' to obtain

$$\frac{\partial}{\partial t}\left(\frac{\partial w'}{\partial x} - \frac{\partial u'}{\partial z}\right) - \frac{g}{\bar{\theta}}\frac{\partial \theta'}{\partial x} = 0 \quad (7.41)$$

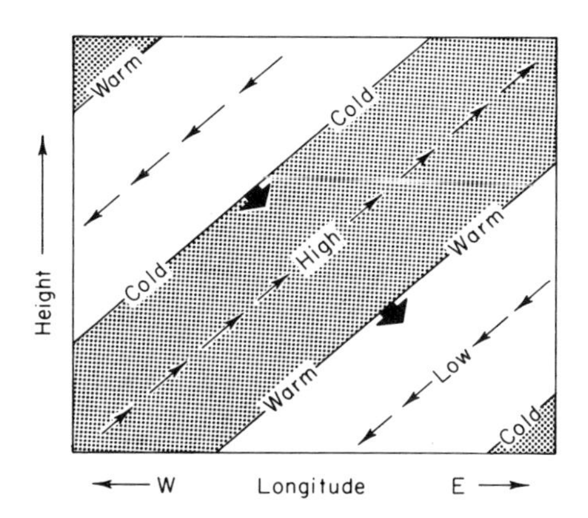

Fig. 7.9 Idealized cross section showing phases of the pressure, temperature, and velocity perturbations for an internal gravity wave. Thin arrows indicate the perturbation velocity field, blunt arrows the phase velocity. (After Wallace and Kousky, 1968. Reproduced with permission of the American Meteorological Society.)

With the aid of (7.39) and (7.40) u' and θ' can be eliminated from (7.41) to yield a single equation for w':

$$\frac{\partial^2}{\partial t^2}\left(\frac{\partial^2 w'}{\partial x^2} + \frac{\partial^2 w'}{\partial z^2}\right) + N^2 \frac{\partial^2 w'}{\partial x^2} = 0 \qquad (7.42)$$

Here $N^2 \equiv g\, d \ln \bar{\theta}/dz$ is the square of the buoyancy frequency, which is assumed to be constant.[3]

We now assume that (7.42) has harmonic wave solutions of the form

$$w' = \mathrm{Re}\{Ae^{i\phi}\} \qquad (7.43)$$

where $\phi = kx + mz - \nu t$ is the phase, which in this example depends linearly on z as well as on x and t. Here the wave number may be regarded as a vector (k, m) whose components, $k = 2\pi/L_x$ and $m = 2\pi/L_z$, are inversely proportional to the horizontal and vertical wavelengths, respectively. Substitution of the assumed solution into (7.42) yields the dispersion relationship

$$\nu^2(k^2 + m^2) - N^2k^2 = 0$$

or

$$\nu = \pm Nk/(k^2 + m^2)^{1/2} \qquad (7.44)$$

If we let $k > 0$ and $m < 0$ then lines of constant phase tilt eastward with respect to height as shown in Fig. 7.9 (i.e., for $\phi = kx + mz$ to remain constant as x increases, z must also increase when $k > 0$ and $m < 0$). The choice of ν positive in (7.44) then corresponds to eastward and downward phase propagation with horizontal and vertical components of the phase velocity given

[3] Strictly speaking, N^2 cannot be exactly constant if ρ_0 is constant. However, for shallow disturbances the variation of N^2 with height is unimportant.

by $c_x = \nu/k$ and $c_z = \nu/m$, respectively. The components of the group velocity U_g and W_g, on the other hand, are given by

$$U_g = \frac{\partial \nu}{\partial k} = \frac{Nm^2}{(k^2 + m^2)^{3/2}} > 0$$

$$W_g = \frac{\partial \nu}{\partial m} = \frac{-Nkm}{(k^2 + m^2)^{3/2}} > 0 \tag{7.45}$$

Thus, recalling that energy propagates with the group velocity, we see that for internal gravity waves *downward* phase propagation implies *upward* energy propagation. In the atmosphere internal gravity waves generated in the troposphere may propagate energy upward many scale heights into the upper atmosphere even though individual fluid parcel oscillations may be confined to vertical distances much less than a kilometer.

Referring again to Fig. 7.9 it is evident that the angle of the phase lines to the local vertical is given by

$$\cos \alpha = L_z/(L_x^2 + L_z^2)^{1/2}$$
$$= k/(k^2 + m^2)^{1/2}$$

Thus, $\nu = \pm N \cos \alpha$ in agreement with the heuristic parcel oscillation model (7.24). The tilt of phase lines for internal gravity waves depends only on the ratio of the wave frequency to the buoyancy frequency, and is independent of wavelength.

7.4.1 Lee Waves

When air is forced to flow over a mountain under statically stable conditions, individual air parcels are displaced from their equilibrium levels and will thus undergo buoyancy oscillations as they move downstream of the mountain. In this manner, as shown in Fig. 7.10 an internal gravity wave system is excited in the lee of the mountain stationary with respect to the ground. If the vertical motion associated with these *lee waves* is strong enough, and the air is moist enough, condensation may occur in the updraft portions of the oscillations giving rise to wave clouds. Such clouds are a common occurrence to the east of large north–south oriented mountain barriers such as the Rockies.

Since lee waves are stationary with respect to the ground, the horizontal phase speed vanishes ($c_x = 0$). However, when viewed in a coordinate system moving at the speed of the mean zonal wind, constant phase lines of lee waves set up by westerly flow appear to progress upstream toward the west. Since the energy source for lee waves is at the ground, such waves must transport energy upward. Hence, the phase velocity relative to the mean

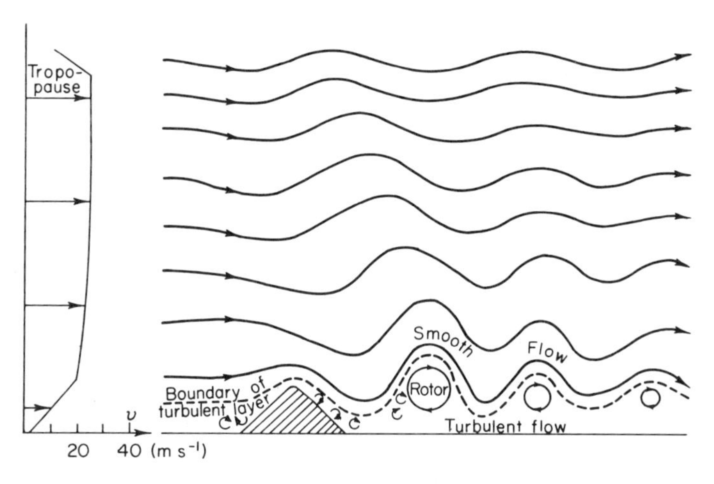

Fig. 7.10 Schematic diagram showing streamlines based on observed lee waves. The upstream velocity profile is indicated on the left. (After Gerbier and Berenger, 1961.)

zonal flow must have a downward component. Consequently, the lines of constant phase must tilt westward with height as shown in Fig. 7.10. For an observer moving with the basic state wind at a constant velocity $\bar{u}$, the frequency of lee waves generated by sinusoidal topographic features of wave number k would be $\nu = -\bar{u}k$. Thus, from (7.44) we see that in this case

$$\bar{u} = N/(k^2 + m^2)^{1/2} \tag{7.46}$$

For given values of N, k, and $\bar{u}$ (7.46) can be solved for m to determine the tilt of phase lines with respect to height. In order that the waves propagate vertically we must have $m^2 > 0$ (i.e., m must be real). Thus, vertical propagation is possible only for $\bar{u} < N/k$. Stable stratification, wide mountains, and comparatively weak zonal flow provide favorable conditions for the formation of vertically propagating lee waves. Of course (7.46) was obtained for conditions of constant basic state flow. In reality both the zonal wind $\bar{u}$ and the stability parameter N generally vary with height. Under certain conditions (e.g., an intense inversion just above the mountain top) large amplitude waves can be formed which may generate severe downslope surface winds and zones of strong clear air turbulence.

7.5 Rossby Waves

The wave type which is of most importance for large-scale meteorological processes is the *Rossby wave*, or *planetary wave*. In a barotropic atmosphere the Rossby wave is an absolute vorticity conserving motion which owes its

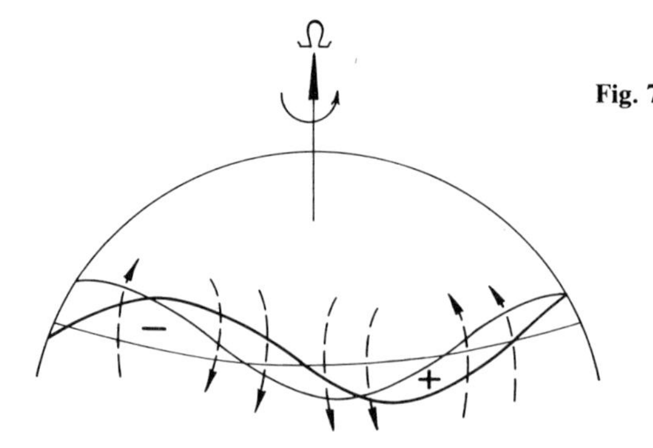

Fig. 7.11 Perturbation vorticity field and induced velocity field (dashed arrows) for a meridionally displaced chain of fluid parcels. Heavy wavy line shows original perturbation position, light line westward displacement of the pattern due to advection by the induced velocity field.

existence to the variation of the Coriolis force with latitude, the so-called β effect.

Rossby wave propagation can be understood in a qualitative fashion by considering a closed chain of fluid parcels initially aligned along a circle of latitude. Recall that the absolute vorticity η is given by $\eta = \zeta + f$, where ζ is the relative vorticity and f is the Coriolis parameter. Assume that $\zeta = 0$ at time t_0. Now suppose that at t_1, δy is the meridional displacement of a fluid parcel from the original latitude. Then at t_1 we have

$$(\zeta + f)_{t_1} = f_{t_0}$$

or

$$\zeta_{t_1} = f_{t_0} - f_{t_1} = -\beta\delta y \tag{7.47}$$

From (7.47) it is evident that if the chain of parcels is subject to a sinusoidal meridional displacement then the induced perturbation vorticity will be positive (i.e., cyclonic) for a southward displacement and negative (anticyclonic) for a northward displacement as indicated schematically in Fig. 7.11.

The meridional velocity field associated with the perturbation vorticity field advects the chain of fluid parcels southward west of the vorticity maximum and northward west of the vorticity minimum, as indicated in Fig. 7.11. Thus, the pattern of vorticity maxima and minima propagates to the west. This westward propagating vorticity field constitutes a Rossby wave.

The dispersion relationship for barotropic Rossby waves may be derived formally by finding wave-type solutions of the linearized barotropic vorticity equation. The barotropic vorticity equation (4.25) states that the vertical component of absolute vorticity is conserved following the horizontal motion:

$$\left(\frac{\partial}{\partial t} + u\frac{\partial}{\partial x} + v\frac{\partial}{\partial y}\right)\zeta + v\frac{df}{dy} = 0 \tag{7.48}$$

We now assume that the motion consists of a basic state zonal velocity plus a small horizontal perturbation:

$$u = \bar{u} + u', \qquad v = v', \qquad \zeta = \zeta'$$

We define a perturbation streamfunction ψ according to

$$u' = -\frac{\partial \psi}{\partial y}, \qquad v' = \frac{\partial \psi}{\partial x}$$

from which $\zeta' = \nabla^2 \psi$. The perturbation form of (7.48) is then

$$\left(\frac{\partial}{\partial t} + \bar{u}\frac{\partial}{\partial x}\right)\nabla^2\psi + \beta\frac{\partial \psi}{\partial x} = 0 \tag{7.49}$$

where we now assume that $\beta \equiv df/dy$ is a constant, and as usual we have neglected terms involving the products of perturbation quantities. We assume a solution exists of the form

$$\psi = \mathrm{Re}\{Ae^{i\phi}\} \tag{7.50}$$

where $\phi = kx + ly - \nu t$. Here k and l are wave numbers in the zonal and meridional directions, respectively. Substituting from (7.50) into (7.49) gives

$$(-\nu + k\bar{u})(-k^2 - l^2) + k\beta = 0$$

which may immediately be solved for ν:

$$\nu = \bar{u}k - \beta k/(k^2 + l^2) \tag{7.51}$$

Recalling that $c_x = \nu/k$ we find that the zonal phase speed relative to the mean wind is

$$c_x - \bar{u} = -\beta/(k^2 + l^2) \tag{7.52}$$

Thus, the Rossby wave propagates *westward* relative to the mean zonal flow. Furthermore the Rossby wave speed depends on the zonal and meridional wave numbers. Therefore, Rossby waves are dispersive waves whose phase speeds increase with wavelength. This result is consistent with the discussion in Section 6.2.2 in which we showed that the advection of planetary vorticity, which tends to make disturbances retrogress, increasingly dominates over relative vorticity advection as the wavelength of a disturbance increases. Equation (7.52) merely provides a quantitative measure of this effect in cases where the disturbance is small enough so that perturbation theory is applicable. For a typical midlatitude synoptic disturbance with zonal wavelength $\sim$6000 km and latitudinal width $\sim$3000 km, the Rossby wave speed relative to the zonal flow calculated from (7.52) is approximately -6 m s^{-1}. Thus, synoptic scale Rossby waves move quite slowly.

It is possible to carry out a less restrictive analysis using the perturbation form of the full primitive equations. The results of such an analysis are mathematically complicated but qualitatively not much different from the pure wave cases discussed here. It turns out that the free oscillations allowed in a hydrostatic gravitationally stable atmosphere consists of eastward and westward moving gravity waves which are slightly modified by the rotation of the earth, and westward moving Rossby waves which are slightly modified by gravitational stability. These free oscillations are the *normal modes* of oscillation of the atmosphere. As such, they are continually excited by the various forces acting on the atmosphere. The planetary scale free oscillations, although they can be detected by careful observational studies, appear to have rather weak amplitudes. Presumably this is because the forcing is quite weak at the large phase speeds characteristic of these long waves.

Problems

1. Show that the Fourier component

$$F(x) = \mathrm{Re}\{Ce^{imx}\}$$

can be written as

$$F(x) = |C| \cos m(x + x_0)$$

where $x_0 = m^{-1} \sin^{-1}[C_i/|C|]$ and C_i stands for the imaginary part of C.

2. In the study of atmospheric wave motions it is often necessary to consider the possibility of amplifying or decaying waves. In such a case we might assume that a solution has the form

$$\psi = A \cos(kx - vt - kx_0)e^{\alpha t}$$

where A is the initial amplitude, α the amplification factor, and x_0 the initial phase. Show that this expression can be written more concisely as

$$\psi = \mathrm{Re}\{Be^{ik(x-ct)}\}$$

where both B and c are complex constants. Determine the real and imaginary parts of B and c in terms of A, α, k, v, and x_0.

3. Several of the wave types discussed in this chapter are governed by equations which are generalizations of the wave equation

$$\frac{\partial^2 \psi}{\partial t^2} = c^2 \frac{\partial^2 \psi}{\partial x^2}$$

This equation can be shown to have solutions corresponding to waves of arbitrary profile moving at the speed c in both the positive and negative x directions. We consider an arbitrary initial profile of the field ψ:

$$\psi = f(x) \qquad \text{at} \quad t = 0$$

If the profile is translated in the positive x direction at speed c without change of shape then

$$\psi = f(x')$$

where x' is a coordinate moving at speed c so that $x = x' + ct$. Thus, in terms of the fixed coordinate x we can write

$$\psi = f(x - ct)$$

corresponding to a profile which moves in the positive x direction at speed c without change of shape. Verify that $\psi = f(x - ct)$ is a solution for any arbitrary continuous profile $f(x - ct)$. *Hint*: Let $x - ct = x'$ and differentiate f using the chain rule.

4. Assuming that the pressure perturbation for a one-dimensional acoustic wave is given by (7.15) find the corresponding solutions of the zonal wind and density perturbations. Express the amplitude and phase for u' and ρ' in terms of the amplitude and phase of p'.

5. Show that for isothermal motion ($dT/dt = 0$) the acoustic wave speed is given by $(gH)^{1/2}$ where $H = RT/g$ is the scale height.

6. If the surface height perturbation in a shallow water gravity wave is given by

$$h' = \text{Re}\{Ae^{ik(x-ct)}\}$$

find the corresponding velocity perturbation $u'(x, t)$. Sketch the phase relationship between h' and u' for an eastward propagating wave.

7. Assuming that the vertical velocity perturbation for a two-dimensional internal gravity wave is given by (7.43), obtain the corresponding solution for the u', p', and θ' fields. Use these results to verify the approximation $|\rho_0 \theta'/\bar{\theta}| \gg |p'/c_s^2|$ which was used in (7.36).

8. For the situation in Problem 7 express the vertical flux of horizontal momentum, $\langle \rho_0 u'w' \rangle$, in terms of the amplitude A of the vertical velocity perturbation. (The angle brackets denote horizontal averaging over one wavelength.) Hence, show that the momentum flux is positive for waves in which phase propagates eastward and downward.

9. Show that if (7.38) is replaced by the hydrostatic equation (i.e., $\partial w'/\partial t$ is neglected) the resulting frequency equation for internal gravity waves is just the asymptotic limit of (7.44) for waves in which $|k| \ll |m|$.

10. (a) Show that the group velocity vector in two-dimensional internal gravity waves is parallel to lines of constant phase.
(b) Show that in the long-wave limit ($|k| \ll |m|$) the magnitude of the group velocity equals the magnitude of the phase velocity so that energy propagates one wavelength per wave period.

***11.** Determine the perturbation horizontal and vertical velocity fields for stationary gravity waves forced by flow over sinusoidally varying topography, given the following conditions: the height of the ground is $h = h_0 \cos kx$, where $h_0 = 50$ m is a constant; $N = 2 \times 10^{-2}\ \mathrm{s}^{-1}$; $\bar{u} = 5\ m\,\mathrm{s}^{-1}$; and $k = 3 \times 10^{-3}\ \mathrm{m}^{-1}$. *Hint*: For small amplitude topography ($h_0 k \ll 1$) we can approximate the lower boundary condition by

$$w' = \frac{dh}{dt} = \bar{u}\frac{\partial h}{\partial x} \qquad \text{at} \quad z = 0.$$

***12.** Derive the Rossby wave speed for a homogeneous incompressible ocean of depth h using the β-plane approximation. Assume a motionless basic state and small perturbations which depend only on x and t,

$$u = u'(x, t), \qquad v = v'(x, t), \qquad h = H + h'(x, t)$$

where H is the mean depth of the ocean. With the aid of the continuity equation for a homogeneous layer (7.21) and the geostrophic wind relationship

$$v' = \frac{g}{f_0}\frac{\partial h'}{\partial x}$$

show that the perturbation vorticity equation can be written

$$\frac{\partial}{\partial t}\left(\frac{\partial^2}{\partial x^2} - \frac{f_0^2}{gH}\right)h' + \beta\frac{\partial h'}{\partial x} = 0$$

and that $h' = Ae^{ik(x-ct)}$ is a solution provided that

$$c = -\frac{\beta}{(k^2 + f_0^2/gH)}$$

If the ocean is 4 km deep, what is the Rossby wave speed at latitude 45° for a wave of 10,000-km zonal wavelength?

***13.** In Section 4.3 we showed that for a homogeneous incompressible fluid a decrease in depth with latitude has the same dynamic effect as a latitudinal dependence of the Coriolis parameter. Thus, Rossby-type waves can be produced in a rotating cylindrical vessel if the depth of the fiuid is dependent on the radial coordinate. To determine the Rossby wave speed formula for this so-called "equivalent β effect," we assume that the flow is confined between rigid lids in an annular region whose distance from the axis of rotation is large enough so that the curvature terms in the equations can be neglected. We can then refer the motion to cartesian coordinates with x directed azimuthally and y directed toward the axis of rotation. If the system is rotating at angular velocity Ω and the depth is linearly dependent on y,

$$H(y) = H_0 - \gamma y$$

show that the perturbation continuity equation can be written

$$H_0\left(\frac{\partial u'}{\partial x} + \frac{\partial v'}{\partial y}\right) - \gamma v' = 0$$

and that the perturbation quasi-geostrophic vorticity is thus

$$\frac{\partial}{\partial t}\nabla^2 h' + \beta\frac{\partial h'}{\partial x} = 0$$

where h' is the deviation of the depth from H, and

$$\beta = \frac{2\Omega}{H_0}$$

Hint: Assume that the velocity field is geostrophic except in the divergence term. What is the Rossby wave speed in this situation for waves of wavelength 100 cm in both the x and y directions if $\Omega = 1\ \mathrm{s}^{-1}$, $H_0 =$ 20 cm, and $\gamma = 0.05$?

14. Show by scaling arguments that if the horizontal wavelength is much greater than the depth of the fluid, two-dimensional surface gravity waves will be hydrostatic so that the "shallow water" approximation applies.

15. Derive an expression for the group velocity of a barotropic Rossby wave and compare with the phase velocity.

16. The linearized form of the quasi-geostrophic vorticity equation (6.10) can be written as follows:

$$\frac{\partial}{\partial t}\nabla^2\psi + \bar{u}\frac{\partial}{\partial x}\nabla^2\psi + \beta\frac{\partial\psi}{\partial x} = -f_0\,\nabla\cdot\mathbf{V} \qquad \text{where} \quad \psi \equiv \Phi/f_0$$

Suppose that the horizontal divergence field is given by $\mathbf{\nabla} \cdot \mathbf{V} = A \cos[k(x - ct)]$ where A is a constant. Find a solution for the corresponding relative vorticity field. What is the phase relationship between vorticity and divergence? For what value of c does the vorticity become infinite?

Suggested References

Hildebrand, *Advanced Calculus for Applications*, is one of many standard textbooks which discuss the mathematical techniques used in this chapter, including the representation of functions in Fourier series and the general properties of the wave equation.

Turner, *Buoyancy Effects in Fluids*, contains an excellent discussion of internal gravity waves.

Gossard and Hooke, *Waves in the Atmosphere*, provides a thorough advanced treatment of both theoretical and observational aspects of gravity and acoustic waves in the atmosphere.

Chapman and Lindzen, *Atmospheric Tides*, thoroughly covers both observational and theoretical aspects of a class of atmospheric motions in which analysis by the linear perturbation method has proved particularly successful.

Scorer, *Natural Aerodynamics*, contains an excellent qualitative discussion on many aspects of waves generated by barriers such as lee waves.

Chapter

8 Numerical Prediction

The most important practical application of dynamic meteorology is in weather prediction by numerical methods. Stated simply, the objective of numerical prediction is to predict the future state of the atmospheric circulation from knowledge of its present state by use of the dynamical equations. To fulfill this objective the following information is required: (1) the initial state of the field variables; (2) a closed set of prediction equations relating the field variables; (3) a method of integrating the equations in time to obtain the future distribution of the field variables.

Numerical prediction is a highly specialized field which is still in a state of rapid development. However, there do exist some fairly standard dynamical methods of forecasting the short-term evolution of synoptic scale circulation patterns whose general description is within the scope of this text. It is these methods, which rely on the scaling considerations introduced in the previous chapters, which will be emphasized in the present discussion. In particular, we will show how the quasi-geostrophic system can be used in numerical forecasting.

8.1 Historical Background

The first attempt to numerically predict the weather was due to the British scientist L. F. Richardson. His book, *Weather Prediction by Numerical*

Process, published in 1922, is the classic treatise in this field. In this work Richardson showed how the differential equations governing atmospheric motions could be written approximately as a set of algebraic difference equations for values of the tendencies of various field variables at a finite number of points in space. Given the observed values of the field variables at these *grid points* the tendencies could be calculated numerically by solving the algebraic difference equations. By extrapolating the computed tendencies ahead a small increment in time an estimate of the fields at a short time in the future could be obtained. These new values of the field variables could then be used to recompute the tendencies. The new tendencies could then be used to extrapolate further ahead in time, etc. Even for short-range forecasting over a small area of the earth, this procedure requires an enormous number of arithmetic operations. Richardson did not foresee the development of high-speed digital computers. He estimated that a work force of 64,000 people would be required just to keep up with the weather on a global basis.

Despite the tedious labor involved, Richardson worked out one example forecast for surface pressure tendencies at two grid points. Unfortunately, the results were very unimpressive. Predicted pressure changes were an order of magnitude larger than those observed. At the time this failure was thought to be due primarily to the poor initial data available—especially the absence of upper-air soundings. However, it is now known that there were other even more serious problems with Richardson's scheme.

After Richardson's failure to obtain a reasonable forecast, numerical prediction was not again attempted for many years. Finally, after World War II interest in numerical prediction revived due partly to the vast expansion of the meteorological observation network, which provided much improved initial data, but even more importantly to the development of digital computers which made the enormous volume of arithmetic operations required in a numerical forecast feasible. At the same time it was realized that Richardson's scheme was not the simplest possible scheme for numerical prediction. Richardson's equations governed not only the slow-moving meteorologically important motions, but also included high-speed sound and gravity waves as solutions. These high-speed sound and gravity waves are in nature very weak in amplitude. However, for reasons that will be explained later, if Richardson had carried his numerical calculation beyond the initial time step, these oscillations would have amplified spuriously, thereby introducing so much "noise" in the solution that the meteorologically relevant disturbances would have been obscured.

J. G. Charney in 1948 showed how the dynamical equations could be simplified by systematic introduction of the geostrophic and hydrostatic assumptions so that the sound and gravity oscillations were filtered out. The equations which resulted from Charney's filtering approximations were essentially those of the quasi-geostrophic model. A special case of this model, the so-called equivalent barotropic model, was used in 1950 to make the first numerical forecast.

The first model provided forecasts only of the geopotential, and hence the geostrophic winds, near 500 mb. Thus, this model did not forecast "weather" in the usual sense, but only the winds at 500 mb. Later multilevel models based on the quasi-geostrophic theory could, however, be used to diagnose the large-scale vertical motion field at the forecast time. Since vertical motion is correlated with precipitation, this information could be used by forecasters as an aid in predicting the local weather associated with large-scale circulations.

With the development of vastly more powerful computers and more sophisticated modeling techniques the emphasis in numerical forecasting has now returned to models which are quite similar to Richardson's formulation.

8.2 Filtering of Sound and Gravity Waves

One difficulty in directly applying the unsimplified equations of motion is that meteorologically important motions are easily lost in the noise introduced by large-amplitude sound and gravity waves which may arise as a result of errors in the initial data, and then spuriously amplify by a process called *computational instability*. As an example of how this problem might arise, we know that on the synoptic scale the pressure and density fields are in hydrostatic balance to a very good approximation. As a consequence. vertical accelerations are extremely small. However, if the pressure and density fields were determined independently by observations, as would be the case if the complete equations were used, small errors in the observed fields would lead to large errors in the computed vertical acceleration simply because the vertical acceleration is the very small difference between two large forces—the vertical pressure gradient and gravity. Such spurious accelerations would appear in the computed solution as high-speed sound waves of very large amplitude. In a similar fashion errors in the initial velocity and pressure fields would lead to spuriously large horizontal accelerations since the horizontal acceleration results from the small difference between the Coriolis and pressure gradient forces. Such spurious horizontal accelerations would excite both sound and gravity waves.

In order to overcome this problem, the simplest procedure is to simplify the governing equations to remove the physical mechanisms responsible for the occurrence of the unwanted oscillations, while still preserving the meteorologically important motions. To see how such a "filtering" of the equations can be accomplished we must refer back to Section 7.3 in which the physical properties of sound and gravity waves were discussed.

If the pipe in Fig. 7.5 is tipped up in the vertical, it can be used to generate vertically traveling sound waves. If we now require that pressure be hydrostatic, the pressure at any point along the pipe must be determined solely by the *weight* of the air above that point. Hence, the vertical pressure gradient cannot be influenced by adiabatic compression. Therefore, sound waves cannot propagate *vertically* if it is assumed that the motions are hydrostatic. Replacement of the vertical momentum equation by the hydrostatic approximation is thus sufficient to filter out ordinary sound waves. However, a hydrostatically balanced atmosphere can still support a special class of *horizontally* propagating acoustic waves. In this type of acoustic wave the vertical velocity is zero (neglecting orographic effects and departures of the basic state temperature from isothermal conditions), but the pressure, horizontal velocity, and density oscillate with the horizontal structure of the simple acoustic waves described in Section 7.3.1. These oscillations have maximum amplitude at the lower boundary and decay away from the boundary with the pressure and density fields remaining in hydrostatic balance everywhere. Since these so-called *Lamb waves* have maximum pressure oscillations at the ground they may be filtered out simply by requiring that $\omega \equiv dp/dt = 0$ at the lower boundary. This boundary condition is most easily applied by formulating the equations in isobaric coordinates. In that case the condition $\omega = 0$ at the lower boundary is a natural first approximation for geostrophically scaled motions as can be seen from the discussion in Section 3.5.1.

Thus, as a result of the minimum simplification necessary to filter out sound waves, the prediction equations become the isobaric horizontal momentum equation, continuity equation, and hydrostatic thermodynamic energy equation subject to the condition $\omega = 0$ at the lower boundary. These may be written using vector notation as

$$\left(\frac{\partial}{\partial t} + \mathbf{V} \cdot \nabla\right)\mathbf{V} + \omega \frac{\partial \mathbf{V}}{\partial p} + f\mathbf{k} \times \mathbf{V} = -\nabla\Phi \tag{8.1}$$

$$\nabla \cdot \mathbf{V} + \frac{\partial \omega}{\partial p} = 0 \tag{8.2}$$

$$\left(\frac{\partial}{\partial t} + \mathbf{V} \cdot \nabla\right)\left(\frac{\partial \Phi}{\partial p}\right) + \sigma\omega = 0 \tag{8.3}$$

where as before $\sigma \equiv -(\alpha/\theta)(\partial\theta/\partial p)$.

The system (8.1)–(8.3) together with the condition $\omega = 0$ at the lower boundary no longer contains the mechanism for sound-wave transmission but is still capable of describing *gravity waves*.

From the discussion of gravity waves in Sections 7.3.2 and 7.4 it should be apparent that a divergent horizontal velocity field which changes in time is essential for the propagation of gravity waves. In fact it turns out that neglecting the local rate of change of the horizontal divergence in computing the relationship between the mass and velocity fields is sufficient to filter out the time-dependent gravity waves. Of course, the momentum equation (8.1) does not explicitly contain the local derivative of $\nabla \cdot \mathbf{V}$. However, the prediction system can be rewritten so that this term does appear if we replace the horizontal equations of motion by the vorticity and divergence equations which are both differentiated forms of the momentum equation. Although we have previously discussed the vorticity equation in Sections 4.4 and 6.2 the derivation will be briefly repeated here in vectorial form for completeness.

Using the vector identity

$$(\mathbf{V} \cdot \nabla)\mathbf{V} = \nabla\left(\frac{\mathbf{V} \cdot \mathbf{V}}{2}\right) + \mathbf{k} \times \mathbf{V}\zeta$$

where $\zeta = \mathbf{k} \cdot \nabla \times \mathbf{V}$ is as before the vertical component of vorticity, we can rewrite (8.1) in the form

$$\frac{\partial \mathbf{V}}{\partial t} = -\nabla\left(\Phi + \frac{\mathbf{V} \cdot \mathbf{V}}{2}\right) - \mathbf{k} \times \mathbf{V}(\zeta + f) - \omega \frac{\partial \mathbf{V}}{\partial p} \tag{8.4}$$

The vorticity and divergence equations are now obtained by operating on (8.4) with the vector operators $\mathbf{k} \cdot \nabla \times (\quad)$ and $\nabla \cdot (\quad)$, respectively. The resulting equations are the vorticity equation,

$$\frac{\partial \zeta}{\partial t} = -\mathbf{V} \cdot \nabla(\zeta + f) - \omega \frac{\partial \zeta}{\partial p} - (\zeta + f)\,\nabla \cdot \mathbf{V} + \mathbf{k} \cdot \left(\frac{\partial \mathbf{V}}{\partial p} \times \nabla\omega\right) \tag{8.5}$$

and the divergence equation,

$$\frac{\partial}{\partial t}(\nabla \cdot \mathbf{V}) = -\nabla^2\left(\Phi + \frac{\mathbf{V} \cdot \mathbf{V}}{2}\right) - \nabla \cdot [\mathbf{k} \times \mathbf{V}(\zeta + f)] - \omega \frac{\partial}{\partial p}(\nabla \cdot \mathbf{V}) - \frac{\partial \mathbf{V}}{\partial p} \cdot \nabla\omega \tag{8.6}$$

Equations (8.5)–(8.6) are independent scalar equations which can be used in place of the horizontal equations of motion in our prediction system. If we now simply set the term on the left-hand side in (8.6) equal to zero, we will eliminate all solutions corresponding to time-dependent gravity waves. This is the minimum simplification required to filter gravity waves.

8.3 Filtered Forecast Equations

Merely neglecting the term on the left-hand side turns (8.6) into a rather intractable diagnostic relationship between ω, $\mathbf{V}$, and Φ. However, further simplifications can be made with the aid of scale analysis. Indeed, for synoptic scale motions the neglect of $\partial(\nabla \cdot \mathbf{V})\ \partial t$ in (8.6) can itself be justified on the basis of scaling arguments. Thus, neglecting terms in (8.6) whose magnitudes are small compared to $\nabla^2\Phi$ is sufficient to filter the gravity waves.

In order to exhibit the connection between scaling and filtering more clearly it is convenient to partition the horizontal velocity field into nondivergent and irrotational components. A theorem of Helmholtz[1] states that any velocity field can be divided into a *nondivergent* part $\mathbf{V}_\psi$ plus an *irrotational* part $\mathbf{V}_e$ such that

$$\mathbf{V} = \mathbf{V}_\psi + \mathbf{V}_e \tag{8.7}$$

where

$$\nabla \cdot \mathbf{V}_\psi = 0 \qquad \text{and} \qquad \nabla \times \mathbf{V}_e = 0$$

If the velocity field is two dimensional, the nondivergent part can be expressed in terms of the *streamfunction* ψ defined by letting

$$\mathbf{V}_\psi = \mathbf{k} \times \nabla\psi \tag{8.8}$$

or in cartesian components,

$$u_\psi = -\frac{\partial\psi}{\partial y}, \qquad v_\psi = \frac{\partial\psi}{\partial x}$$

from which it is easily verified that

$$\nabla \cdot \mathbf{V}_\psi = 0 \qquad \text{and} \qquad \zeta \equiv \mathbf{k} \cdot \nabla \times \mathbf{V} = \nabla^2\psi$$

Because the isolines of ψ correspond to streamlines for the nondivergent velocity and the distance separating the isolines of ψ is inversely proportional to the magnitude of the nondivergent velocity, the spatial distribution of $\mathbf{V}_\psi$ can be easily pictured by plotting lines of constant ψ on a map.

In our scale analysis of the vorticity equation in Chapter 4 we showed that for midlatitude synoptic scale motions $\mathbf{V}$ must be quasi-nondivergent, that is,

$$|\mathbf{V}_\psi| \gg |\mathbf{V}_e|$$

Thus, if (8.6) is scaled in a similar fashion we find that the terms involving ω and $\mathbf{V}_e$ may all be neglected. The remaining terms imply a relationship

[1] See, for example, Bourne and Kendall (1968, p. 190).

between $\mathbf{V}_\psi$ and Φ which is known as the *balance equation.* Using the definition (8.8) the balance equation may be written in terms of ψ and Φ as

$$\nabla^2[\Phi + \tfrac{1}{2}(\nabla\psi)^2] = \nabla \cdot [(f + \nabla^2\psi)\,\nabla\psi] \tag{8.9}$$

The system (8.2), (8.3), (8.5), and (8.9) may then be used as the basis of a forecast model which filters out the gravity wave solutions.[2] However, (8.9) expresses a rather complicated nonlinear relationship between ψ and Φ which is closely related to the gradient wind balance (3.10). Because of its complexity the balance equation (8.9) has not been widely employed in numerical forecasting.

A simpler filtered model, which is still quite accurate outside the tropics, can be obtained by noting that for synoptic scale motions the nonlinear terms in (8.9) are small compared to the linear terms so that to a reasonable approximation (8.9) can be replaced by the so-called *linear balance equation*

$$\nabla^2\Phi = \nabla \cdot (f\,\nabla\psi) \tag{8.10}$$

In an analogous fashion we may neglect small terms in (8.5) to obtain the approximate vorticity equation

$$\frac{\partial\zeta}{\partial t} + \mathbf{V}_\psi \cdot \nabla(\zeta + f) + \mathbf{V}_e \cdot \nabla f + f\,\nabla \cdot \mathbf{V}_e = 0 \tag{8.11}$$

The advection of the planetary vorticity by the divergent wind, $-\mathbf{V}_e \cdot \nabla f$, is actually small compared to the remaining terms in (8.11) but must be retained for energetic consistency when the linear balance equation (8.10) is used to relate Φ and ψ. If we wish to neglect this small term it is necessary at the same time to replace f by a constant mean value f_0 in (8.10) and in the last term in (8.11). The resulting approximate vorticity and divergence equations, valid for midlatitude synoptic scale motions, are

$$\frac{\partial\zeta}{\partial t} = -\mathbf{V}_\psi \cdot \nabla(\zeta + f) - f_0\,\nabla \cdot \mathbf{V}_e \tag{8.12}$$

and

$$\nabla^2\Phi = -f_0\,\nabla \cdot (\mathbf{k} \times \mathbf{V}_\psi) = f_0\,\nabla^2\psi \tag{8.13}$$

Equation (8.13) simply states that to a first approximation the vorticity is the vorticity of the geostrophic wind computed using a constant midlatitude value of the Coriolis parameter. Thus, for midlatitude synoptic scale motions the streamfunction is given approximately by the relation $\psi = \Phi/f_0$. Hence, the geopotential field on a constant pressure chart is

[2] Lorenz (1960) has shown that for energetic consistency in this case $\mathbf{V}$ must be replaced by $\mathbf{V}_\psi$ in the advection term in (8.3) and in the last term (tilting term) in (8.5).

approximately proportional to the streamfunction field, and to the same order of approximation

$$\mathbf{V}_\psi = \frac{\mathbf{k} \times \nabla\Phi}{f_0} \tag{8.14}$$

The geostrophic vorticity equation and hydrostatic thermodynamic energy equation can now be written in terms of ψ and ω as

$$\frac{\partial}{\partial t}\nabla^2\psi = -\mathbf{V}_\psi \cdot \nabla(\nabla^2\psi + f) + f_0\frac{\partial\omega}{\partial p} \tag{8.15}$$

$$\frac{\partial}{\partial t}\left(\frac{\partial\psi}{\partial p}\right) = -\mathbf{V}_\psi \cdot \nabla\left(\frac{\partial\psi}{\partial p}\right) - \frac{\sigma}{f_0}\omega \tag{8.16}$$

Aside from the change of notation this set of equations is identical to the quasi-geostrophic system derived in Chapter 6.

Differentiating (8.16) with respect to p after multiplying through by f_0^2/σ, and adding the result to (8.15) gives the quasi-geostrophic *potential vorticity equation*,

$$\left(\frac{\partial}{\partial t} + \mathbf{V}_\psi \cdot \nabla\right)q = 0 \tag{8.17}$$

where

$$q = \nabla^2\psi + f + f_0^2\frac{\partial}{\partial p}\left(\frac{1}{\sigma}\frac{\partial\psi}{\partial p}\right)$$

and we have again assumed that σ is a function of pressure only.[3] This equation states that the geostrophic potential vorticity q is *conserved following the nondivergent wind* in pressure coordinates. This conservation law is the basis for most of the numerical prediction schemes discussed in this chapter. Comparing with Eq. (4.12) we see that this law is similar to, but not identical to, the more general potential vorticity conservation law of Chapter 4.

Again, as in Chapter 6, if the time derivatives are eliminated between (8.15) and (8.16), we obtain the diagnostic omega equation

$$\left(\nabla^2 + \frac{f_0^2}{\sigma}\frac{\partial^2}{\partial p^2}\right)\omega = \frac{f_0}{\sigma}\frac{\partial}{\partial p}[\mathbf{V}_\psi \cdot \nabla(\nabla^2\psi + f)] - \frac{f_0}{\sigma}\nabla^2\left[\mathbf{V}_\psi \cdot \nabla\left(\frac{\partial\psi}{\partial p}\right)\right] \tag{8.18}$$

Equation (8.18) can be used to diagnose the ω field at any instant provided that the ψ field is known.

[3] Equation (8.17) is, except for the inclusion of pressure dependence in σ, identical to (6.18) if we use $\psi = \Phi/f_0$.

Now since according to (8.17) q is conserved following the nondivergent wind, prediction with the quasi-geostrophic system in principle merely involves advecting the q field with the nondivergent wind at each point. Of course the nondivergent wind itself changes in time as q is advected along. Thus, after extrapolating ahead for a short time, we must invert the operator

$$\nabla^2 + f_0^2 \frac{\partial}{\partial p}\left(\frac{1}{\sigma}\frac{\partial}{\partial p}\right)$$

to obtain ψ from the computed field of q. We may then use the new ψ field to recompute the nondivergent wind and advect q along for another short time. By continually repeating this process a forecast for the distribution of ψ at any time in the future can be prepared.

In the usual method of preparing a numerical prediction using the quasi-geostrophic system, $\psi(x, y, p)$ is represented at a finite number of points on a three-dimensional *grid* network. The differential equation (8.17) is then replaced by its finite difference analog. The result is a set of simultaneous algebraic equations which can be solved to obtain predictions for the time evolution of ψ at each gridpoint. To adequately resolve synoptic scale systems it is necessary that the grid points be separated by distances of no more than a few hundred kilometers in the horizontal. Ideally grid points should be spaced every 2 or 3 km in the vertical. However, the use of many data levels in the vertical requires storage and computation of ψ at an enormous number of points. Fortunately most synoptic scale systems have vertical scales comparable to the depth of the troposphere. Thus, it is possible to obtain useful forecasts with models which have very crude vertical resolution.

8.4 One-Parameter Models

The simplest type of model is one which refers to only one data level in the vertical. Such a model is called a one-parameter model. One-parameter models are inherently barotropic for the inclusion of thickness advection, which is crucial to baroclinic processes, requires knowledge of Φ at more than a single level.

8.4.1 The Barotropic Model

The simplest one-parameter model is obtained by assuming that there exists a level of nondivergence in the atmosphere. This level is usually taken to be the 500-mb surface. Thus, we assume that

$$\nabla \cdot \mathbf{V} = -\frac{\partial \omega}{\partial p} = 0 \qquad \text{at} \quad 500 \text{ mb}.$$

This assumption, although certainly not true everywhere, is partly justified observationally since computed vertical motion patterns generally show a maximum in ω near 500 mb. (This fact was used in Section 6.2.4 in our qualitative discussion of the omega equation.) Assuming that 500 mb is a level of nondivergence, the vorticity equation (8.12) simplifies to

$$\frac{\partial}{\partial t}\nabla^2\psi = -\mathbf{V}_\psi\cdot\nabla(\nabla^2\psi + f) \tag{8.19}$$

This equation allows one to compute the evolution of the flow at the level of nondivergence only. No vertical coupling is involved. Thus, no predictions for levels above or below the nondivergent level are possible. The barotropic vorticity equation (8.19) is an exact model only for a homogeneous incompressible fluid confined between rigid, frictionless horizontal boundaries. Clearly the atmosphere does not meet these basic requirements. However, because midtropospheric synoptic scale flow is in general quasi-nondivergent, (8.19) is a useful approximate forecast equation. Furthermore, unlike the potential vorticity equation (8.17), the barotropic vorticity equation does not require that the motion be quasi-geostrophic. It is only necessary that the horizontal velocity be quasi-nondivergent. Thus, as we shall see in Chapter 12, (8.19) remains a useful approximation even in equatorial regions. In fact, outside areas of active precipitation, (8.19) tends to be even a better approximation in the tropics than it is at middle latitudes.

8.4.2 A Modified Barotropic Model

The linearized form of (8.19) has free solutions (i.e., resonant oscillations) corresponding to westward propagating Rossby waves. As indicated by the dispersion relation (7.52), for long wavelengths Rossby waves retrograde (i.e., propagate westward) with large phase speeds. Although such free oscillations do exist in the atmosphere, most of the energy of the planetary scale flow is in quasi-stationary motions in which the planetary vorticity advection is balanced not by the local vorticity change as in Rossby waves, but by the divergence term in the vorticity equation (8.15). Thus, neglect of the horizontal divergence as in (8.19) seriously misrepresents the vorticity balance for the planetary scale motions. Although proper physical treatment of the divergence effect requires a multilevel baroclinic model, it is possible to improve the performance of the barotropic forecast model by including an empirical correction for the long-wave divergence.

Comparing (8.17) and (8.19) we see that the barotropic model can be obtained from the quasi-geostrophic potential vorticity equation by assuming that the "stretching" term (the term involving partial differentiation of ψ with respect to p) vanishes at the 500-mb level. A more realistic approximation may be obtained by "parameterizing" this baroclinic term as follows:

$$f_0^2 \frac{\partial}{\partial p}\left(\frac{1}{\sigma}\frac{\partial \psi}{\partial p}\right) = -\lambda^2 \psi \tag{8.20}$$

where λ^{-1} is an empirical constant with units of length.[4] Substituting from (8.20) into (8.17) we obtain the modified barotropic vorticity equation:

$$\frac{\partial}{\partial t}(\nabla^2\psi - \lambda^2\psi) = -\mathbf{V}_\psi \cdot \nabla(\nabla^2\psi - \lambda^2\psi + f) \tag{8.21}$$

Extensive operational forecast experience has shown that (8.21) provides much better short-range predictions for the long-wave components of the 500-mb flow than does (8.19), provided that we choose $\lambda \simeq 10^{-6}\ \mathrm{m}^{-1}$. In fact for many situations forecasts based on (8.21) have as much skill in the range of 24–48 h as do multilevel models of much greater physical sophistication. It must, of course, be kept in mind that (8.21) can only forecast the evolution of the geostrophic streamfunction in the midtroposphere. The output from this type of model must be used in conjunction with other techniques to forecast surface weather elements such as temperature and precipitation.

8.5 A Two-Parameter Baroclinic Model

In Chapter 6 we showed that thermal advection processes are essential to the development of synoptic systems. The barotropic and modified barotropic models do not allow temperature advection; therefore they cannot forecast the development of new systems. In fact barotropic models are really merely extrapolation formulas which state that the vertical vorticity distribution at any instant is advected isobarically by the windfield. The fact that barotropic prognoses are quite effective in predicting the evolution of midtropospheric flow for periods of up to two or three days indicates that in the short range, barotropic vorticity advection is the primary mechanism governing the flow. This fact simply reflects the quasi-horizontal and quasi-nondivergent character of midlatitude synoptic scale flows.

However, mere advection of the initial circulation field is clearly not satisfactory if we wish to produce forecasts which are consistently reliable. It is necessary, in addition, to predict the development of new systems. To include thermal advection processes which are essential for baroclinic

[4] λ^{-1} is essentially the *radius of deformation* discussed in Section 8.5.

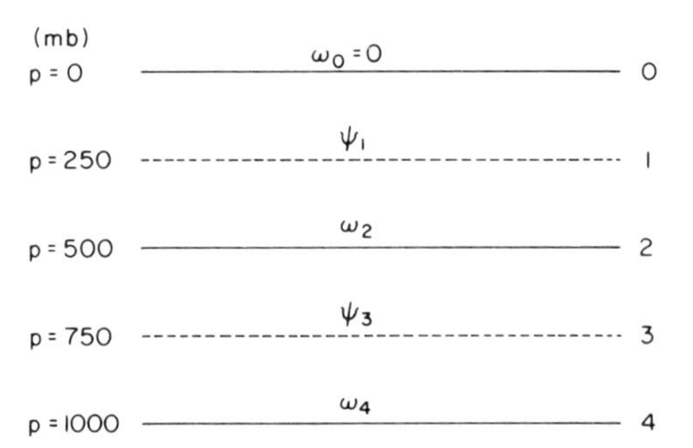

Fig. 8.1 Arrangement of variables in the vertical for the two-parameter baroclinic model.

development we must use a model which involves more than one data level in the atmosphere. We must also explicitly use the thermodynamic energy equation. More than a single data level is required because to compute temperature advection we must know the thickness, which in turn requires measurement of the difference in geopotential between two levels in the vertical.

The simplest model which can incorporate the baroclinic temperature advection process is one in which the geopotential is predicted at *two* levels. The thickness or mean temperature can then be represented as the difference in geopotential between those two levels. To derive this model we divide the atmosphere into two discrete layers bounded by surfaces numbered 0, 2, and 4 as shown in Fig. 8.1.

We now apply the vorticity equation (8.15) at the two levels designated as 1 and 3, which are at the midlevels of the two layers. To do this we must evaluate the divergence term $\partial\omega/\partial p$ at each level. Using *finite difference* approximations to the vertical derivatives we obtain

$$\left(\frac{\partial \omega}{\partial p}\right)_1 \simeq \frac{\omega_2 - \omega_0}{\Delta p}, \qquad \left(\frac{\partial \omega}{\partial p}\right)_3 \simeq \frac{\omega_4 - \omega_2}{\Delta p}$$

where Δp is the pressure interval between levels 0–2 and 2–4, and subscript notation is used to designate the vertical level (0–4) for each dependent variable. The resulting vorticity equations are

$$\frac{\partial}{\partial t}\nabla^2\psi_1 = -(\mathbf{k} \times \nabla\psi_1)\cdot\nabla(\nabla^2\psi_1 + f) + \frac{f_0}{\Delta p}\omega_2 \tag{8.22}$$

$$\frac{\partial}{\partial t}\nabla^2\psi_3 = -(\mathbf{k} \times \nabla\psi_3)\cdot\nabla(\nabla^2\psi_3 + f) - \frac{f_0}{\Delta p}\omega_2 \tag{8.23}$$

where for simplicity we have assumed that $\omega_4 = 0$, which will be approximately true for flat ground.

We next write the thermodynamic energy equation (8.16) for level 2. Here we evaluate $\partial\psi/\partial p$ using the difference formula

$$\left(\frac{\partial\psi}{\partial p}\right)_2 \simeq \frac{\psi_3 - \psi_1}{\Delta p}$$

The result is

$$\frac{\partial}{\partial t}(\psi_1 - \psi_3) = -(\mathbf{k} \times \nabla\psi_2)\cdot\nabla(\psi_1 - \psi_3) + \frac{\sigma\,\Delta p}{f_0}\omega_2 \tag{8.24}$$

The first term on the right-hand side in (8.24) is the advection of the 250–750-mb thickness by the wind at 500 mb. However, ψ_2, the 500-mb streamfunction, is not a predicted quantity in this model. Therefore, ψ_2 must be obtained by linearly interpolating between 250 and 750 mb:

$$\psi_2 = \tfrac{1}{2}(\psi_1 + \psi_3) \tag{8.25}$$

If this interpolation formula is used, (8.22)–(8.24) become a closed set of prediction equations in the variables ψ_1, ψ_3, and ω_2. We now eliminate ω_2 between (8.22)–(8.24) to obtain two equations in ψ_1 and ψ_3 alone. We first simply add (8.22) and (8.23) to obtain

$$\frac{\partial}{\partial t}\nabla^2(\psi_1 + \psi_3) = -(\mathbf{k} \times \nabla\psi_1)\cdot\nabla(\nabla^2\psi_1 + f) - (\mathbf{k} \times \nabla\psi_3)\cdot\nabla(\nabla^2\psi_3 + f) \tag{8.26}$$

We now introduce a length scale λ^{-1}, called the *radius of deformation*, by letting

$$\lambda^2 \equiv \frac{f_0^2}{\sigma(\Delta p)^2}$$

We next subtract (8.23) from (8.22) and add the result to $-2\lambda^2$ times (8.24) to get

$$\begin{aligned}\frac{\partial}{\partial t}[(\nabla^2 - 2\lambda^2)(\psi_1 - \psi_3)] = {} & -(\mathbf{k} \times \nabla\psi_1)\cdot\nabla(\nabla^2\psi_1 + f)\\ & + (\mathbf{k} \times \nabla\psi_3)\cdot\nabla(\nabla^2\psi_3 + f)\\ & + 2\lambda^2(\mathbf{k} \times \nabla\psi_2)\cdot\nabla(\psi_1 - \psi_3)\end{aligned} \tag{8.27}$$

Essentially, (8.26) states that the local rate of change of the vertically averaged vorticity (that is, the average of the 250- and 750-mb vorticities) is equal to the average of the 250- and 750-mb vorticity advections. Thus, (8.26) governs the barotropic part of the flow.

On the other hand, (8.27) may be regarded as an equation for the thickness tendency. Equation (8.27) states that the local rate of change of the 250–750-mb thickness is proportional to the difference between the vorticity advections at 250 and 750 mb plus the thermal advection. Thus, the physical mechanisms expressed by the terms on the right-hand side in (8.27) are identical to those contained in the diagnostic tendency equation (6.14).

We can also get an omega equation for the two-parameter model by combining (8.22)–(8.24) so that the time derivatives are eliminated. It is merely necessary to operate on (8.24) with ∇^2 then add (8.23) and subtract (8.22) from the result. After rearranging terms we get

$$(\nabla^2 - 2\lambda^2)\omega_2 = \frac{f_0}{\sigma\,\Delta p}\{\nabla^2[(\mathbf{k}\times\nabla\psi_2)\cdot\nabla(\psi_1 - \psi_3)] + (\mathbf{k}\times\nabla\psi_3)\cdot\nabla(\nabla^2\psi_3 + f) - (\mathbf{k}\times\nabla\psi_1)\cdot\nabla(\nabla^2\psi_1 + f)\} \tag{8.28}$$

Equation (8.28) is simply a two-level finite difference approximation for the omega equation (6.21) previously discussed. In fact (8.28) could have been obtained directly from (6.21) by approximating the partial derivatives with respect to p in (6.21) using finite differences centered at $p = 500$ mb.

The two-level model has not proved to be a very successful forecast model, primarily because it tends to produce stronger baroclinic development than is observed in many cases. However, the simplicity of this model does make it a useful tool for analysis of the physical processes occurring in baroclinic disturbances, as we shall see in Chapter 9.

8.6 Numerical Solution of the Barotropic Vorticity Equation

So far in this chapter we have discussed three different filtered models, all of which are based on the quasi-geostrophic system, and all requiring the numerical solution of some type of vorticity equation in order to produce a forecast. In this section we use the simplest prototype model, the barotropic vorticity equation, to illustrate the procedures and problems involved in preparing a numerical prediction with a filtered model.

The simplest approach to solving a nonlinear advection equation like (8.19) is to use *finite differences* to approximate the differential equation on a grid of points in space and time.

There are a number of problems associated with deriving finite difference analogs to partial differential equations such as the vorticity equation. Some of the most serious problems relate to the *computational stability* of the finite difference scheme. In order that a finite difference scheme be computationally stable in the sense that a solution of the difference equations will approximate a solution of the original system, it turns out that the ratio of the time and space increments must satisfy certain conditions. Only if the difference scheme is computationally stable will solutions of the finite difference equations converge to the solutions of the original difference equations as the space and time differencing increments approach zero. However, for finite difference solutions stability alone does not guarantee accurate solutions because all such solutions are subject to *truncation error* due to the approximate nature of the finite difference estimates of space and time derivatives.

8.6.1 Finite Differencing

The barotropic vorticity equation (8.19) can be rewritten in the following form:

$$\nabla^2 \chi + F(x, y, t) = 0 \tag{8.29}$$

where

$$F(x, y, t) = \mathbf{V}_\psi \cdot \nabla(\nabla^2 \psi + f) = \frac{\partial \psi}{\partial x}\frac{\partial}{\partial y}\nabla^2 \psi - \frac{\partial \psi}{\partial y}\frac{\partial}{\partial x}\nabla^2 \psi + \beta \frac{\partial \psi}{\partial x}$$

and

$$\chi = \frac{\partial \psi}{\partial t}$$

The advection of absolute vorticity $F(x, y, t)$ may be calculated at any point in space provided that we know the field of $\psi(x, y, t)$. Equation (8.29) is a *Poisson equation* in the variable χ with $F(x, y, t)$ a known source function. It may be solved for χ by several standard methods. The most common method is to write (8.29) in finite difference form and solve for χ approximately using the iterative method called *relaxation.*

Suppose that the horizontal x,y space is divided into a grid of $(M + 1) \times (N + 1)$ points separated by distance increments d. Then we can write the coordinate distances as $x = md$ and $y = nd$ where $m = 0, 1, 2, \ldots, M$ and $n = 0, 1, 2, \ldots, N$. Thus any point on the grid is uniquely identified by the indices (m, n). A portion of such a grid space is shown in Fig. 8.2.

Using centered difference formulas, derivatives of ψ at the point (m, n) may be expressed in terms of the values of ψ at surrounding points. For example,

$$\left(\frac{\partial \psi}{\partial x}\right)_{m,n} \simeq \frac{\psi_{m+1,n} - \psi_{m-1,n}}{2d}$$

$$\left(\frac{\partial \psi}{\partial y}\right)_{m,n} \simeq \frac{\psi_{m,n+1} - \psi_{m,n-1}}{2d}$$

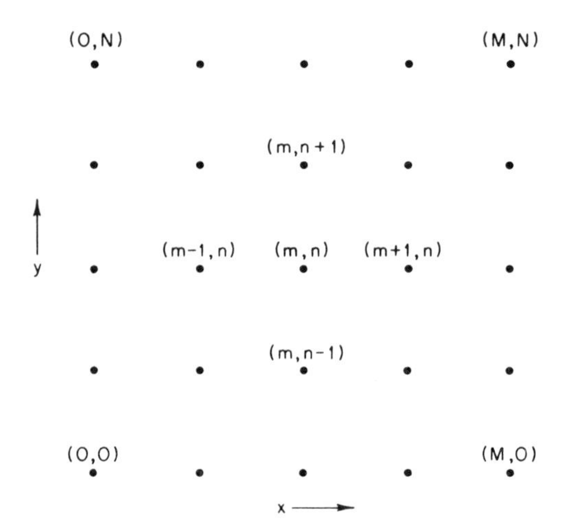

Fig. 8.2 Identification of mesh points on a finite difference grid of $(M + 1) \times (N + 1)$ points.

In a similar fashion second derivatives at any point can be evaluated by taking the difference between the first derivatives evaluated at half grid increments on each side of the point:

$$\left(\frac{\partial^2 \psi}{\partial x^2}\right)_{m,n} \simeq \frac{(\partial \psi/\partial x)_{m+1/2,n} - (\partial \psi/\partial x)_{m-1/2,n}}{d}$$

$$\simeq \frac{\psi_{m+1,n} - \psi_{m,n}}{d^2} - \frac{\psi_{m,n} - \psi_{m-1,n}}{d^2}$$

$$= \frac{\psi_{m+1,n} - 2\psi_{m,n} + \psi_{m-1,n}}{d^2}$$

Similarly,

$$\left(\frac{\partial^2 \psi}{\partial y^2}\right)_{m,n} \simeq \frac{\psi_{m,n+1} - 2\psi_{m,n} + \psi_{m,n-1}}{d^2}$$

Thus, we may write as the finite difference approximation to the horizontal Laplacian:

$$\nabla^2 \psi \simeq \frac{\psi_{m+1,n} + \psi_{m,n+1} + \psi_{m-1,n} + \psi_{m,n-1} - 4\psi_{m,n}}{d^2}$$

$$\equiv \nabla^2 \psi_{m,n} \tag{8.30}$$

The finite difference form of the Laplacian is simply proportional to the difference between the value of ψ at the central point and the average value at the four surrounding points.

We next write the term $F(x,y)$ in (8.29) in finite difference form. However, before doing so we observe that if $F(x, y)$ is integrated over a closed region bounded by a line along which ψ is constant, it is easily verified that the average value of $F(x, y)$ over that region is zero. This would be the case, for example, if we integrated over the entire Northern Hemisphere and assumed that there was no flow across the equator. Hence, from (8.29) the average vorticity over the region is conserved.

Since $F(x, y)$ has an average value of zero, it is desirable that any finite difference analog to $F(x, y)$ also have zero average value. Otherwise, the average vorticity will not be conserved in the finite difference form of the vorticity equation. In fact there are other integral constraints which motions governed by (8.29) must satisfy. In addition to conservation of average vorticity, kinetic energy and mean square vorticity are also conserved. Finite difference schemes which conserve all these integral invariants have been designed. However, such treatments are beyond the scope of this discussion. It is sufficient here to indicate that by finite differencing $F(x, y)$ using the form

$$F(x, y) = \frac{\partial}{\partial y}\left(\frac{\partial \psi}{\partial x}\nabla^2\psi\right) - \frac{\partial}{\partial x}\left(\frac{\partial \psi}{\partial y}\nabla^2\psi\right) + \beta\frac{\partial \psi}{\partial x} \tag{8.31}$$

it is possible to obtain a finite difference analog in which both the energy and average vorticity are conserved on the grid mesh. Thus, as a finite difference form for $F(x, y)$ we take centered differences in (8.31) to get

$$\begin{aligned} F_{m,n} = \frac{1}{4d^2}\big[&(\psi_{m+1,n+1} - \psi_{m-1,n+1})\nabla^2\psi_{m,n+1} \\ &- (\psi_{m+1,n-1} - \psi_{m-1,n-1})\nabla^2\psi_{m,n-1} \\ &- (\psi_{m+1,n+1} - \psi_{m+1,n-1})\nabla^2\psi_{m+1,n} \\ &+ (\psi_{m-1,n+1} - \psi_{m-1,n-1})\nabla^2\psi_{m-1,n}\big] \\ &+ \frac{\beta}{2d}(\psi_{m+1,n} - \psi_{m-1,n}) \end{aligned} \tag{8.32}$$

It is readily verified that

$$\sum_{m=1}^{M-1}\sum_{n=1}^{N-1} F_{m,n} = 0$$

provided that ψ is constant on the boundaries. Therefore, the form of the advection term given in (8.32) conserves the average vorticity. In a similar fashion it is possible to demonstrate that (8.32) also conserves the average kinetic energy. Thus, the finite difference form of (8.29) may be written as

$$\nabla^2\chi_{m,n} + F_{m,n} = 0 \tag{8.33}$$

8.6.2 Map Factors

Before discussing the relaxation method of inverting the Laplacian in (8.33) to obtain $\chi_{m,n}$ it is appropriate to digress slightly and discuss the question of map factors. It has been assumed above that the points on a finite difference grid represent points in space separated by equal distances along the surface of the earth. In reality finite difference grids are often square grids set on a map which is a flat projection of part of the curved earth. Usually meteorologists use *conformal* maps which preserve the shape in mapping any small area element of the earth, but may distort the relative size of the element. Thus, mesh points of a square grid on a conformal map will not in general correspond to uniform spacing on the real earth. An example of such a map is shown in Fig. 8.3 which shows a *polar stereographic* projection.

If the grid distance d_s on the map is measured in units which correspond to the actual grid distance at the latitude where the map projection is true, then the *map factor* μ may be defined as the ratio of the map grid distance to the actual grid distance at any other latitude. For a polar stereographic projection true at 90°N it can be shown that

$$\mu = \frac{d_s}{d} = \frac{2}{(1 + \sin \phi)}$$

In this case a given grid distance d_s will represent twice as much real distance at the pole as at the equator. Therefore, in order to preserve the accuracy of finite difference formulas, we must replace d by d_s/μ in (8.30) and (8.32) when the grid is referred to a map projection.

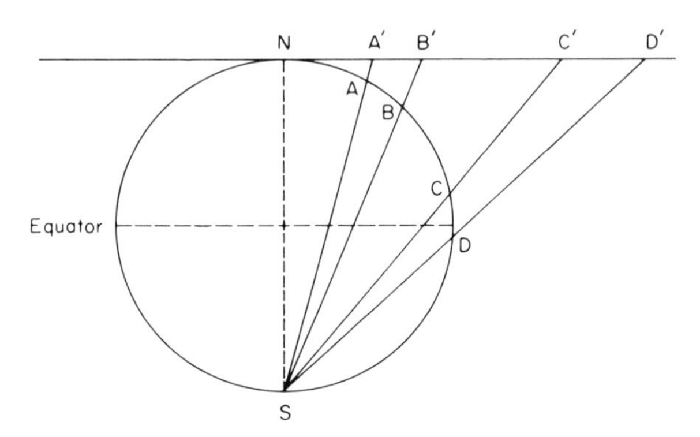

Fig. 8.3 Geometry of the polar stereographic projection. The arcs AB and CD on the surface of the earth map onto the lines $A'B'$ and $C'D'$, respectively.

8.6.3 Relaxation

We now consider the problem of solving (8.33) for $\chi_{m,n}$. Formally, we can write the solution as $\chi_{m,n} = -\nabla^{-2}F_{m,n}$ merely by inverting the finite difference Laplacian operator. However, in order to actually invert this operator by any standard method, we must first specify suitable boundary conditions on χ along a closed curve. For our purposes, it is sufficient to assume that ψ is zero for all time at all boundary points.

A simple scheme for solving (8.33) on a large grid mesh is an iterative technique known as *relaxation*. Relaxation schemes may be divided into two classes: (1) simultaneous relaxation, and (2) successive relaxation. These two methods will be discussed in turn below.

Since relaxation is an iterative procedure, some method is needed to identify the order of approximation. We thus label χ with the superscript v to indicate the vth guess χ^v. If the method is convergent, χ^v should approach the true solution χ at all points as $v \to \infty$. Using this notation we write the initial guess for the χ field as $\chi^0_{m,n}$. Substituting the initial guess into (8.33) we obtain

$$\chi^0_{m+1,n} + \chi^0_{m-1,n} + \chi^0_{m,n+1} + \chi^0_{m,n-1} - 4\chi^0_{m,n} = -d^2F_{m,n} + R^0_{m,n} \tag{8.34}$$

where $R^0_{m,n}$ is the *residual*, which is a measure of the difference between the initial guess and the true solution. The residual at any point (m, n) may be reduced to zero by altering the guess $\chi^0_{m,n}$ at that point to $\chi^1_{m,n}$ defined by

$$\chi^1_{m,n} = \chi^0_{m,n} + \tfrac{1}{4}R^0_{m,n} \tag{8.35}$$

while leaving the guesses at all surrounding points unchanged. This can be seen by substituting back into (8.34). If the correction formula (8.35) is applied at all interior grid points, the result will be a new estimate of the field, designated as $\chi^1_{m,n}$. The residuals can now be computed again by generalizing (8.34) to any order of approximation v: Thus,

$$R^v_{m,n} = d^2F_{m,n} + \nabla^2\chi^v_{m,n} \tag{8.36}$$

The $(v + 1)$th guess can then be computed according to

$$\chi^{v+1}_{m,n} = \chi^v_{m,n} + \tfrac{1}{4}R^v_{m,n} \tag{8.37}$$

This scheme is called simultaneous relaxation because the entire new field $\chi^{v+1}_{m,n}$ is guessed using residuals computed from the old field.

However, it is clear that once a new guess has been made at a given point, the new values can be used to modify the residuals at the surrounding points. Thus, the residuals can be computed sequentially starting from grid point (1, 1) and working to the right along the grid to point $(M - 1, 1)$, then skipping to the second interior row of points and working from point (1, 2)

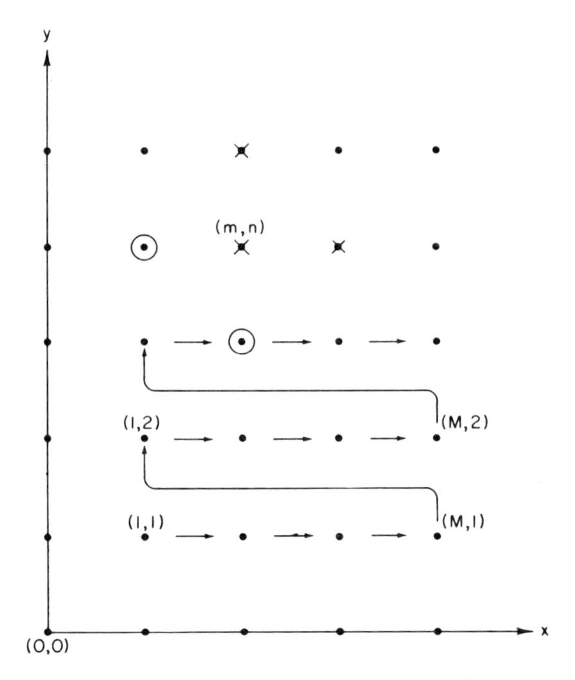

Fig. 8.4 Arrangement of grid for successive relaxation. Arrows show order of progression through the grid. The residual at (m, n) is computed using old guesses at the points indicated with $\times$ and new guesses at the circled points.

to $(M - 1, 2)$, etc., as shown in Fig. 8.4. The formula for computing residuals then becomes

$$R^{\nu}_{m,n} = d^2 F_{m,n} + \chi^{\nu}_{m+1,n} + \chi^{\nu}_{m,n+1} + \chi^{\nu+1}_{m-1,n} + \chi^{\nu+1}_{m,n-1} - 4\chi^{\nu}_{m,n} \quad (8.38)$$

Thus as shown in Fig. 8.4 at each point we use two new guesses and three old guesses to obtain the residual.

Although a general discussion of the conditions for convergence and rates of convergence for relaxation techniques is quite complicated, a simple example will suffice to indicate the superiority of successive relaxation to simultaneous relaxation. We consider the example shown in Fig. 8.5 in

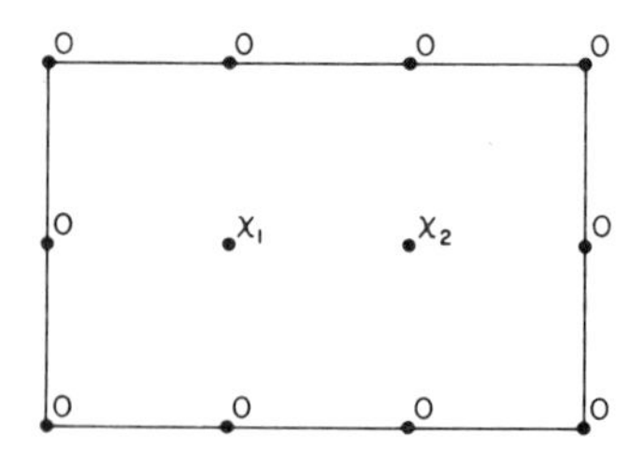

Fig. 8.5 A grid with two interior points.

which we solve the finite difference equation

$$\nabla^2\chi + 1 = 0 \tag{8.39}$$

for a mesh on which there are only two interior grid points labeled 1 and 2. The value of χ is taken to be zero on all boundary points. Applying formula (8.39) we find that simultaneous relaxation gives for the $(\nu + 1)$th guess at the interior points

$$\begin{aligned} -4\chi_1^{\nu+1} + \chi_2^{\nu} &= -1 \\ -4\chi_2^{\nu+1} + \chi_1^{\nu} &= -1 \end{aligned} \tag{8.40}$$

Now since χ^ν is the νth approximation to the true solution χ, we can express the error in the νth approximation as $\varepsilon^\nu = \chi^\nu - \chi$. Substituting this form into (8.40) and observing that χ satisfies (8.40) exactly we get

$$\begin{aligned} -4\varepsilon_1^{\nu+1} + \varepsilon_2^{\nu} &= 0 \\ -4\varepsilon_2^{\nu+1} + \varepsilon_1^{\nu} &= 0 \end{aligned} \tag{8.41}$$

Since (8.41) is a recursive formula valid for all ν, we can write

$$-4\varepsilon_1^{\nu+2} + \varepsilon_2^{\nu+1} = 0$$

which, upon substituting for $\varepsilon_2^{\nu+1}$ from (8.41), gives

$$\varepsilon_1^{\nu+2} = \tfrac{1}{16}\varepsilon_1^{\nu}$$

Thus, the error in this simple example decreases to $\frac{1}{16}$ of its initial value in two iterations of simultaneous relaxation.

In the successive method, on the other hand, the $(\nu + 1)$th guess is given by

$$\begin{aligned} -4\chi_1^{\nu+1} + \chi_2^{\nu} &= 1 \\ -4\chi_2^{\nu+1} + \chi_1^{\nu+1} &= 1 \end{aligned}$$

In this case the errors satisfy the equations

$$\begin{aligned} -4\varepsilon_1^{\nu+1} + \varepsilon_2^{\nu} &= 0 \\ \varepsilon_1^{\nu+1} - 4\varepsilon_2^{\nu+1} &= 0 \end{aligned}$$

from which we find

$$\varepsilon_1^{\nu+1} = \tfrac{1}{16}\varepsilon_1^{\nu}$$

Hence, the error decreases twice as fast as in the case of simultaneous relaxation. For more realistic cases involving a larger number of grid points, the convergence of the relaxation procedure would not be so rapid. However, in all cases it is found that using the successive method will improve the rate of convergence. Further improvement is possible by systematically *over-relaxing*. Over-relaxation merely involves multiplying the residual by a factor somewhat greater than $\frac{1}{4}$ in computing the new guess. In practice, only a few iterations are usually required for adequate convergence, provided that the initial guess is reasonably good.

8.6.4 Time Integration: Linear Computational Stability

In the previous subsections we have shown how to write the space derivatives in the vorticity equation in finite difference form and how to use relaxation to obtain the $\chi_{m,n}$ field at any time given the $F_{m,n}$ field at that time. To forecast the future circulation we must now extrapolate ahead in time using a finite difference approximation. Choosing a centered differencing scheme we may write

$$\psi(t_0 + \delta t) = \psi(t_0 - \delta t) + 2\,\delta t\,\chi(t_0)$$

This scheme requires knowledge of ψ at two time levels, $\psi(t_0 - \delta t)$ and $\psi(t_0)$ in order to compute $\psi(t_0 + \delta t)$. Since at the first time step of the forecast ($t_0 = 0$) only $\psi(t_0)$ is known, we must use a forward time step to initiate the forecast. Thus,

$$\psi(\delta t) = \psi(0) + \delta t\,\chi(0)$$

Specification of the time increment δt turns out to be a matter of crucial importance. Obviously, a larger value of δt will tend to give a poor approximation because in effect the tendency is linearly extrapolated over the entire interval $2\,\delta t$. Errors of this type which arise from the fact that finite differences are only approximations to the actual derivatives are called *truncation errors*. Such errors can be controlled by making δt sufficiently small. Of far greater significance, however, is the phenomenon of *computational instability*.

The problem of computational instability can be illustrated simply by examining finite difference solutions for the one-dimensional advection equation

$$\frac{\partial \zeta}{\partial t} = -U\frac{\partial \zeta}{\partial x} \tag{8.42}$$

Here U is a constant advecting velocity and $\zeta(x, t)$ is the advected field. If we assume that ζ is periodic in x and that initially

$$\zeta(x, 0) = e^{ikx} \tag{8.43}$$

then the exact analytic solution for (8.42) is

$$\zeta(x, t) = e^{ik(x - Ut)}$$

Using centered differences in space and time, (8.42) can be approximated in finite difference form by

$$\hat{\zeta}_{m,r+1} - \hat{\zeta}_{m,r-1} = -\mu(\hat{\zeta}_{m+1,r} - \hat{\zeta}_{m-1,r}) \tag{8.44}$$

where $\mu \equiv U\,\delta t/d$ and the space and time increments are defined by

$$x = md, \quad m = 0, 1, 2\ldots, \qquad t = r\,\delta t, \quad r = 0, 1, 2\ldots$$

We now assume that solutions of (8.44) exist in the form

$$\hat{\zeta}_{m,r} = V_r e^{ikmd} \tag{8.45}$$

where the amplitude coefficient V_r depends only on the time step r. If we choose $V_0 = 1$ then (8.45) exactly satisfies the initial condition (8.43). Substituting from (8.45) into (8.44) gives a three-term recursion formula for the coefficients V_r:

$$V_{r+1} - V_{r-1} = -\mu V_r(2i \sin kd) \tag{8.46}$$

We now let $V_r = \omega^r$, where ω is a constant to be determined. Substituting into (8.46) and dividing through by ω^{r-1} we obtain a quadratic equation for ω:

$$\omega^2 + 2i \sin\theta\,\omega - 1 = 0 \tag{8.47}$$

where

$$\sin\theta \equiv \mu \sin kd. \tag{8.48}$$

Eq. (8.47) yields two solutions,

$$\omega_1 = e^{-i\theta} \qquad \text{and} \qquad \omega_2 = -e^{i\theta}$$

Thus the complete solution (8.45) can be expressed as

$$\begin{aligned}\hat{\zeta}_{m,r} &= (A\omega_1^r + B\omega_2^r)e^{ikmd} \\ &= Ae^{i(kmd-\theta r)} + (-1)^r Be^{i(kmd+\theta r)}\end{aligned} \tag{8.49}$$

where A and B are constants. From inspection it is clear that $\hat{\zeta}_{m,r}$ will remain finite for $r \to \infty$ provided that θ is real. If θ is complex then one term in (8.49) will grow exponentially and the solution will be unbounded for $r \to \infty$. Referring back to (8.48) it is clear that for θ to be real we must require that $\mu \sin kd \leq 1$. But $\sin kd \leq 1$ so that the *Von Neumann necessary condition* for stability is

$$\frac{U\,\delta t}{d} \leq 1$$

Thus, for a given space increment d the time step δt must be chosen so that the dependent field will be advected a distance less than one grid length per time step.

The complete solution (8.49) can also be used to examine the truncation error of the finite difference approximation. Recalling that initially $\hat{\zeta}_{m,0} = e^{ikmd}$, we find from (8.49) with $r = 0$

$$1 = A + B$$

so that B may be replaced by $1 - A$ in (8.49). The remaining constant A must be determined from the first time step which, as has already been explained, is generally taken to be a forward step:

$$\hat{\zeta}_{m,1} = \hat{\zeta}_{m,0} - \frac{\mu}{2}(\hat{\zeta}_{m+1,0} - \hat{\zeta}_{m-1,0})$$

Substituting from the initial condition we obtain

$$\hat{\zeta}_{m,1} = e^{ikmd}(1 - i\mu \sin kd)$$

But from (8.49)

$$\hat{\zeta}_{m,1} = e^{ikmd}[Ae^{-i\theta} - (1 - A)e^{i\theta}]$$

Thus,

$$A = \frac{1 + \cos\theta}{2\cos\theta}, \qquad B = -\frac{1 - \cos\theta}{2\cos\theta}$$

Now from (8.48) we see that for $\delta t \to 0$ and d held constant $\theta \to 0$. Thus, $A \to 1$ and $B \to 0$ for $\delta t \to 0$. Examining the argument of the first term on the right in (8.49) we see that

$$kmd - \theta r = k[md - \theta r\,\delta t/(k\,\delta t)] = k(x - U't)$$

where

$$U' = \theta/(k\,\delta t) = \frac{U \sin^{-1}(\mu \sin kd)}{\mu kd}$$

Thus for $\mu \to 0$, $U' \to U$ and the finite difference solution approaches the true analytic solution. For this reason the first term on the right in (8.49) is referred to as the *physical mode*. The second term in the finite difference solution has no equivalent in the solution to the original differential equation. This term, called the *computational mode*, appears only because the centered difference time scheme produces a difference equation which is second order in time while the original equation (8.42) was first order in time. The computational mode propagates in a direction opposite to that of the physical mode and changes sign with each successive time step.

Table 8.1 *Accuracy of the Centered Difference Scheme*

$L/d = 2\pi/(kd)$	θ	$(U - U')/U$	B/A
2	180°	1	∞
3	139°	0.55	7.3
4	49°	0.28	0.2
6	41°	0.10	0.14
8	32°	0.05	0.08
12	22°	0.02	0.04
24	11°	0.005	0.01

Clearly, if (8.49) is to be a satisfactory approximation to the true solution, it is necessary that $B/A \ll 1$ and $(U - U')/U \ll 1$. These ratios (which measure the amplitude and phase error, respectively, for the numerical solution) are shown for disturbances of various wavelengths in Table 8.1 for the case $\mu = 0.75$. Only disturbances with wavelengths greater than four times the grid spacing are reasonably approximated by the finite difference solution. The numerically computed phase speed U' decreases with wavelength. This *numerical dispersion* is a computational artifact which can cause serious errors in the forecast of short-wavelength disturbances.

The results in Table 8.1 are for a computationally stable case. If we let $\mu = 1.1$, corresponding to a computationally unstable condition, it turns out that the computational mode would amplify by a factor of $\simeq 1.5$ per time step. In such a case the numerical solution would quickly lose all resemblance to the true solution. Thus, linear computational stability is an absolutely essential requirement for any numerical prediction model. Although we have here explicitly considered only a simple one-dimensional advection model, the necessary condition for stability is easily extended to more general cases. For a two-dimensional grid with uniform grid spacing d, in the x and y directions it can be shown that δt and d must satisfy

$$\frac{c\,\delta t}{d} \leq \frac{1}{\sqrt{2}} \tag{8.50}$$

where c is the propagation speed of the fastest moving disturbance generated by the model.

The existence of computational instability is one of the prime motivations for using filtered equations. In the quasi-geostrophic system no gravity or sound waves occur. Thus, the speed c in (8.50) is just the maximum wind speed or Rossby wave speed, whichever is greater. Typically, $c < 50$ m s^{-1} so that for a grid interval of 200 km a time increment of over one hour would be permissible. On the other hand, if we use the complete equations, c must be set equal to the speed of sound which is the fastest wave described by the equations. Thus, $c \simeq 300 \text{ m s}^{-1}$ and for a 200-km grid interval a time step of only a few minutes is permitted.

8.6.5 Summary of the Solution Procedure

The procedure for preparing a numerical forecast with the barotropic vorticity equation can now be summarized as follows:

1. Use the observed geopotential field at time $t = 0$ to compute $\psi_{m,n}(0)$.
2. Evaluate $F_{m,n}$ at all grid points.
3. Use relaxation to solve (8.29) for the tendency field $\chi_{m,n}$.
4. Extrapolate ahead a time increment δt to obtain $\psi_{m,n}(t + \delta t)$ using centered differencing except at the first step when a forward difference must be used.
5. Use the predicted field $\psi_{m,n}(t + \delta t)$ as data and repeat steps 2 through 4 until the desired forecast period is reached. Thus for a 24-hr forecast 24 time steps would be required with a one-hour time increment.
6. Use the forecast $\psi_{m,n}$ field to compute the predicted geopotential and nondivergent wind fields.

8.7 Primitive Equation Models

The filtered models discussed in the previous sections of this chapter have all involved the various approximations of the quasi-geostrophic system. Although these approximations are generally valid to within about 10–20% in middle latitudes, they do place a definite limit on the accuracy of the forecast. Furthermore, if these models are to be used in low latitudes, the streamfunction cannot be determined from the geostrophic vorticity field but must be obtained from the more complicated balance equation (8.9). This is a nonlinear equation which is difficult and time consuming to solve at each time step. Therefore, it is worthwhile to investigate the possibility of using less restrictive modeling assumptions.

Our scale analysis (Section 2.4) indicated that the hydrostatic approximation is far more accurate than the other filtering approximations. Thus, we will continue to assume that the motions are hydrostatic. The equations can then be written in pressure coordinates. The resulting system then consists of three prognostic equations (the x and y components of the momentum equation, and the thermodynamic energy equation) and three diagnostic equations (the continuity equation, the hydrostatic approximation, and the equation of state). These constitute a closed set in the dependent variables u, v, ω, Φ, α, and θ which is referred to as the *primitive equations.*

8.7.1 Sigma Coordinates

The isobaric coordinate system has many advantages which have been previously mentioned. Among these are the following: (1) meteorological data is normally referred to isobaric surfaces; (2) the continuity equation has a simple form; (3) density does not explicitly appear; (4) sound waves are completely filtered. However, these advantages are partially offset by the fact that the boundary condition at the ground is difficult to handle in the isobaric system. In fact we have generally used the approximate condition

$$\omega(p_0) \simeq -\rho_0 g w(z_0)$$

as the lower boundary condition. Here we have assumed that the height of the ground z_0 is coincident with the pressure surface p_0 (where p_0 is usually set equal to 1000 mb). These assumptions are of course not strictly valid even when the ground is level. Pressure does change at the ground. But more importantly, the height of the ground generally varies so that even if the pressure tendency were zero everywhere, the lower boundary condition should not be applied at a constant p_0. Rather, we should set $p_0 = p_0(x, y)$. However, it is very inconvenient for mathematical analysis to have a boundary condition which must be applied at a surface which is a function of the horizontal variables. Therefore, a modification of the pressure coordinate system, called the *sigma* system, is usually used in numerical modeling. In the sigma system the vertical coordinate is the pressure normalized with the surface pressure:[5]

$$\sigma \equiv \frac{p}{p_s}$$

where $p_s(x, y, t)$ is the pressure at the surface. Thus, σ is a *nondimensional* independent vertical coordinate which decreases upward from a value $\sigma = 1$ at the ground to $\sigma = 0$ at the top of the atmosphere. Hence, in sigma coordinates the lower boundary condition will always apply exactly at $\sigma = 1$. Furthermore, the vertical σ velocity

$$\dot{\sigma} \equiv \frac{d\sigma}{dt}$$

will always be zero at the ground even in the presence of sloping terrain. Thus, the lower boundary condition in the σ system is merely

$$\dot{\sigma} = 0 \qquad \text{at} \quad \sigma = 1$$

[5] It is important not to confuse the independent variable σ in this section with the static stability parameter of the quasi-geostrophic system which has also been labeled σ.

It is now necessary to transform the primitive equations from the (x, y, p) system to the (x, y, σ) system. In the (x, y, p) system, the basic primitive equations can be written in vector notation as

$$\frac{d\mathbf{V}}{dt} + f\mathbf{k} \times \mathbf{V} = -\nabla\Phi \tag{8.51}$$

$$\nabla \cdot \mathbf{V} + \frac{\partial \omega}{\partial p} = 0 \tag{8.52}$$

$$\frac{\partial \Phi}{\partial p} + \alpha = 0 \tag{8.53}$$

$$c_p \frac{d \ln \theta}{dt} = \frac{ds}{dt} \tag{8.54}$$

$$\theta = \left(\frac{p_0}{p}\right)^{R/c_p} \left(\frac{p\alpha}{R}\right) \tag{8.55}$$

where

$$\frac{d}{dt} = \frac{\partial}{\partial t} + \mathbf{V} \cdot \nabla + \omega \frac{\partial}{\partial p}$$

and the horizontal ∇ operator refers to differentiation at constant pressure. Since in general the σ and p surfaces will not coincide, partial derivatives evaluated at constant σ will not be equal to the analogous derivatives evaluated at constant p. The transformation from p to σ coordinates can be derived in a manner analogous to the generalized vertical coordinate transformation given in Section 1.6.3. Applying (1.23) with p replaced by Φ, s replaced by σ, and z replaced by p, we find that

$$\left(\frac{\partial \Phi}{\partial x}\right)_\sigma = \left(\frac{\partial \Phi}{\partial x}\right)_p + \frac{\sigma}{p_s} \frac{\partial p_s}{\partial x} \frac{\partial \Phi}{\partial \sigma} \tag{8.56}$$

Since any other variable will transform in an analogous way we can write the general transformation as

$$\nabla_p(\ ) = \nabla_\sigma(\ ) - \frac{\sigma}{p_s} \nabla p_s \frac{\partial(\ )}{\partial \sigma} \tag{8.57}$$

Applying the transformation formula (8.57) to the momentum equation (8.51) we get

$$\frac{d\mathbf{V}}{dt} + f\mathbf{k} \times \mathbf{V} = -\nabla\Phi + \frac{\sigma}{p_s} \nabla p_s \frac{\partial \Phi}{\partial \sigma} \tag{8.58}$$

where ∇ is now applied holding σ *constant*, and the total differential is

$$\frac{d}{dt} = \frac{\partial}{\partial t} + \mathbf{V} \cdot \nabla + \dot{\sigma} \frac{\partial}{\partial \sigma}$$

The equation of continuity can be transformed to the σ system as follows: From (8.57) we write the horizontal divergence as

$$(\nabla \cdot \mathbf{V})_p = (\nabla \cdot \mathbf{V})_\sigma - \frac{\sigma}{p_s} \frac{\partial \mathbf{V}}{\partial \sigma} \cdot \nabla p_s \tag{8.59}$$

To transform the term $\partial \omega / \partial p$ we first note that since p_s does not depend on σ:

$$\frac{\partial}{\partial p} = \frac{\partial}{\partial(\sigma p_s)} = \frac{1}{p_s} \frac{\partial}{\partial \sigma}$$

Thus, the continuity equation can be written

$$p_s(\nabla \cdot \mathbf{V})_p + \frac{\partial \omega}{\partial \sigma} = 0 \tag{8.60}$$

Now the sigma vertical velocity $\dot{\sigma}$ can be written as

$$\begin{aligned} \dot{\sigma} \equiv \frac{d\sigma}{dt} &= \left(\frac{\partial \sigma}{\partial t} + \mathbf{V} \cdot \nabla \sigma \right)_p + \omega \frac{\partial \sigma}{\partial p} \\ &= -\frac{\sigma}{p_s} \left(\frac{\partial p_s}{\partial t} + \mathbf{V} \cdot \nabla p_s \right) + \frac{\omega}{p_s} \end{aligned}$$

Differentiating the above with respect to σ and rearranging terms yields

$$\frac{\partial \omega}{\partial \sigma} = p_s \frac{\partial \dot{\sigma}}{\partial \sigma} + \left(\frac{\partial p_s}{\partial t} + \mathbf{V} \cdot \nabla p_s \right)_\sigma + \sigma \frac{\partial \mathbf{V}}{\partial \sigma} \cdot \nabla p_s \tag{8.61}$$

Substituting from (8.59) and (8.61) into (8.60) we obtain the transformed continuity equation

$$\nabla \cdot (p_s \mathbf{V}) + p_s \frac{\partial \dot{\sigma}}{\partial \sigma} + \frac{\partial p_s}{\partial t} = 0 \tag{8.62}$$

With the aid of the equation of state and Poisson's equation (2.44) the hydrostatic approximation (8.53) may be written in the sigma system as

$$\frac{\partial \Phi}{\partial \sigma} = \frac{-RT}{\sigma} = \frac{-R\theta}{\sigma} (p/p_0)^{R/c_p} \tag{8.63}$$

where $p_0 = 10^5$ Pa (1000 mb).

Expanding the total derivative in (8.54) we may write the thermodynamic energy equation in the form

$$\frac{\partial\theta}{\partial t} + \mathbf{V}\cdot\nabla\theta + \dot{\sigma}\,\frac{\partial\theta}{\partial\sigma} = \frac{\theta}{c_p}\frac{ds}{dt} \tag{8.64}$$

If we now multiply (8.62) through by θ, multiply (8.64) by p_s, and add the results we obtain the thermodynamic energy equation in *flux* form:

$$\frac{\partial}{\partial t}(p_s\theta) + \mathbf{V}\cdot(p_s\theta\mathbf{V}) + \frac{\partial}{\partial\sigma}(p_s\theta\dot{\sigma}) = \frac{p_s\theta}{c_p}\frac{ds}{dt} \tag{8.65}$$

A similar transformation of the x and y components of the momentum equation yields

$$\frac{\partial}{\partial t}(p_s u) + \mathbf{V}\cdot(p_s u\mathbf{V}) + \frac{\partial}{\partial\sigma}(p_s u\dot{\sigma}) - fp_s v = -p_s\frac{\partial\Phi}{\partial x} - R\theta\left(\frac{p}{p_0}\right)^{R/c_p}\frac{\partial p_s}{\partial x} \tag{8.66}$$

$$\frac{\partial}{\partial t}(p_s v) + \mathbf{V}\cdot(p_s v\mathbf{V}) + \frac{\partial}{\partial\sigma}(p_s v\dot{\sigma}) + fp_s u = -p_s\frac{\partial\Phi}{\partial y} - R\theta\left(\frac{p}{p_0}\right)^{R/c_p}\frac{\partial p_s}{\partial y} \tag{8.67}$$

The set of equations (8.62), (8.63), (8.65), (8.66), and (8.67) contains the six independent scalar variables u, v, $\dot{\sigma}$, θ, Φ, p_s. We, therefore, need an additional equation to make the system complete. This relation is the surface pressure tendency equation which can be readily obtained by integrating the continuity equation (8.62) vertically and using the boundary conditions that

$$\dot{\sigma} = 0 \qquad \text{at} \quad \sigma = 0, 1$$

The result is

$$\frac{\partial p_s}{\partial t} = -\int_0^1 \mathbf{V}\cdot(p_s\mathbf{V})\,d\sigma \tag{8.68}$$

Equation (8.68) simply states that the rate of increase of the surface pressure at a given point equals the mass convergence in a unit cross-section column above the point. With the inclusion of (8.68) we now have a complete set of prediction equations which can be written in finite difference form and numerically integrated.

8.7.2 A Two-Layer Primitive Equation Model

In the two-level quasi-geostrophic model discussed in Section 8.5 the static stability was a specified constant. In reality, however, static stability varies in space and time in response to various dynamical and thermodynamical processes. This variability is usually incorporated in primitive

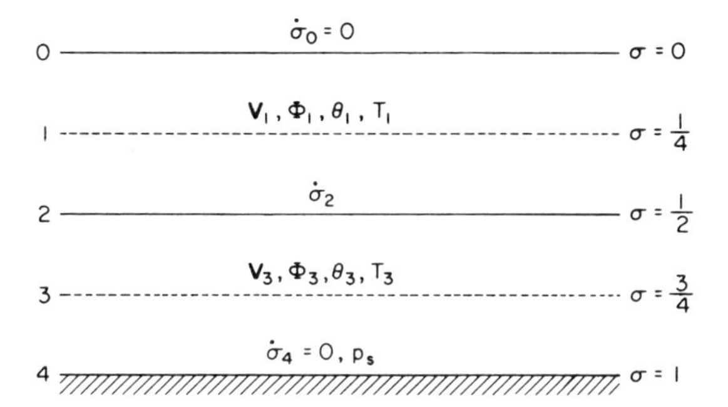

Fig. 8.6 Vertical differencing scheme for a two-level primitive equation model.

equation prediction models by computing $\partial\theta/\partial p$ explicitly for each grid point at each time step. Temperature must then be predicted at a minimum of *two* data levels rather than carried at a single level as in the two parameter quasi-geostrophic model described in Section 8.5. A simple vertical differencing scheme which satisfies this requirement is shown in Fig. 8.6.

As in the two-parameter model we divide the atmosphere into two layers separated by surfaces labeled 0, 2, and 4. The momentum equation and thermodynamic energy equation are applied at levels 1 and 3, which are at the centers of the two layers. Thus, for example, the thermodynamic energy equation takes the form

$$\frac{\partial}{\partial t}(p_s\theta_1) + \mathbf{V}\cdot(p_s\theta_1\mathbf{V}_1) + 2p_s\theta_2\dot{\sigma}_2 = \frac{p_s\theta_1}{c_p}\frac{ds_1}{dt} \tag{8.69}$$

$$\frac{\partial}{\partial t}(p_s\theta_3) + \mathbf{V}\cdot(p_s\theta_3\mathbf{V}_3) - 2p_s\theta_2\dot{\sigma}_2 = \frac{p_s\theta_3}{c_p}\frac{ds_3}{dt} \tag{8.70}$$

Analogous forms hold for the momentum equations (8.66) and (8.67). The vertical differencing in the thermodynamic energy equation involves θ_2 and $\dot{\sigma}_2$, while vertical differencing of the momentum equations involves u_2, v_2, and $\dot{\sigma}_2$. The variables u_2, v_2, and θ_2 are all obtained by linear interpolation; e.g., $\theta_2 = (\theta_1 + \theta_3)/2$. The field of $\dot{\sigma}_2$ may be obtained diagnostically through use of the continuity equation as follows: Writing the continuity equation for levels 1 and 3 with vertical derivatives replaced by centered differences, we obtain

$$\frac{\partial p_s}{\partial t} + \mathbf{V}\cdot(p_s\mathbf{V}_1) + 2p_s\dot{\sigma}_2 = 0 \tag{8.71}$$

$$\frac{\partial p_s}{\partial t} + \mathbf{V}\cdot(p_s\mathbf{V}_3) - 2p_s\dot{\sigma}_2 = 0 \tag{8.72}$$

Adding these two equations gives a finite difference form of the surface pressure tendency equation (8.68),

$$\frac{\partial p_s}{\partial t} + \frac{1}{2}\nabla\cdot[p_s(\mathbf{V}_1 + \mathbf{V}_3)] = 0 \tag{8.73}$$

while subtracting (8.72) from (8.71) yields the desired diagnostic equation for $\dot{\sigma}_2$,

$$\dot{\sigma}_2 = \frac{-1}{4p_s}\nabla\cdot[p_s(\mathbf{V}_1 - \mathbf{V}_3)] \tag{8.74}$$

In order to complete the specification of the dependent variables Φ_1 and Φ_3 must be determined in terms of θ_1 and θ_3 using the hydrostatic relationship. It turns out (Arakawa, 1972) that the finite difference equations will have energy conservation properties similar to the original differential equation system only if Φ_1 and Φ_3 are obtained by using special forms of the hydrostatic equation. Thus, we rewrite (8.63) in the form

$$\frac{\partial \Phi}{\partial p^\kappa} = -c_p\theta(p_0)^{-\kappa} \tag{8.75}$$

Applying (8.75) at level 2 we obtain

$$\Phi_1 - \Phi_3 = -c_p\theta_2\left[\left(\frac{p_1}{p_0}\right)^\kappa - \left(\frac{p_3}{p_0}\right)^\kappa\right] \tag{8.76}$$

Next we use the ideal gas law to rewrite (8.63) as

$$\frac{\partial(\sigma\Phi)}{\partial\sigma} = (\Phi - p_s\sigma\alpha) \tag{8.77}$$

Observing that $\sigma\Phi = 0$ for $\sigma = 0$, we integrate (8.77) with respect to σ from $\sigma = 1$ to $\sigma = 0$ to obtain

$$\Phi_4 = \tfrac{1}{2}[(\Phi_3 - p_s\sigma_3\alpha_3) + (\Phi_1 - p_s\sigma_1\alpha_1)] \tag{8.78}$$

Here Φ_4 is the geopotential at the surface which is a known function of x and y. Combining (8.76) and (8.78) we obtain

$$\Phi_1 = \Phi_4 + \frac{1}{2}c_p\theta_2\left[\left(\frac{p_3}{p_0}\right)^\kappa - \left(\frac{p_1}{p_0}\right)^\kappa\right] + \frac{p_s}{2}(\sigma_3\alpha_3 + \sigma_1\alpha_1) \tag{8.79}$$

$$\Phi_3 = \Phi_4 - \frac{1}{2}c_p\theta_2\left[\left(\frac{p_3}{p_0}\right)^\kappa - \left(\frac{p_1}{p_0}\right)^\kappa\right] + \frac{p_s}{2}(\sigma_3\alpha_3 + \sigma_1\alpha_1) \tag{8.80}$$

This completes the formulation of the two-level model.

To summarize, the procedure for forecasting with a two-level primitive equation model is as follows:

1. Write suitable finite difference analogues of the momentum and thermodynamic energy equations at levels 1 and 3, and the surface pressure tendency equation.
2. Use these prognostic equations to obtain the tendencies of the $\mathbf{V}_1$, $\mathbf{V}_3$, θ_1, θ_3, and p_s fields.
3. Extrapolate the tendencies ahead using a suitable time differencing scheme.
4. Use the new values of the $\mathbf{V}_1$, $\mathbf{V}_3$, θ_1, θ_3, and p_s fields to diagnostically determine $\dot{\sigma}_2$, Φ_1, and Φ_3.
5. Repeat steps 2–4 until the desired forecast time is reached.

In applying this scheme, it must be noted that the primitive equations in sigma coordinates contain the mechanisms for both gravity-wave and horizontal acoustic-wave propagation. It is thus necessary that the time increment be kept small enough to satisfy the condition

$$\frac{c\,\Delta t}{d} = \frac{1}{\sqrt{2}}$$

where c is the speed of a horizontally propagating acoustic wave which is the fastest moving wave solution for these equations. For this reason, time steps in primitive equation forecasts must be considerably shorter than those allowed for a quasi-geostrophic model with equal horizontal resolution.

8.7.3 Initial Data for the Primitive Equations

In discussing Richardson's early attempt at numerical forecasting, it was mentioned that one reason for his poor results was the lack of suitable initial data. Although data coverage has vastly improved since Richardson's time, measured wind velocities still are not accurate enough to allow us to use observed winds and pressures as initial data without some adjustments.

The basic problem in specifying initial data for primitive equation models can be illustrated by considering the relative magnitudes of the various terms in the momentum equation in pressure coordinates

$$\frac{\partial \mathbf{V}}{\partial t} + (\mathbf{V}\cdot\nabla)\mathbf{V} + \omega\frac{\partial \mathbf{V}}{\partial p} = -f\mathbf{k}\times\mathbf{V} - \nabla\Phi \tag{8.81}$$

For synoptic scale motions the wind and pressure fields are in approximate geostrophic balance. Thus, the acceleration following the motion is measured by the small difference between the two nearly equal terms $f\mathbf{k}\times\mathbf{V}$ and $-\nabla\Phi$. Although the Φ field can be determined observationally with quite good accuracy, observed winds are often 10–20% in error. This means that by using observed winds in the initial data the estimated Coriolis force may be 10–20% in error at the initial time. Since the acceleration is normally only about 10% of the Coriolis force in magnitude, an acceleration computed using the observed wind and geopotential fields will generally be 100% in error. Such spurious accelerations not only lead to poor estimates of the initial pressure and velocity tendencies but also may produce large-amplitude gravity-wave oscillations as the flow attempts to adjust from the initial unbalanced state back toward a state of geostrophic balance. These strong gravity waves are not present in nature, but their presence in the solution of the model equations will quickly spoil any chance of a reasonable forecast.

One possible approach to avoid this problem might be to neglect the observed wind data, and derive a wind field from the observed Φ field. The simplest scheme of this type would be to assume that the initial wind field was in geostrophic balance. However, it can be shown that the error in the computed, initial *local* acceleration would still be $\sim 100\%$. This can be seen by considering the example of stationary flow about a circular pressure system. In that case $\partial \mathbf{V}/\partial t = 0$ so that if we neglect the effects of vertical advection the flow will be in gradient wind balance (see Section 3.2.4). In vectorial form this balance is simply

$$(\mathbf{V} \cdot \nabla)\mathbf{V} + f\mathbf{k} \times \mathbf{V} = -\nabla\Phi \tag{8.82}$$

However, if $\mathbf{V}$ were replaced by $\mathbf{V}_g$ in (8.81) the Coriolis and pressure gradient forces would identically balance. Thus, the inertial force would be unbalanced and

$$\left(\frac{\partial \mathbf{V}}{\partial t}\right)_{t=0} \simeq -(\mathbf{V}_g \cdot \nabla)\mathbf{V}_g$$

Therefore, by assuming that initially the wind is in geostrophic balance rather than gradient balance, we compute a local acceleration which is completely erroneous.

It should now be clear that in order to avoid large errors in the initial acceleration, the initial wind field should be determined by the gradient wind balance, not by the geostrophic approximation or direct observations. The gradient wind formula (3.15) is not in itself a suitable balance condition to use, because the radius of curvature must be computed for parcel trajectories. However, a similar balance condition (which is equivalent to the gradient wind for stationary circularly symmetric flows) can be obtained from the divergence equation (8.6). If we assume that *initially*

$$\frac{\partial}{\partial t}(\nabla \cdot \mathbf{V}) = 0 \qquad \text{and} \qquad \nabla \cdot \mathbf{V} = 0$$

then the initial wind field is nondivergent, and is related to the geopotential field by the balance equation (8.9). The balance equation is linear in Φ, but nonlinear in ψ. Thus if ψ is known (8.9) can be solved by relaxation to determine Φ. In the usual midlatitude case, however, Φ is known and more sophisticated iterative techniques must be used to solve for ψ. In fact, since (8.9) is quadratic in ψ, there will generally exist two solutions for the ψ field corresponding to a given Φ field. One solution will correspond to the normal gradient wind balance, the other to the anomalous balance. Once ψ is determined from (8.9), the initial velocity field is computed using (8.8). The velocity field determined in this manner will be in gradient balance with the geopotential field. The initial local accelerations are thus very small. But more importantly, the initial local rate of change of divergence is zero which assures that large amplitude gravity oscillations are not excited by the initial conditions.

8.7.4 The NMC Six-Layer Model

For operational numerical forecasting the United States National Meteorological Center (NMC) uses a model with six prediction levels in the vertical. This model employs a modification of the sigma coordinate system in which four separate σ domains are defined. The vertical structure of the model is shown in Fig. 8.7. The coordinates σ_B, σ_T, and σ_S refer to the planetary boundary layer, the middle troposphere, and the stratosphere, respectively. The choice of separate coordinates for each of these regions allows better vertical resolution both in the boundary layer and in the stratosphere. In this model the tropopause is considered to be a material surface. Thus, an additional dependent variable p_2, the pressure at the tropopause, must be predicted. However the basic scheme is similar to that described for the two-level primitive equation model. The prognostic equations are solved at levels $k = 0, 1, \ldots, 5$. The constant-θ layer at the top is included for computational reasons, but has no meteorological significance.

Initial data in this model must be derived by interpolating geopotential values from constant-pressure maps. Since this interpolation involves some errors, it is necessary to determine the initial velocity field using the balance equation referred to sigma coordinates. The initial mass and velocity fields will then be in gradient balance on the sigma surfaces. For a grid distance of $\sim$400 km, this model requires a time step of about ten minutes or less to avoid computational instability.

The standard NMC primitive equation model uses a grid which covers almost the entire Northern Hemisphere with an average grid spacing of 380-km. In addition NMC has an operational regional limited-area fine mesh model (LFM) which covers the continental United States and the

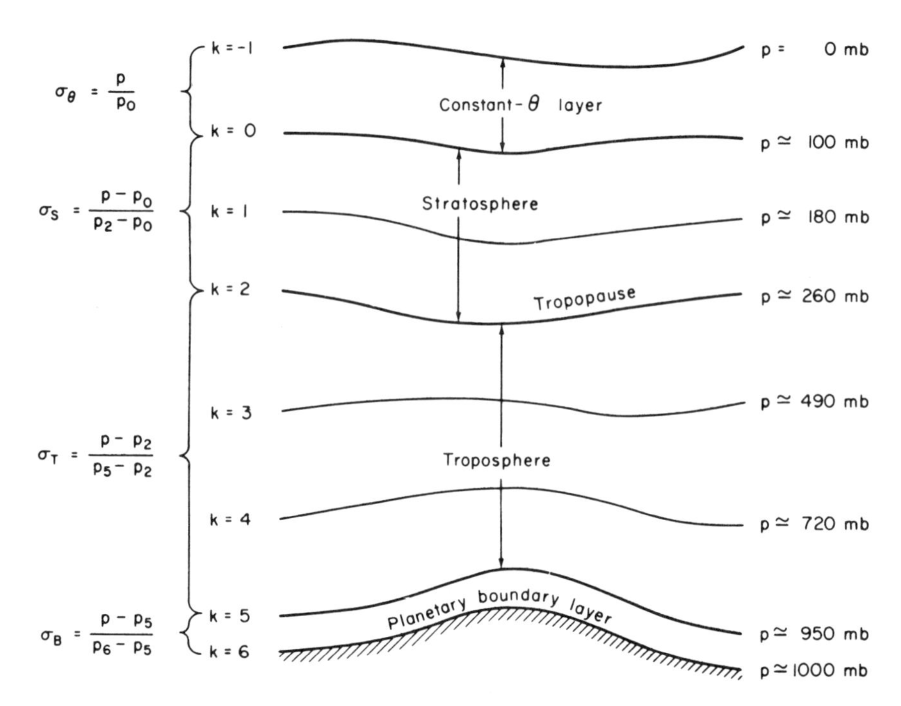

Fig. 8.7 Schematic diagram of the vertical structure of a six-layer σ coordinate model. (After Shuman and Hovermale, 1968. Reproduced with permission of the American Meteorological Society.)

adjacent ocean regions using a grid spacing of 190 km. Forecasts out to 48 h are prepared with both these models twice daily using initial fields based on the 0000 GMT and 1200 GMT observations. In addition the hemispheric model is run out to 84 h once daily using the 0000 GMT observations.

Since the introduction of numerical forecasting techniques in the late 1950s there has been a steady and significant improvement in forecast of the 50-kPa (500-mb) height field. Numerical forecasts of surface weather elements, especially precipitation, still have rather limited reliability. However experience with the regional model (LFM) indicates that finer horizontal grid spacing can lead to significant improvement in forecasts. There is little doubt that as models and initial data continue to improve in the future, there will be further improvement in forecast reliability.

8.7.5 Atmospheric Predictability

For short-range forecasts (one or two days) of the midlatitude 500-mb flow, it is possible to neglect diabatic heating and frictional dissipation. However, it is important to have good initial data in the region of interest

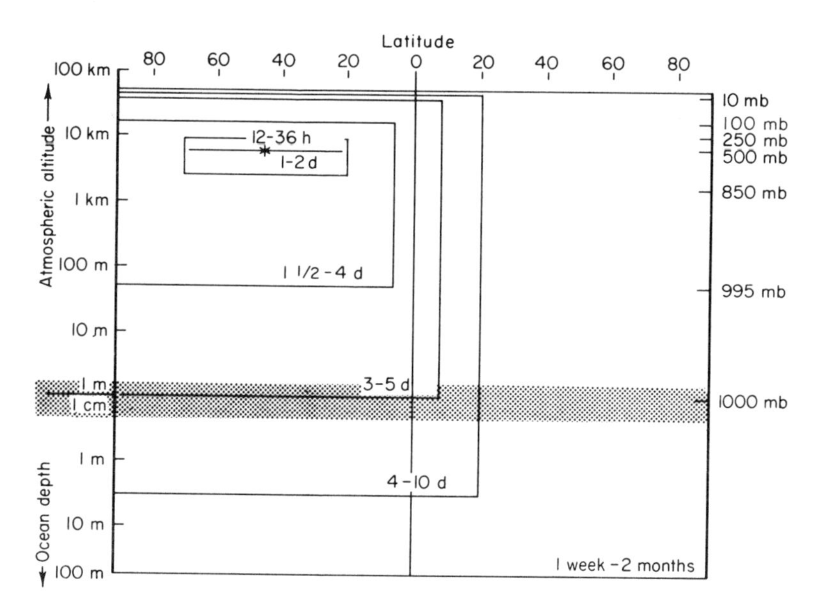

Fig. 8.8 A schematic diagram of the domain of initial data dependence for a prediction point in midtroposphere at midlatitudes (denoted by a star) as a function of forecast timespan. The atmospheric and ocean elevations are given on a logarithmic scale, increasing upward and downward, respectively. The stripped area is the interface zone. (After Smagorinsky, 1967. Reproduced with permission of the American Meteorological Society.)

because on this short time scale forecasting ability depends primarily on the proper advection of the initial vorticity field. As the length of the forecast period increases, the effects of various sources and sinks of energy become increasingly important. Therefore, the flow at a point in the middle-latitude troposphere will depend on initial conditions for an increasing domain as the period of the forecast is extended. In fact, according to the estimate of Smagorinsky shown in Fig. 8.8, for periods greater than a week it is necessary to know the initial state of the entire global atmosphere from the stratosphere to the surface as well as the state of the upper layers of the oceans.

However, even if the ideal data network were available to specify the initial state on a global scale, there still would be a limit beyond which a useful forecast would not be possible. This limit to the predictability of the atmosphere arises because the atmosphere is a continuum with a continuous spectrum of scales of motion. No matter how fine the grid resolution is made, there will always be motions whose scales are too small to be properly represented in the model. Thus, there is an unavoidable level of error in the determination of the initial state. As a forecast proceeds, the nonlinearity and instability of atmospheric flow will cause the inherent errors in the initial data to grow and gradually affect the larger scales of motion so that the forecast flow field will evolve differently from the actual flow field.

Estimates of how this error growth limits the inherent predictability of the atmosphere have been made by a number of groups using primitive equation forecast models. In these *predictability* experiments a "control" run is made using initial data corresponding to the observed flow at a given time. The initial data is then perturbed by introducing small random errors and the model is run again. The growth of inherent error can then be estimated by comparing the second "forecast" run with the control run. Results from a number of such studies indicate that the doubling time for the root mean square geopotential height error is about two to three days for small errors and somewhat greater for large errors. Thus, the theoretical limit for predictability on the synoptic scale is probably about one to two weeks.

Actual predictive skill with present models is, however, much less than the theoretical limit imposed by inherent error growth. Observational and analysis errors in the initial data both limit forecast accuracy; but the primary

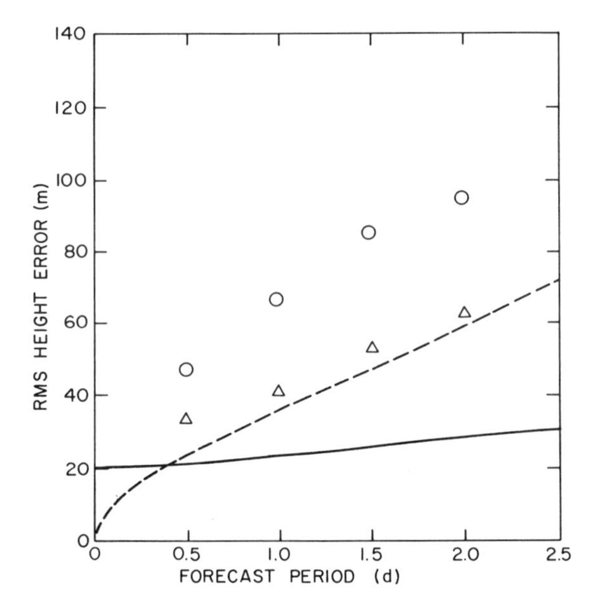

Fig. 8.9 500-mb RMS height error vs. forecast period. The initial height error is 20 m. Solid line shows perfect model forecast, triangles primitive equation forecast, circles persistence forecast. Dashed line shows primitive equation forecast starting with zero initial error. The RMS height error is computed by averaging the square of the height error for all grid points then taking the square root. (After Leith, 1978. Reproduced with permission from *Ann. Rev. Fluid Mechan.* **10**. © 1978 by Annual Reviews Inc.)

sources of error in present numerical predictions are believed to be due to imperfections in the numerical models, rather than to inadequacies in the initial data. An indication of the extent to which present forecasts fall short of the perfect model forecasts is given in Fig. 8.9, in which two day forecasts with a multilevel primitive equation model are compared with persistence (a forecast of no change) and a theoretical perfect model forecast. A number of aspects of the model may contribute to forecast error. In the previous subsection we have already touched on the importance of horizontal resolution. In addition, inadequate representation of mountain effects, precipitation, vertical resolution, and boundary layer processes may all contribute to forecast error. In summary, forecast error cannot be attributed to any single cause, but apparently results from the combined effects of a number of model aspects—both physical and computational.

Problems

1. Show that the nonlinear terms in the balance equation

$$G(x, y) \equiv -\nabla^2\left(\frac{\nabla\psi \cdot \nabla\psi}{2}\right) + \nabla \cdot (\nabla\psi\, \nabla^2\psi)$$

may be written in cartesian coordinates as

$$G(x, y) = 2\left[\frac{\partial^2\psi}{\partial x^2}\frac{\partial^2\psi}{\partial y^2} - \left(\frac{\partial^2\psi}{\partial x\, \partial y}\right)^2\right]$$

2. Express $G(x, y)$ from the previous problem in finite difference form using centered differences.

3. Using the results of Problem 1 show that the balance equation (8.9) is equivalent to the gradient wind equation (3.15) for a circularly symmetric geopotential perturbation given by

$$\Phi = \Phi_0(x^2 + y^2)/L^2$$

where Φ_0 is a constant geopotential and L a constant length scale. *Hint*: Let $\Psi = \Psi_0(x^2 + y^2)/L^2$ and solve for Ψ_0. You must assume that f is constant in this problem.

4. Use relaxation to solve the Poisson equation

$$\nabla^2\chi_{m,n} + F_{m,n} = 0$$

for $\chi_{m,n}$ on the mesh points of the grid shown in Fig. 8.2 assuming that $d = 1$, $M = N = 4$, and that $\chi_{m,n}$ is zero on all boundary points. For an initial guess let $\chi^0_{m,n} = F_{m,n}$ where the values of $F_{m,n}$ at the interior points are as follows:

(m, n)	$F_{m,n}$	(m, n)	$F_{m,n}$
(1, 1)	1	(3, 2)	2
(2, 1)	4	(1, 3)	1
(3, 1)	1	(2, 3)	4
(1, 2)	2	(3, 3)	1
(2, 2)	6		

5. Show that the advection of absolute vorticity by the nondivergent wind

$$\mathbf{V}_\psi \cdot \nabla(\zeta + f)$$

can be written in the form given in (8.31).

6. Use scale analysis to show that (8.10) may be approximated by (8.13) provided that the horizontal scale of the motion is much less than the radius of the earth.

7. Suppose that the streamfunction ψ is given by a single sinusoidal wave $\psi = A \sin kx$. Find the error in the finite difference approximation

$$\frac{\partial^2 \psi}{\partial x^2} \simeq \frac{\psi_{m+1} - 2\psi_m + \psi_{m-1}}{d^2}$$

for $kd = \pi/4, \pi/2, \pi$. Here $x = md$ with $m = 0, 1, 2, \ldots$.

8. Derive the computational stability criterion for a centered difference approximation to the so-called "inertial oscillation" equations

$$\frac{\partial u}{\partial t} = -fv, \qquad \frac{\partial v}{\partial t} = +fu$$

where f is a constant Coriolis parameter. *Hint*: let $V = u + iv$ and combine the two equations to obtain a single equation for V. Assume solutions of the form $V_r = \omega^r$ and reduce the equation to a quadratic form analogous to (8.47).

***9.** For the situation in Problem 8 obtain an analytic solution for $u(t)$ and $v(t)$ given the initial conditions $u = u_0$ and $v = 0$. Compare this solution with the solution to the centered difference approximation obtained in Problem 8. Use a forward time difference at the initial step. Compare the amplitudes of the computational and physical modes and the error in the computed period for $f\,\Delta t = 0.2, 0.4, 0.6, 0.8, 1$.

***10.** Using the method given in Section 8.6.4 evaluate the computational stability of the following two finite difference approximations to the one-dimensional linear advection equation:

(a) $$\hat{\zeta}_{m,r+1} - \hat{\zeta}_{m,r} = -\mu(\hat{\zeta}_{m,r} - \hat{\zeta}_{m-1,r})$$

(b) $$\hat{\zeta}_{m,r+1} - \hat{\zeta}_{m,r} = -\mu(\hat{\zeta}_{m+1,r} - \hat{\zeta}_{m,r})$$

where $\mu \equiv U\,\delta t/d > 0$.

(Schemes (a) and (b) are referred to as *upstream* and *downstream* differencing, respectively.) Show that scheme (a) damps the $\hat{\zeta}_m$ field and compute the percent damping per time step for $\mu = 0.25$ and $kd = \pi/8$ for a field with the initial form (8.43).

11. Show by a formal transformation of variables from the (x, y, p) to the (x, y, σ) system that

$$\left(\frac{\partial}{\partial t} + \mathbf{V}\cdot\nabla\right)_p + \omega\frac{\partial}{\partial p} = \left(\frac{\partial}{\partial t} + \mathbf{V}\cdot\nabla\right)_\sigma + \dot{\sigma}\frac{\partial}{\partial\sigma}$$

12. Show that (8.65) can be rewritten in terms of temperature as

$$\frac{\partial}{\partial t}(p_s T) + \nabla\cdot(p_s T\mathbf{V}) + \frac{\partial}{\partial\sigma}(p_s T\dot{\sigma}) - \frac{p_s\alpha}{c_p}\left(\sigma\frac{\partial p_s}{\partial t} + \sigma\mathbf{V}\cdot\nabla p_s + p_s\dot{\sigma}\right) = \frac{p_s}{c_p}\dot{q}$$

where $\dot{q}$ is the diabatic heating rate per unit mass.

***13.** Show that for stationary circularly symmetric flow on an f plane (i.e., f = const.) the balance equation (8.9) is equivalent to the gradient wind equation. *Hint*: use cylindrical coordinates.

Suggested References

Haltiner, *Numerical Weather Prediction*, presents an excellent treatment of the entire subject at graduate level.

Platzman (1967) presents an interesting historical review of Richardson's work and its relation to modern-day forecasting methods.

Richtmeyer and Morton, *Difference Methods for Initial Value Problems*, is an advanced text which thoroughly treats the questions of truncation error and computational stability for a wide variety of problems.

Shuman and Hovermale (1968) present a complete description of the NMC operational six-layer primitive equation forecast model.

Phillips (1973) provides an excellent account of many important dynamical aspects of large-scale numerical weather prediction including dynamic balance of initial data, and spectral modeling techniques.

Fawcett (1977) discusses the operational forecasting models of the National Meteorological Center.

Leith (1978) presents an interesting discussion of the predictive skill of numerical forecasting models.

Chapter

9 The Development and Motion of Midlatitude Synoptic Systems

In Chapter 6 we discussed the observed structure of midlatitude synoptic disturbances and showed that simple diagnostic relationships based on the quasi-geostrophic system can qualitatively account for the observed relationships among the pressure, temperature, and velocity fields. In particular, we found that horizontal temperature advection plays an essential role in the amplification of quasi-geostrophic disturbances. However, such diagnostic studies, although useful for interpreting the structure of baroclinic systems, are in themselves inadequate to determine the *origin* of these disturbances.

The presently accepted view is that baroclinic synoptic scale disturbances in the middle latitudes are initiated as the result of a *hydrodynamic instability* of the basic zonal current with respect to small perturbations of the flow. In this chapter we will examine this instability hypothesis for the origin of baroclinic waves and the energy conversions involved in the development of such waves. We will also consider briefly the development of fronts in association with synoptic disturbances.

9.1 Hydrodynamic Instability

The concept of hydrodynamic instability may be qualitatively understood by considering the motion of an individual fluid parcel in a steady zonal current. In Chapter 2 the parcel method was used to derive the criterion for static stability by examining the conditions under which a parcel displaced vertically without disturbing its surroundings would either return to its original position or be accelerated further from the initial level. The parcel method can be generalized by assuming that the parcel displacement is in an arbitrary direction. If the basic state is stably stratified and the zonal current has horizontal and vertical shear, the analysis becomes quite complicated. However, if the parcel is displaced horizontally across the basic flow, then the buoyancy force plays no role. This special case, called *inertial stability*, can be analyzed very simply if the basic flow is assumed to be geostrophic.

If we designate the basic state flow by

$$u_g = -\frac{1}{f}\frac{\partial \Phi}{\partial y}$$

and assume that the parcel displacement does not perturb the pressure field, the approximate equations of motion become

$$\frac{du}{dt} = fv = f\frac{dy}{dt} \tag{9.1}$$

$$\frac{dv}{dt} = f(u_g - u) \tag{9.2}$$

We consider a parcel which is moving with the geostrophic basic state motion at a position $y = y_0$. If the parcel is displaced across stream by a distance δy, we can obtain its new zonal velocity by integrating (9.1):

$$u(y_0 + \delta y) = u_g(y_0) + f\,\delta y \tag{9.3}$$

The geostrophic wind at $y_0 + \delta y$ can be approximated as

$$u_g(y_0 + \delta y) = u_g(y_0) + \frac{\partial u_g}{\partial y}\delta y \tag{9.4}$$

Substituting from (9.3) and (9.4) into (9.2) we obtain

$$\frac{dv}{dt} = \frac{d^2(\delta y)}{dt^2} = -f\left(f - \frac{\partial u_g}{\partial y}\right)\delta y \tag{9.5}$$

This equation is mathematically of the same form as (2.52), the equation for the motion of a vertically displaced particle in a stable atmosphere. Depending on the sign of the coefficient on the right-hand side in (9.5) the parcel will either be forced to return to its original position or will accelerate further from that position. In the Northern Hemisphere where f is positive the inertial stability condition thus becomes

$$f - \frac{\partial u_g}{\partial y} \begin{cases} > 0 & \text{stable} \\ = 0 & \text{neutral} \\ < 0 & \text{unstable} \end{cases} \tag{9.6}$$

Since $f - \partial u_g/\partial y$ is the absolute vorticity of the basic flow, the inertial stability condition is simply that the absolute vorticity be positive. Observations indicate that on the synoptic scale the absolute vorticity is nearly always positive. The occurrence of a negative absolute vorticity over any large area would be expected to trigger immediately inertially unstable motions which would mix the fluid laterally and reduce the shear until the absolute vorticity was again positive. This mechanism is called *inertial instability* since when viewed in an absolute reference frame the instability results from an imbalance between the pressure gradient and Coriolis (that is, inertial) forces for a parcel displaced radially in an axisymmetric vortex.

Inertial and static instability are merely two of many possible types of hydrodynamic instability. In general, a basic flow subject to arbitrary perturbations may be subject to a variety of modes of instability which depend on the horizontal and vertical shear, the static stability, the variation of the Coriolis parameter, the influence of friction, etc. In only a few cases can the simple parcel method give satisfactory stability criteria. Usually a more rigorous approach is required in which a linearized version of the governing equations is analyzed to determine the conditions under which the solutions describe amplifying disturbances. As indicated in Problem 2 of Chapter 7, the usual approach is to assume a wave type solution of the form

$$e^{ik(x-ct)}$$

and to determine the conditions for which the phase velocity c has an imaginary part. This technique, which is called the *normal modes* method, will be applied in the next section to analyze the stability of a baroclinic current.

9.2 Baroclinic Instability: Cyclogenesis

The development of synoptic scale weather disturbances is often referred to as *cyclogenesis*. The process of cyclogenesis will be regarded here as a manifestation of the amplification of an infinitesimal perturbation superposed on an unstable zonal current. For a perturbation to amplify it is clear that the basic flow must give up potential and (or) kinetic energy to the perturbation. In middle latitudes baroclinic instability is the most important cyclogenetic process. In baroclinic instability it turns out, as we shall show later, that the potential energy of the basic state flow is converted to potential and kinetic energy of the perturbation.

In this section we will derive the conditions for baroclinic instability using the two-level quasi-geostrophic model discussed in Chapter 8. This model certainly oversimplifies the vertical structure of baroclinic systems, but it does nevertheless contain the essential features necessary for a qualitative understanding of baroclinic instability.

We recall from (8.22)–(8.24) that the basic equations of the two-level model are

$$\frac{\partial}{\partial t}\nabla^2\psi_1 + \mathbf{V}_1\cdot\nabla(\nabla^2\psi_1 + f) = \frac{f_0}{\Delta p}\omega_2 \tag{9.7}$$

$$\frac{\partial}{\partial t}\nabla^2\psi_3 + \mathbf{V}_3\cdot\nabla(\nabla^2\psi_3 + f) = -\frac{f_0}{\Delta p}\omega_2 \tag{9.8}$$

$$\frac{\partial}{\partial t}(\psi_1 - \psi_3) + \mathbf{V}_2\cdot\nabla(\psi_1 - \psi_3) = \frac{\sigma\,\Delta p}{f_0}\omega_2 \tag{9.9}$$

where

$$\mathbf{V}_j = \mathbf{k}\times\nabla\psi_j \qquad \text{for} \quad j = 1, 2, 3$$

The arrangement of these variables in the vertical was shown in Fig. 8.1. To keep the analysis as simple as possible we now assume that the stream functions ψ_1 and ψ_3 consist of basic state parts which depend linearly on y alone, plus perturbations which depend only on x and t. Thus we let

$$\begin{aligned}\psi_1 &= -U_1 y + \psi_1'(x, t)\\ \psi_3 &= -U_3 y + \psi_3'(x, t)\\ \omega_2 &= \omega_2'(x, t)\end{aligned} \tag{9.10}$$

The zonal velocities at levels 1 and 3 are then U_1 and U_3, respectively. Hence, the perturbation field has meridional and vertical velocity components only.

Substituting from (9.10) into (9.7)–(9.9) and linearizing we obtain the perturbation equations

$$\left(\frac{\partial}{\partial t} + U_1\frac{\partial}{\partial x}\right)\frac{\partial^2\psi_1'}{\partial x^2} + \beta\frac{\partial\psi_1'}{\partial x} = \frac{f_0}{\Delta p}\omega_2' \tag{9.11}$$

$$\left(\frac{\partial}{\partial t} + U_3\frac{\partial}{\partial x}\right)\frac{\partial^2\psi_3'}{\partial x^2} + \beta\frac{\partial\psi_3'}{\partial x} = -\frac{f_0}{\Delta p}\omega_2' \tag{9.12}$$

$$\left(\frac{\partial}{\partial t} + \frac{U_1 + U_3}{2}\frac{\partial}{\partial x}\right)(\psi_1' - \psi_3') - \frac{U_1 - U_3}{2}\frac{\partial}{\partial x}(\psi_1' + \psi_3') = \frac{\sigma\,\Delta p}{f_0}\omega_2' \tag{9.13}$$

where we have used the β-plane approximation, $\beta \equiv df/dy$, and have linearly interpolated to express $\mathbf{V}_2$ in terms of ψ_1 and ψ_3. Equations (9.11)–(9.13) are a linear set in ψ'_1, ψ'_3, and ω'_2. As in Chapter 7 we assume wave-type solutions

$$\psi'_1 = Ae^{ik(x-ct)}, \qquad \psi'_3 = Be^{ik(x-ct)}, \qquad \omega'_2 = Ce^{ik(x-ct)} \tag{9.14}$$

Substituting these assumed solutions into (9.11)–(9.13) we find that the amplitude factors A, B, and C must satisfy the following set of simultaneous, homogeneous, linear algebraic equations:

$$ik[(c - U_1)k^2 + \beta]A - \frac{f_0}{\Delta p} C = 0 \tag{9.15}$$

$$ik[(c - U_3)k^2 + \beta]B + \frac{f_0}{\Delta p} C = 0 \tag{9.16}$$

$$-ik(c - U_3)A + ik(c - U_1)B - \frac{\sigma\, \Delta p}{f_0} C = 0 \tag{9.17}$$

Since this set is homogeneous, nontrivial solutions will exist only if the determinant of the coefficients of A, B, and C is zero. Thus we require that the phase speed c satisfy

$$\begin{vmatrix} ik[(c - U_1)k^2 + \beta] & 0 & -f_0/\Delta p \\ 0 & ik[(c - U_3)k^2 + \beta] & f_0/\Delta p \\ -ik(c - U_3) & ik(c - U_1) & -\sigma\, \Delta p/f_0 \end{vmatrix} = 0$$

Multiplying out the terms in the determinant we obtain a quadratic equation in c:

$$\begin{aligned}(k^4 + 2\lambda^2 k^2)c^2 &+ [2\beta(k^2 + \lambda^2) - (U_1 + U_3)(k^4 + 2\lambda^2 k^2)]c \\ &+ [k^4 U_1 U_3 + \beta^2 - (U_1 + U_3)(k^2 + \lambda^2)\beta + \lambda^2 k^2 (U_3^2 + U_1^2)] = 0\end{aligned} \tag{9.18}$$

where we have let $\lambda^2 \equiv f_0^2/(\sigma\, \Delta p^2)$. Alternatively, (9.18) could also have been obtained by eliminating any two of the variables A, B, or C between (9.15)–(9.17). Solving (9.18) for the phase speed we obtain

$$c = U_{\mathrm{m}} - \frac{\beta(k^2 + \lambda^2)}{k^2(k^2 + 2\lambda^2)} \pm \delta^{1/2} \tag{9.19}$$

where

$$\delta \equiv \frac{\beta^2 \lambda^4}{k^4(k^2 + 2\lambda^2)^2} - \frac{U_{\mathrm{T}}^2(2\lambda^2 - k^2)}{(k^2 + 2\lambda^2)}$$

and

$$U_m \equiv \frac{U_1 + U_3}{2}, \qquad U_T \equiv \frac{U_1 - U_3}{2}$$

Thus U_m and U_T are, respectively, the vertically averaged zonal wind and the basic state thermal wind for the interval $\Delta p/2$.

We have now shown that (9.14) is a solution for the system (9.11)–(9.13) only if the phase speed satisfies (9.19). Equation (9.19) is a rather complicated expression. However, we can note immediately that if $\delta < 0$ the phase speed will have an imaginary part and the perturbations will amplify exponentially. Before discussing the general properties of (9.19) in detail, we will consider two special cases.

As the first special case, we consider a barotropic basic state. If we let $U_T = 0$ so that the basic state thermal wind vanishes, the phase speeds

$$c_1 = U_m - \frac{\beta}{k^2} \tag{9.20}$$

and

$$c_2 = U_m - \frac{\beta}{(k^2 + 2\lambda^2)} \tag{9.21}$$

both are solutions. These are real quantities which correspond to the free (normal mode) oscillations for the two-level model with a barotropic basic state current. The phase speed c_1 is simply the dispersion relationship for a barotropic Rossby wave with no y dependence (see Section 7.5). Substituting the expression (9.20) in place of c in (9.15)–(9.17) we see that in this case $A = B$ and $C = 0$ so that the *perturbation is barotropic.* The expression (9.21), on the other hand, may be interpreted as the phase speed for an internal baroclinic Rossby wave. Note that c_2 is a dispersion relationship analogous to the Rossby wave speed for a homogeneous ocean with a free surface which was given in Problem 12 of Chapter 7. But, in the two-level model, the factor $2\lambda^2 \equiv 2f_0^2/\sigma\,\Delta p^2$ appears in the denominator in place of the f_0^2/gH for the oceanic case. In each of these cases there is vertical motion associated with the Rossby wave so that static stability modifies the wave speed. It is left as a problem for the reader to show that if c_2 is substituted into (9.15)–(9.17), the resulting fields of ψ_1 and ψ_3 are 180° out of phase so that the perturbation is baroclinic, although the basic state is barotropic. Furthermore, the ω_2' field leads the 250-mb geopotential field by 90° phase (that is, the maximum upward motion occurs west of the 250-mb trough). This vertical motion pattern may be understood if we note that $c_2 - U_m < 0$, so that the disturbance pattern moves westward *relative to the mean wind.*

Now, viewed in a coordinate system moving with the mean wind the vorticity changes are due only to the planetary vorticity advection and the convergence terms, while the thickness changes must be caused solely by the adiabatic heating or cooling due to vertical motion. Hence, there must be rising motion west of the 250-mb trough in order to produce the thickness changes required by the westward motion of the system.

Comparing (9.20) and (9.21) we see that the phase speed of the baroclinic mode is generally much less than that of the barotropic mode since for average midlatitude tropospheric conditions $\lambda^2 \simeq 2 \times 10^{-12}\ \mathrm{m}^{-2}$, which is comparable in magnitude to k for zonal wavelength of ~4500 km.[1]

As the second special case, we assume that $\beta = 0$. This case corresponds, for example, to a laboratory situation in which the fluid is bounded above and below by rotating horizontal planes so that the gravity and rotation vectors are everywhere parallel. In such a situation

$$c = U_{\mathrm{m}} \pm U_{\mathrm{T}}\left(\frac{k^2 - 2\lambda^2}{k^2 + 2\lambda^2}\right)^{1/2} \tag{9.22}$$

For waves with zonal wave numbers satisfying $k^2 < 2\lambda^2$, (9.22) has an imaginary part. Thus, all waves longer than the critical wavelength $L_{\mathrm{c}} = \sqrt{2}\pi/\lambda$ will amplify. From the definition of λ we can write

$$L_{\mathrm{c}} = \Delta p \pi (2\sigma)^{1/2}/f_0$$

Now for typical tropospheric conditions $(2\sigma)^{1/2} \simeq 2 \times 10^{-3}\ \mathrm{N}^{-1}\ \mathrm{m}^3\ \mathrm{s}^{-1}$. Thus with $\Delta p = 50$ kPa and $f_0 = 10^{-4}\ \mathrm{s}^{-1}$ we find that $L_{\mathrm{c}} \simeq 3000$ km. It is also clear from this formula that the critical wavelength for baroclinic instability increases with the static stability. The role of static stability in stabilizing the shorter waves can be understood qualitatively as follows: For a sinusoidal perturbation, the relative vorticity, and hence the differential vorticity advection, increases with the square of the wave number. But, as shown in Chapter 6, a secondary vertical circulation is required to maintain hydrostatic temperature changes and geostrophic vorticity changes in the presence of differential vorticity advection. Thus, for a geopotential perturbation of fixed amplitude the relative strength of the accompanying vertical circulation must increase as the wavelength of the disturbance decreases. Since static stability tends to resist vertical displacements, the shortest wavelengths will thus be stabilized.

[1] The presence of the free internal Rossby wave should actually be regarded as a weakness of the two-level model. Lindzen *et al.* (1968) have shown that this mode does not correspond to any free oscillation of the real atmosphere but is, rather, a spurious mode resulting from the use of the upper boundary condition $\omega = 0$ at $p = 0$ which formally turns out to be equivalent to putting a lid at the top of the atmosphere.

It is also of interest that with $\beta = 0$ the criterion for instability does not depend on the magnitude of the basic state thermal wind U_T. All wavelengths longer than L_c are unstable even for very small vertical shear. However, the growth rate of the perturbation does depend on U_T. From (9.14) we see that the exponential growth rate is $\alpha = kc_i$, where c_i designates the imaginary part of the phase speed. In the present case

$$\alpha = kU_T\left(\frac{2\lambda^2 - k^2}{2\lambda^2 + k^2}\right)^{1/2} \tag{9.23}$$

so that the growth rate increases linearly with the mean thermal wind.

Returning to the general case where all terms are retained in (9.19), the stability criterion is most easily understood by computing the so-called *neutral curve* which connects all values of U_T and k for which $\delta = 0$ so that the flow is *marginally stable*. From (9.19), the condition $\delta = 0$ implies that

$$\frac{\beta^2\lambda^4}{k^4(k^2 + 2\lambda^2)} = U_T^2(2\lambda^2 - k^2) \tag{9.24}$$

This complicated relationship between U_T and k can best be displayed by solving (9.24) for $k^4/2\lambda^4$, yielding

$$\frac{k^4}{2\lambda^4} = 1 \pm \left[1 - \frac{\beta^2}{4\lambda^4 U_T^2}\right]^{1/2}$$

In Fig. 9.1 the nondimensional quantity $k^2/2\lambda^2$, which is a measure of the zonal wavelength, is plotted against the nondimensional parameter $2\lambda^2 U_T/\beta$, which is proportional to the thermal wind. As indicated in the figure the neutral curve separates the unstable region of the U_T, k plane from the stable region. It is clear that the inclusion of the β effect serves to stabilize the flow, for now unstable roots exist only for

$$|U_T| > \frac{\beta}{2\lambda^2}$$

In addition the minimum value of U_T required for unstable growth depends strongly on k. Thus, the β effect strongly stabilizes the long-wave end of the wave spectrum ($k \to 0$). Again the flow is always stable for waves shorter than the critical wavelength $L_c = \sqrt{2}\pi/\lambda$.

This long wave stabilization associated with the β effect is caused by the rapid westward propagation of long waves (i.e., Rossby wave propagation) which occurs only when the β effect is included in the model. It can be shown that baroclinically unstable waves always propagate at a speed which lies between the maximum and minimum mean zonal wind speeds. Thus, for our two-level model in the usual midlatitude case where $U_1 > U_3 > 0$,

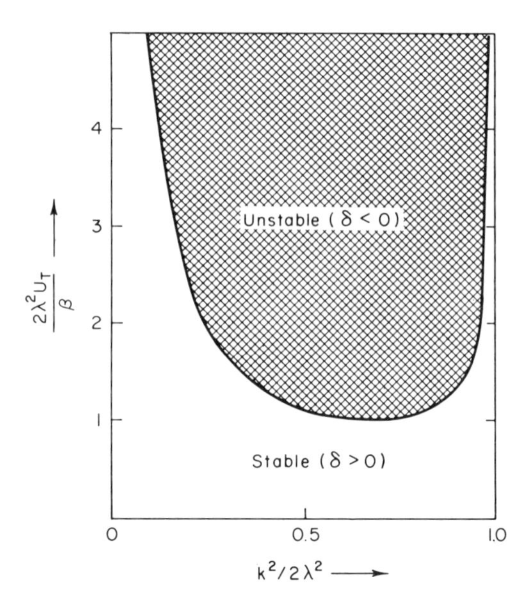

Fig. 9.1 Neutral stability curve for the two-level baroclinic model.

$U_3 < c_r < U_1$ for unstable waves. For long waves and weak basic state wind shear, solutions will have $c_r < U_3$ and unstable growth cannot then occur.

Differentiating (9.24) with respect to k and setting $dU_T/dk = 0$ we find that the minimum value of U_T for which unstable waves may exist occurs when $k^2 = \sqrt{2}\lambda^2$. This wave number corresponds to the wave of *maximum instability*. It seems likely that the wave numbers of observed disturbances should be close to the wave number of maximum instability, for if U_T were gradually raised from zero the flow would first become unstable for perturbations of wave number $k = 2^{1/4}\lambda$. Those perturbations would then amplify and in the process remove energy from the mean thermal wind, thereby decreasing U_T and stabilizing the flow. Under normal conditions of static stability the wavelength of maximum instability is about 4000 km, which is close to the average wavelength for midlatitude synoptic systems. Furthermore, the thermal wind required for marginal stability at this wavelength is only about $U_T \simeq 4 \text{ m s}^{-1}$, which implies a shear of 8 m s^{-1} between 250 and 750 mb. Shears greater than this are certainly common in middle latitudes for the zonally averaged flow. Therefore, the observed behavior of midlatitude synoptic systems is consistent with the hypothesis that such systems can originate from infinitesimal perturbations of a baroclinically unstable basic current. Of course in the real atmosphere many other factors may influence the development of synoptic systems, for example, instabilities due to lateral shear in the jet stream, nonlinear interactions of finite amplitude perturbations, and the release of latent heat. However, the evidence from observational studies, laboratory simulations, and numerical models all suggests that baroclinic instability is the primary mechanism for synoptic scale wave development in middle latitudes.

9.2.1 Vertical Motion in Baroclinically Unstable Waves

Since the two-level model has been derived from the equations of the quasi-geostrophic system, the physical mechanisms responsible for forcing vertical motions should be those which were discussed in Section 6.2.4. In this subsection we will verify that this is indeed the case. We also will show that the simplicity of the two-level model helps elucidate certain aspects of the divergent secondary circulations which were not previously discussed.

To review briefly the properties of the quasi-geostrophic system, we recall first that this system requires that the atmosphere simultaneously satisfy two constraints: (1) vorticity changes are geostrophic, and (2) temperature changes are hydrostatic. Thus both vorticity and temperature are proportional to derivatives of the geopotential field. In order that these constraints both be satisfied the omega field must at every instant be adjusted so that the divergent motions keep the vorticity changes geostrophic and the vertical motions keep the temperature changes hydrostatic.

These properties of the quasi-geostrophic model are clearly revealed in the linearized two-level model by examining the omega equation for the model. An omega equation for our linearized model can be obtained easily by taking the second derivative of (9.13) with respect to x, then eliminating the time derivatives with the aid of (9.11) and (9.12). If for simplicity we neglect the β effect, the resulting omega equation may be written as

$$\left(\frac{\partial^2}{\partial x^2} - 2\lambda^2\right)\omega_2' = \frac{f_0}{\sigma\,\Delta p}\left\{\frac{\partial^2}{\partial x^2}\left[U_{\mathrm{m}}\left(\frac{\partial\psi_1'}{\partial x} - \frac{\partial\psi_3'}{\partial x}\right)\right]\right.$$
$$\left. - \frac{\partial^2}{\partial x^2}\left[U_{\mathrm{T}}\left(\frac{\partial\psi_1'}{\partial x} + \frac{\partial\psi_3'}{\partial x}\right)\right] - \left[U_1\frac{\partial}{\partial x}\left(\frac{\partial^2\psi_1'}{\partial x^2}\right) - U_3\frac{\partial}{\partial x}\left(\frac{\partial^2\psi_3'}{\partial x^2}\right)\right]\right\} \tag{9.25}$$

The first term on the right in (9.25) represents the Laplacian of the advection of the perturbation thickness by the basic state mean (vertically averaged) wind. The second term is proportional to the Laplacian of the advection of the basic state thickness by the vertically averaged perturbation meridional wind. The final term (enclosed by the square brackets) represents the differential advection of perturbation vorticity by the basic state wind. Thus, it appears that three separate physical processes force vertical motion in this model. However, the first and third terms can be combined to yield an expression identical to the second term so that (9.25) becomes

$$\left(\frac{\partial^2}{\partial x^2} - 2\lambda^2\right)\omega_2' = -\frac{2f_0}{\sigma\,\Delta p}\frac{\partial^2}{\partial x^2}\left[U_T\left(\frac{\partial\psi_1'}{\partial x} + \frac{\partial\psi_3'}{\partial x}\right)\right] \tag{9.26}$$

Observing that

$$\left(\frac{\partial^2}{\partial x^2} - 2\lambda^2\right)\omega_2' \propto -\omega_2'$$

and that from geostrophy

$$U_T \propto -\frac{\partial\overline{T}}{\partial y}$$

we have from (9.26)

$$\omega_2' \propto -\frac{\partial^2}{\partial x^2}\left[v_2'\frac{\partial\overline{T}}{\partial y}\right] \propto v_2'\frac{\partial\overline{T}}{\partial y}$$

Therefore, in the linearized two-level model with $\beta = 0$, the *net* forcing of the vertical motion is proportional to the advection of the basic state temperature field by the vertically averaged perturbation meridional wind. Alternatively, we may write

$$\frac{\partial^2}{\partial x^2}\left[U_T\left(\frac{\partial\psi_1'}{\partial x} + \frac{\partial\psi_3'}{\partial x}\right)\right] = U_T\frac{\partial}{\partial x}(\zeta_1' + \zeta_3')$$

where $\zeta_1' = \partial^2\psi_1'/\partial x^2$ and $\zeta_3' = \partial^2\psi_3'/\partial x^2$. We then have the linearized version of the *Sutcliffe development formula*

$$\omega_2' \propto U_T\frac{\partial\zeta_2'}{\partial x}$$

which states that vertical motion is forced by the advection of the vertically averaged perturbation vorticity by the basic state thermal wind. Therefore, we may summarize as follows:

(a) cold (warm) advection forces sinking (rising) motion, or
(b) negative (positive) vorticity advection by the thermal wind forces sinking (rising) motion.

In the lower part of Fig. 9.2 we show schematically the phase relationship between the geopotential field and the divergent secondary motion field for a developing baroclinic wave in the two-level model for the usual mid-

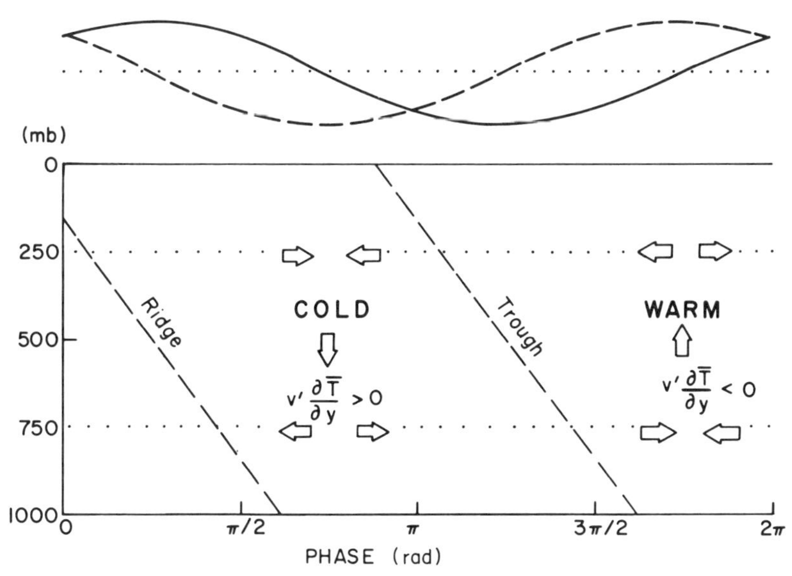

Fig. 9.2 Upper part: relative phases of the 500-mb geopotential perturbation (solid line) and temperature perturbation (dashed line) for an unstable baroclinic wave. Lower part: vertical cross section showing phase relationships between geopotential, vertical motion, and temperature fields for an unstable baroclinic wave in the two-level model.

latitude situation where $U_T > 0$. Linear interpolation has been used between levels so that the trough and ridge axes are straight lines tilted back toward the west with height. In this example the ψ_1 field lags the ψ_3 field by about 65° in phase so that the trough at 250 mb lies 65° in phase to the west of the 750-mb trough. At 500 mb the thickness field lags the geopotential field by one-quarter wavelength as shown in the top part of Fig. 9.2 and the thickness and vertical motion fields are in phase. Note that the temperature advection by the perturbation meridional wind v_2' is in phase with the 500-mb thickness field so that the advection of the basic state temperature by the perturbation wind acts to intensify the perturbation thickness field. In addition the divergence pattern associated with the subsiding branch of the vertical circulation contributes a positive vorticity tendency at the 250-mb trough and a negative vorticity tendency at the 750-mb ridge. Conversely, in the region of ascending motion the divergence field contributes a negative vorticity tendency at the 250-mb ridge and a positive tendency at the 750-mb trough. Since in all cases these vorticity tendencies tend to increase the extreme values of vorticity at the troughs and ridges, this secondary circulation system will act to increase the strength of the disturbances.

The role of differential vorticity advection in developing baroclinic waves can be elucidated with the aid of Figs. 9.3 and 9.4. In Fig. 9.3 the pattern of

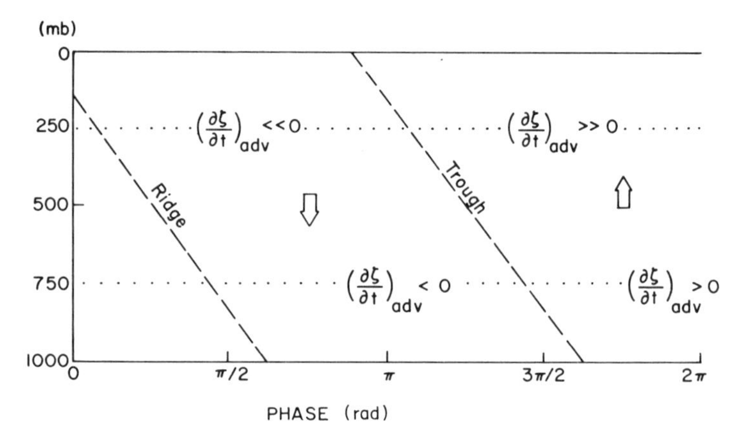

Fig. 9.3 Vertical cross section showing phase of vorticity change due to vorticity advection for an unstable baroclinic wave in the two-level model.

local vorticity change due to advection is shown schematically. Vorticity advection leads the vorticity field by one-quarter wavelength. Since in this case the basic state wind increases with height, the vorticity advection at 250 mb is larger than that at 750 mb. If no other processes influenced the vorticity field, the effect of this differential vorticity advection would be to move the upper-level trough and ridge pattern eastward more rapidly than the lower-level pattern. Thus, the westward tilt of the trough–ridge pattern would quickly be destroyed. The maintenance of this tilt in the presence of differential vorticity advection is due to the vorticity generation through

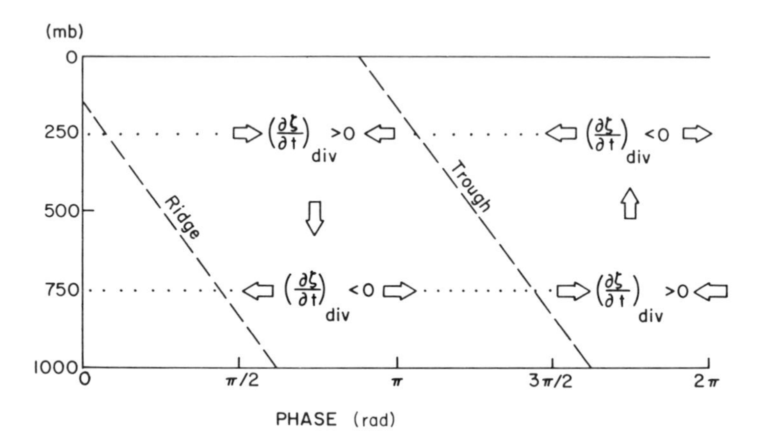

Fig. 9.4 Vertical cross section showing phase of vorticity change due to divergence–convergence for an unstable baroclinic wave in the two-level model.

vortex stretching associated with the divergent secondary circulation. Referring to Fig. 9.4, we see that vorticity generation by the divergence effect lags the vorticity field by 65.5° at 250 mb and leads the vorticity field by 65.5° at 750 mb. As a result the net vorticity tendencies ahead of the vorticity maxima and minima are less than the advective tendencies at the upper level and greater than the advective tendencies at the lower level:

$$\left|\frac{\partial \zeta_1}{\partial t}\right| < \left|\frac{\partial \zeta_1}{\partial t}\right|_{\mathrm{adv}}, \qquad \left|\frac{\partial \zeta_3}{\partial t}\right| > \left|\frac{\partial \zeta_3}{\partial t}\right|_{\mathrm{adv}}$$

Furthermore, vorticity generation by the divergence effect will tend to amplify the vorticity perturbations in the troughs and ridges at both the 250-mb and 750-mb levels as is required for a growing disturbance.

We can now also understand the cancellation that occurs between terms one and three on the right-hand side in (9.25). From Fig. 9.2 we see that the advection of the perturbation thickness field by U_{m} is in phase with the trough and ridge pattern at 500 mb (i.e., cold advection is a maximum at the 500-mb trough). We have previously seen that cold advection forces subsidence and hence leads to vortex tube stretching at 250 mb which keeps the vorticity in geostrophic balance in the presence of negative thickness tendencies. However, in the present case there is positive differential vorticity advection in phase with the 500-mb trough. This differential vorticity advection is just strong enough to provide the vorticity tendency required to maintain geostrophic balance. Thus, there is no net forcing of vertical motion in phase with the 500-mb troughs and ridges. In the region one-quarter wavelength behind the 500-mb trough, however, differential vorticity advection forces a subsidence equal to that forced by the thickness advection. Therefore, the net forcing of vertical motion in this simple model can be expressed in terms of temperature advection alone as in (9.26).

9.3 The Energetics of Baroclinic Waves

We have seen in the previous section that under suitable conditions a basic state current which contains vertical shear will be unstable to small perturbations. Such perturbations can then amplify exponentially by drawing potential and/or kinetic energy from the mean flow. In this section we analyze the energetics of linearized baroclinic disturbances and show that these perturbations grow by converting potential energy of the mean flow.

9.3.1 Available Potential Energy

Before discussing the energetics of baroclinic waves, it is necessary to consider the energy of the atmosphere from a more general point of view.

For all practical purposes, the total energy of the atmosphere is the sum of the internal energy, gravitational potential energy, and kinetic energy. However, it is not necessary to consider separately the variations of internal and gravitational potential energy because in a hydrostatic atmosphere these two forms of energy are proportional and may be combined into a single term called the *total potential energy*. The proportionality of internal and gravitational potential energy may be shown simply by considering these forms of energy for a column of air of unit horizontal cross section which extends from the surface to the top of the atmosphere.

If we let dE_{I} be the internal energy in a vertical section of the column of height dz, then from the definition of internal energy

$$dE_{\mathrm{I}} = \rho c_v \, T \, dz$$

so that the internal energy for the entire column is

$$E_{\mathrm{I}} = c_v \int_0^{\infty} \rho T \, dz \tag{9.27}$$

On the other hand, the gravitational potential energy for a slab of thickness dz at a height z is just

$$dE_{\mathrm{p}} = \rho g z \, dz$$

so that the gravitational potential energy in the entire column is

$$E_{\mathrm{p}} = \int_0^{\infty} \rho g z \, dz = -\int_{p_0}^{0} z \, dp \tag{9.28}$$

where we have substituted from the hydrostatic equation to obtain the last integral in (9.28). Integrating (9.28) by parts and using the ideal gas law we obtain

$$E_{\mathrm{p}} = \int_0^{\infty} p \, dz = R \int_0^{\infty} \rho T \, dz \tag{9.29}$$

Comparing (9.27) and (9.29) we see that

$$c_v E_{\mathrm{p}} = R E_{\mathrm{I}}$$

Thus, the total potential energy may be expressed as

$$E_{\mathrm{p}} + E_{\mathrm{I}} = \frac{c_p}{c_v} E_{\mathrm{I}} = \frac{c_p}{R} E_{\mathrm{p}} \tag{9.30}$$

Therefore, in a hydrostatic atmosphere the total potential energy can be obtained by computing either E_{I} or E_{p} alone.

The total potential energy is not a very suitable measure of energy in the atmosphere because only a very small fraction of the total potential energy

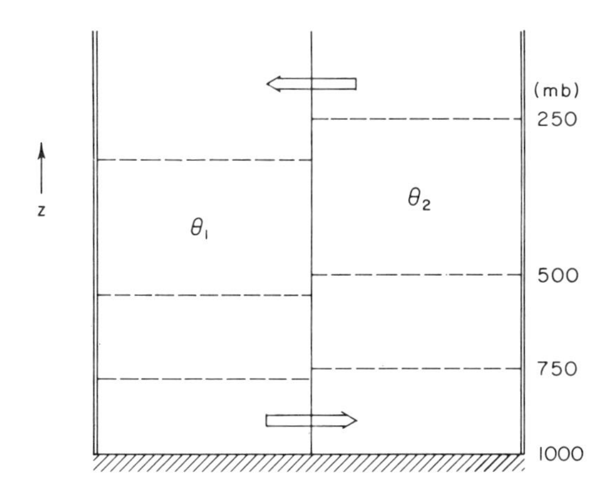

Fig. 9.5 Two air masses of differing potential temperature separated by a vertical partition. Dashed lines indicate isobaric surfaces. Arrows show direction of motion when partition is removed.

is available for conversion to kinetic energy in storms. To understand qualitatively why most of the total potential energy is unavailable we consider a simple model atmosphere which initially consists of two equal masses of dry air separated by a vertical partition as shown in Fig. 9.5. The two air masses are at uniform potential temperatures θ_1 and θ_2, respectively, with $\theta_1 < \theta_2$. The ground level pressure on each side of the partition is taken to be 1000 mb. We now wish to compute the maximum kinetic energy which can be realized by an *adiabatic* rearrangement of mass within the same volume when the partition is removed. Now for an adiabatic process, total energy is conserved:

$$E_{\mathrm{k}} + E_{\mathrm{p}} + E_{\mathrm{I}} = \mathrm{const}$$

where E_{k} denotes the kinetic energy. If the air masses are initially at rest $E_{\mathrm{k}} = 0$. Thus, if we let primed quantities denote the final state

$$E'_{\mathrm{k}} + E'_{\mathrm{p}} + E'_{\mathrm{I}} = E_{\mathrm{p}} + E_{\mathrm{I}}$$

so that with the aid of (9.30) we find that the kinetic energy realized by removal of the partition is

$$E'_{\mathrm{k}} = \frac{c_p}{c_v}(E_{\mathrm{I}} - E'_{\mathrm{I}})$$

Since θ is conserved for an adiabatic process, no mixing is allowed. It is clear that E'_{I} will be a minimum (designated by E''_{I}) when the masses are rearranged so that the air at θ_1 lies entirely beneath the θ_2 air with the 500-mb surface as the horizontal boundary between the two masses. In that case the total potential energy

$$\frac{c_p}{c_v} E_{\text{I}}''$$

is not available for conversion to kinetic energy, because no adiabatic process can further reduce E_{I}''.

The *available potential energy* (often abbreviated as APE) can now be defined as the difference between the total potential energy of a closed system and the minimum total potential energy which could result from an adiabatic redistribution of mass. Thus, for the idealized model given above

$$P \equiv \frac{c_p}{c_v}(E_{\text{I}} - E_{\text{I}}'') \tag{9.31}$$

which is equivalent to the maximum kinetic energy which can be realized by an adiabatic process.

Lorenz (1960) has shown that for the earth's atmosphere, the available potential energy is given approximately by the volume integral over the entire atmosphere of the variance of potential temperature on isobaric surfaces. Thus, letting $\bar{\theta}$ designate the average potential temperature for a given pressure surface, and θ' the local deviation from the average, the average available potential energy per unit volume satisfies the proportionality

$$\bar{P} \propto \frac{1}{V}\int \frac{\overline{\theta'^2}}{\bar{\theta}^2}\, dV$$

where V designates the total volume. For the quasi-geostrophic model, this proportionality is an *exact* measure of the available potential energy, as we will show in the following subsection.

Observations indicate that for the atmosphere as a whole

$$\frac{\bar{P}}{(c_p/c_v)\bar{E}_{\text{I}}} \sim \frac{1}{200}, \quad \frac{\bar{K}}{\bar{P}} \sim \frac{1}{10}$$

Thus only about 0.5% of the total potential energy of the atmosphere is available, and of the portion available only about 10% is actually converted to kinetic energy. From this point of view the atmosphere is a rather inefficient heat engine.

9.3.2 Energy Equations for the Two-Level Linearized Quasi-geostrophic Model

In our two-level model the perturbation temperature field is proportional to $\psi_1' - \psi_3'$, the 250–750-mb thickness. Thus, in view of the discussion in the previous section, we anticipate that the available potential energy in this case is proportional to $(\psi'_1 - \psi'_3)^2$. To show that this in fact must be the case, we derive the energy equations for the system in the following manner: We first multiply (9.11) by $-\psi'_1$, (9.12) by $-\psi'_3$, and (9.13) by $\psi'_1 - \psi'_3$. We then integrate the resulting equations over one wavelength of the perturbation in the zonal direction. The resulting zonally averaged[2] terms will be denoted by the angle bracket notation

$$\langle (\ ) \rangle = \frac{1}{L} \int_0^L (\ )\, dx$$

where L is the wavelength of the perturbation. Thus, for the first term in (9.11) we have after multiplying by $-\psi'_1$

$$-\left\langle \psi'_1 \frac{\partial}{\partial t}\left(\frac{\partial^2 \psi'_1}{\partial x^2}\right)\right\rangle = -\left\langle \psi'_1 \frac{\partial^2}{\partial x^2}\left(\frac{\partial \psi'_1}{\partial t}\right)\right\rangle$$

$$= \underbrace{-\left\langle \frac{\partial}{\partial x}\left[\psi'_1 \frac{\partial}{\partial x}\left(\frac{\partial \psi'_1}{\partial t}\right)\right]\right\rangle}_{\text{A}} + \underbrace{\left\langle \frac{\partial \psi'_1}{\partial x}\frac{\partial}{\partial t}\left(\frac{\partial \psi'_1}{\partial x}\right)\right\rangle}_{\text{B}}$$

Term A vanishes because it is the integral of a perfect differential in x over a complete cycle. Term B can be rewritten as

$$\left\langle \frac{1}{2}\frac{\partial}{\partial t}\left(\frac{\partial \psi'_1}{\partial x}\right)^2\right\rangle$$

which is just the rate of change of the perturbation kinetic energy per unit mass averaged over a wavelength. Similarly, $-\psi'_1$ times the advection term on the left in (9.11) can be written after integration in x as

$$-U_1\left\langle \psi'_1 \frac{\partial^2}{\partial x^2}\left(\frac{\partial \psi'_1}{\partial x}\right)\right\rangle = -U_1\left\langle \frac{\partial}{\partial x}\left[\psi'_1 \frac{\partial}{\partial x}\left(\frac{\partial \psi'_1}{\partial x}\right)\right]\right\rangle + U_1\left\langle \frac{\partial \psi'_1}{\partial x}\frac{\partial^2 \psi'_1}{\partial x^2}\right\rangle$$

$$= +\frac{U_1}{2}\left\langle \frac{\partial}{\partial x}\left[\frac{\partial \psi'_1}{\partial x}\right]^2\right\rangle = 0$$

Thus, the advection of kinetic energy vanishes when integrated over a wavelength. Evaluating the various terms in (9.12) and (9.13) in the same manner

[2] A *zonal* average generally designates the average around an entire circle of latitude. However, for a disturbance consisting of a single sinusoidal wave of wave number $k = m/(a \cos \phi)$, where m is an integer, the average over a wavelength is identical to a zonal average.

after multiplying through by $-\psi'_3$ and $(\psi'_1 - \psi'_3)$, respectively, we obtain the following set of perturbation energy equations:

$$\frac{1}{2}\left\langle \frac{\partial}{\partial t}\left(\frac{\partial \psi'_1}{\partial x}\right)^2 \right\rangle = -\frac{f_0}{\Delta p}\langle \omega'_2 \psi'_1 \rangle \tag{9.32}$$

$$\frac{1}{2}\left\langle \frac{\partial}{\partial t}\left(\frac{\partial \psi'_3}{\partial x}\right)^2 \right\rangle = \frac{f_0}{\Delta p}\langle \omega'_2 \psi'_3 \rangle \tag{9.33}$$

$$\frac{1}{2}\left\langle \frac{\partial}{\partial t}(\psi'_1 - \psi'_3)^2 \right\rangle = U_{\mathrm{T}}\left\langle (\psi'_1 - \psi'_3)\frac{\partial}{\partial x}(\psi'_1 + \psi'_3)\right\rangle + \frac{\sigma\,\Delta p}{f_0}\langle \omega'_2(\psi'_1 - \psi'_3)\rangle \tag{9.34}$$

where as before $U_{\mathrm{T}} \equiv (U_1 - U_3)/2$.

Defining the perturbation kinetic energy to be the sum of the kinetic energies of the 250- and 750-mb levels,

$$K' \equiv \frac{1}{2}\left\langle\left(\frac{\partial \psi'_1}{\partial x}\right)^2\right\rangle + \frac{1}{2}\left\langle\left(\frac{\partial \psi'_3}{\partial x}\right)^2\right\rangle$$

we find by adding (9.32) and (9.33) that

$$\frac{dK'}{dt} = -\frac{f_0}{\Delta p}\langle \omega'_2(\psi'_1 - \psi'_3)\rangle \tag{9.35}$$

Thus, the rate of change of perturbation kinetic energy is proportional to the correlation between the perturbation thickness and vertical motion. If we now define the perturbation available potential energy as

$$P' = \frac{\lambda^2\langle(\psi'_1 - \psi'_3)^2\rangle}{2}$$

we obtain from (9.34)

$$\frac{dP'}{dt} = \lambda^2 U_{\mathrm{T}}\left\langle (\psi'_1 - \psi'_3)\frac{\partial}{\partial x}(\psi'_1 + \psi'_3)\right\rangle + \frac{f_0}{\Delta p}\langle \omega'_2(\psi'_1 - \psi'_3)\rangle \tag{9.36}$$

The last term in (9.36) is just equal and opposite to the kinetic energy source term in (9.35). This term clearly must represent a conversion between potential and kinetic energy. If on the average the vertical motion is positive ($\omega'_2 < 0$) where the thickness is greater than average ($\psi'_1 - \psi'_3 > 0$) and vertical motion is negative where thickness is less than average, we have

$$\langle \omega'_2(\psi'_1 - \psi'_3)\rangle < 0 \tag{9.37}$$

perturbation potential energy is then being converted to kinetic energy. Physically, this correlation represents an overturning in which warm air is rising and cold air sinking, a situation which clearly tends to lower the center of mass and hence the potential energy of the perturbation. However, the available potential energy and kinetic energy of a disturbance can still grow simultaneously, provided that the potential energy generation due to the first term in (9.36) exceeds the rate of potential energy conversion to kinetic energy.

The potential energy generation term in (9.36) depends on the correlation between the perturbation thickness $(\psi'_1 - \psi'_3)$ and meridional velocity $\partial(\psi'_1 + \psi'_3)/\partial x$ at 500 mb. In order to understand the role of this term, it is helpful to consider a particular sinusoidal wave disturbance. Suppose that the barotropic and baroclinic parts of the disturbance can be written respectively as

$$\begin{aligned} \psi'_1 + \psi'_3 &= A_M \cos k(x - ct) \\ \psi'_1 - \psi'_3 &= A_T \cos k(x + x_0 - ct) \end{aligned} \tag{9.38}$$

where x_0 designates the phase difference. Since $\psi'_1 + \psi'_3$ is proportional to the 500-mb geopotential, and $\psi'_1 - \psi'_3$ is proportional to the 500-mb temperature, the phase angle kx_0 gives the phase difference between the geopotential and temperature fields at 500 mb. Furthermore, A_M and A_T are measures of the amplitudes of the 500-mb disturbance geopotential and temperature fields, respectively. Using the expressions in (9.38) we obtain

$$\begin{aligned} &\left\langle (\psi'_1 - \psi'_3) \frac{\partial}{\partial x} (\psi'_1 + \psi'_3) \right\rangle \\ &\quad = -\frac{k}{L} \int_0^L A_T A_M \cos k(x + x_0 - ct) \sin k(x - ct)\, dx \\ &\quad = \frac{k A_T A_M \sin kx_0}{L} \int_0^L [\sin k(x - ct)]^2\, dx \\ &\quad = \frac{A_T A_M k \sin kx_0}{2} \end{aligned} \tag{9.39}$$

From (9.36) we see that for the usual midlatitude case of a westerly thermal wind ($U_T > 0$) the correlation in (9.39) must be positive if the perturbation potential energy is to increase. Thus, x_0 must satisfy

$$0 < kx_0 < \pi$$

Furthermore, the correlation will be a positive maximum for $kx_0 = \pi/2$, that is, when the *temperature wave lags the geopotential wave by* $90°$ at 500 mb.

This case is shown schematically in Fig. 9.2. Clearly, when the temperature wave lags the geopotential by one-quarter cycle, the northward advection of warm air by the geostrophic wind east of the 500-mb trough, and the southward advection of cold air west of the 500-mb trough are both maximized. As a result, the cold advection in the lower troposphere is strong below the 250-mb trough and the warm advection in the lower troposphere is strong below the 250-mb ridge. In that case, as discussed previously in Section 6.2.2, the upper-level disturbance will intensify. It should also be noted here that if the temperature wave lags the geopotential wave, the trough and ridge axes will tilt westward with height, which, as mentioned in Section 6.1, is observed to be the case for amplifying midlatitude synoptic systems.

Referring again to Fig. 9.2 and recalling the vertical motion pattern implied by the omega equation (9.26), we see that the signs of the two terms on the right in (9.36) cannot be the same. For in the westward tilting perturbation of Fig. 9.2, the vertical motion must be downward in the cold air behind the trough at 500 mb and upward in the warm air ahead of the trough in order that the divergent motions can produce the necessary geostrophic vorticity change at 250 mb. Hence, the correlation between temperature and vertical velocity must be positive in this situation. Thus, for quasi-geostrophic perturbations, a westward tilt of the perturbation with height implies *both* that the horizontal temperature advection will increase the available potential energy of the perturbation and that the vertical circulation will convert perturbation available potential energy to perturbation kinetic energy. Conversely, an eastward tilt of the system with height would change the direction of both terms on the right in (9.36).

Although the signs of the potential energy generation term and the potential energy conversion term in (9.36) are always opposite for a developing baroclinic wave, it is only the potential energy generation rate which determines the growth of the total energy $P' + K'$ of the disturbance. This may be proved by adding (9.35) and (9.36) to obtain

$$\frac{d}{dt}(P' + K') = \lambda^2 U_T \left\langle (\psi'_1 - \psi'_3) \frac{\partial}{\partial x}(\psi'_1 + \psi'_3) \right\rangle$$

Thus, provided the correlation between the meridional velocity and temperature is positive and $U_T > 0$, the total energy of the perturbation will increase. Note that the vertical circulation merely converts disturbance energy between the available potential and kinetic forms without affecting the total energy of the perturbation.

The rate of increase of the total energy of the perturbation depends on the magnitude of U_T, that is, on the zonally averaged meridional temperature gradient. Since the generation of perturbation energy requires systematic

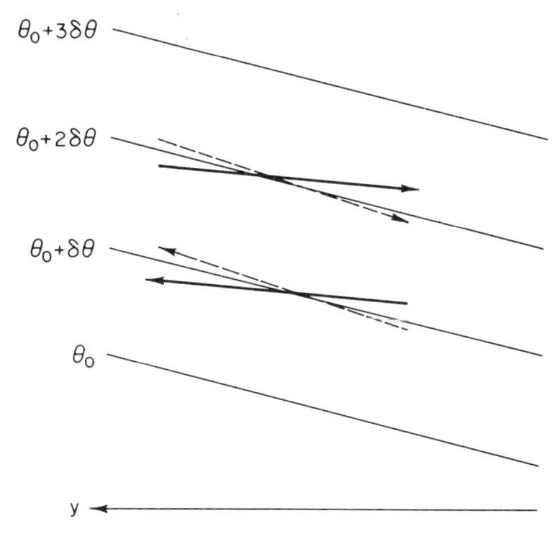

Fig. 9.6 Slopes of parcel trajectories relative to the zonal mean potential temperature surfaces for a baroclinically unstable disturbance (solid arrows) and for a baroclinically stable disturbance (dashed arrows).

northward transport of warm air and southward transport of cold air, it is clear that these disturbances will tend to reduce the meridional temperature gradient and hence the available potential energy of the mean flow. This latter process cannot be mathematically described in terms of the linearized equations. However, from Fig. 9.6 we can see qualitatively that parcels which move poleward (equatorward) and upward (downward) with slopes less than the slope of the zonal mean potential temperature surface will become warmer (colder) than their surroundings. For such parcels the correlations between meridional velocity and temperature, and between vertical velocity and temperature will both be positive as required for baroclinically unstable disturbances. Parcels which have trajectory slopes greater than the mean potential temperature slope will, on the other hand, have negative meridional velocity—temperature and vertical velocity—temperature correlations. Such parcels must then convert disturbance kinetic energy to disturbance available potential energy, which is in turn converted to zonal mean available potential energy. Therefore in order that perturbations be able to extract potential energy from the mean flow the perturbation parcel trajectories in the meridional plane must have slopes less than the slopes of the potential temperature surfaces. Since we have previously seen that northward-moving air must rise and southward-moving air must sink, it is clear that the rate of energy generation can be greater for an atmosphere in which the meridional slope of the potential temperature surfaces is large. We can also see more clearly why there is a short-wave cutoff from baroclinic instability. As previously mentioned, the intensity of the vertical circulation must increase as

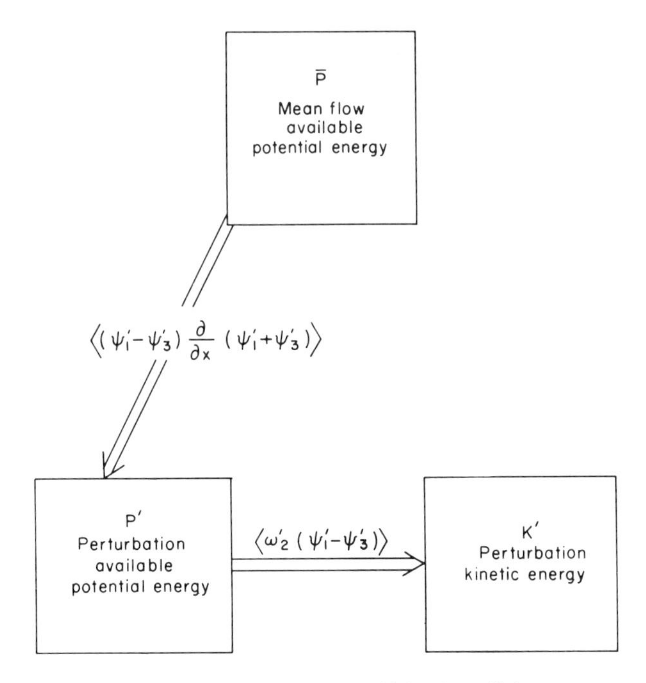

Fig. 9.7 Energy flow in an amplifying baroclinic wave.

the wavelength decreases. Thus, the slopes of the parcel trajectories must increase with decreasing wavelength, and for some critical wavelength the trajectory slopes will become greater than the slopes of the potential temperature surfaces.

The energy flow for quasi-geostrophic perturbations is summarized in Fig. 9.7 by means of a block diagram. In this type of energy diagram each block represents a reservior of a particular type of energy and the arrows designate the direction of energy flow. The complete energy cycle cannot be derived in terms of linear perturbation theory, but will be discussed qualitatively in Chapter 10.

9.4 Fronts and Frontogenesis

In the above discussion of baroclinic instability, we assumed that the mean thermal wind U_T was a constant independent of the y coordinate. This assumption was necessary to obtain a mathematically simple model which could still illustrate the basic instability mechanism. However, it was pointed out in Section 6.1 that baroclinicity is not uniformly distributed, but rather the horizontal temperature gradients tend to be concentrated in narrow frontal zones associated with the tropospheric jet streams. Thus in the real atmosphere the thermal wind is strongly latitudinally dependent with a maximum at the latitude of the jet core. This lateral dependence of the thermal wind must be included in any complete theory of baroclinic wave development. However, the fact that the simple theory of the previous section was able to qualitatively reproduce the essential features of developing baroclinic waves indicates that the fundamental process in the generation of midlatitude storms is baroclinic instability due to vertical shear, not some process (e.g., barotropic instability) associated with the lateral shear of the zonal wind across the jet stream.

We have shown in the previous section that the energetics of baroclinic waves require that the waves remove available potential energy from the mean flow. Thus, baroclinic wave development will tend to weaken the mean meridional temperature gradient (that is, reduce the mean thermal wind). The mean pole-to-equator temperature gradient is of course continually restored by differential solar heating. However, differential heating cannot account for the tendency of the meridional temperature gradient to be concentrated along the polar front. Clearly, some dynamic process is required which can continuously reestablish the strong gradients characteristic of the frontal zone. Such a process is called *frontogenetic*. Frontogenesis usually occurs in association with developing baroclinic waves. Thus, even though baroclinic disturbances transport heat down the mean temperature gradient and tend to reduce the temperature difference between the polar and tropical regions, locally, in the vicinity of the jet stream, the flow associated with baroclinic disturbances acts to enhance the temperature gradient.

9.4.1 THE KINEMATICS OF FRONTOGENESIS

A complete dynamical model of frontogenesis is beyond the scope of this text. However, a qualitative description of frontogenesis can be obtained by a strictly *kinematic* analysis, that is, a description of the geometry of the flow without reference to the underlying physical forces.

There are at least four basic flow configurations which can change horizontal temperature gradients in the atmosphere: (1) horizontal deformation; (2) horizontal shear; (3) vertical deformation; (4) differential vertical motion. These are illustrated schematically in Fig. 9.8.

Horizontal shearing will tend to stretch a fluid parcel along the shear axis (the y axis in Fig. 9.8a) and contract the parcel along the horizontal direction perpendicular to the shear axis. Horizontal shear is an important frontogenetic mechanism in cold front situations. For example, in the schematic surface pressure chart of Fig. 9.9 the geostrophic wind has a northerly component west of point B and a southerly component east of point B. The

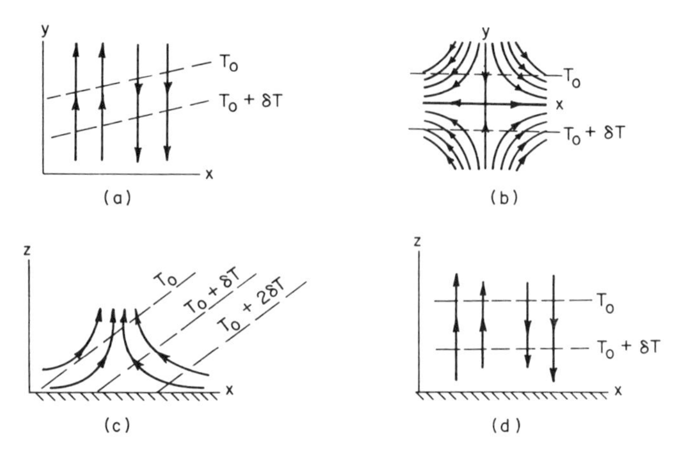

Fig. 9.8 Four flow configurations which can intensify horizontal temperature gradients: (a) horizontal shear; (b) horizontal deformation; (c) vertical deformation; (d) differential vertical advection. (Adapted from Hoskins and Bretherton, 1972.)

resulting cyclonic shear will tend to concentrate the isotherms along the line of maximum shear passing through *B*. (Note the strong cold advection northwest of *B* and the weak thermal advection southeast of *B*.)

A horizontal deformation field will tend to advect the temperature field so that the isotherms become concentrated along the axis of *dilation* (the *x* axis in Fig. 9.8b), provided that the initial temperature field has a finite gradient along the axis of *contraction* (the *y* axis in Fig. 9.8b). The velocity

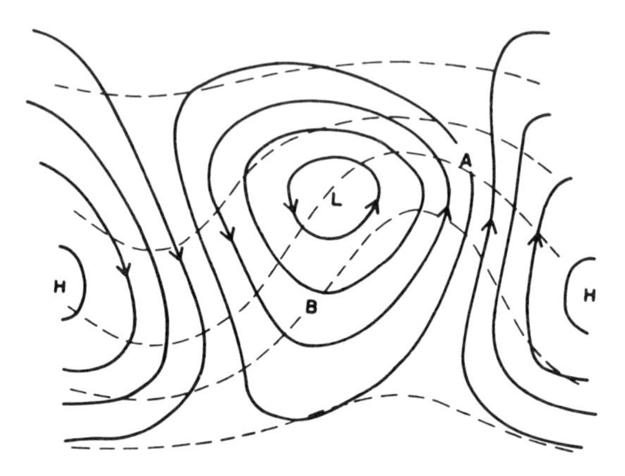

Fig. 9.9 Schematic surface isobars (solid lines) and isotherms (dashed lines) for a baroclinic wave disturbance. Arrows show direction of geostrophic wind. Horizontal deformation intensifies the temperature gradient at *A*, horizontal shear intensifies the gradient at *B*. (After Hoskins and Bretherton, 1972. Reproduced with permission of the American Meteorological Society.)

field shown in Fig. 9.8b is a pure deformation field which has a streamfunction given by $\psi = -Kxy$, where K is a constant. It is easily verified that a pure deformation field is both irrotational and nondivergent. A parcel advected by a pure deformation field will merely have its shape changed in time, without any change in horizontal area. Thus, in the flow field shown in Fig. 9.8b, a square parcel with sides parallel to the x and y axes would be deformed into a rectangle.

Horizontal deformation at low levels is an important mechanism for the development of warm fronts. In the example of Fig. 9.9 there is a deformation field centered near point A with its axis of contraction nearly orthogonal to the isotherms. This deformation field leads to strong warm advection south of point A and weak warm advection north of point A. The result is to concentrate the temperature gradient in the vicinity of point A.

Although, as shown in Fig. 9.9 the low level flow in the vicinity of a developing warm front may resemble a pure deformation field, the flow in the upper troposphere seldom has the characteristics of a pure deformation field. However, if a component of horizontal deformation is added to a mean zonal flow the result is a *confluent* flow as shown in Fig. 9.10. Such confluent regions are always present in the tropospheric jet stream due to the influence of quasi-stationary planetary scale waves on the position and intensity of the jet. In fact, even a monthly mean 500-mb chart (see Fig. 6.3) reveals two regions of large-scale confluence immediately to the east of the continents of Asia and North America. Observationally, these two regions are known to be regions of intense baroclinic wave development and frontogenesis.

The mechanisms of horizontal shear and horizontal deformation operate primarily to concentrate the pole–equator temperature gradient on the synoptic scale (~ 1000 km). These processes alone can not cause the rapid

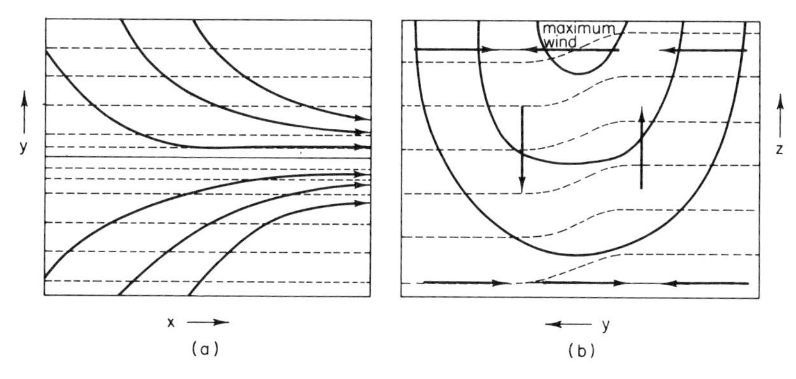

Fig. 9.10 (a) Horizontal streamlines and isotherms in a frontogenetic confluence. (b) Vertical section across the confluence, showing isotachs (solid), isotherms (dashed), and vertical and transverse motions (arrows). (After Sawyer, 1956.)

frontogenesis often observed in extratropical systems in which the temperature gradient can become concentrated in a zone of characteristic width ~ 50 km on a time scale of 1–2 d. This rapid reduction in scale is due primarily to the frontogenetic character of the vertical deformation field (Fig. 9.8c) associated with the secondary circulation driven by the quasi-geostrophic synoptic scale flow. The nature of the secondary flow may be understood by referring to Fig. 9.10. The role of the confluent flow is to increase the horizontal temperature gradient perpendicular to the jet axis. This concentration of the isotherms in turn requires an acceleration of the zonal flow in order to maintain geostrophic balance. The required zonal acceleration can only be produced by the Coriolis force due to a poleward flowing ageostrophic wind in the jet stream core as shown in Fig. 9.10b. As the jet accelerates cyclonic vorticity must be generated north of the jet axis and anticyclonic vorticity to the south. These vorticity changes require that the horizontal flow at the jet stream level be convergent north of the jet axis and divergent south of the jet axis. The vertical circulation and low-level secondary ageostrophic motion required by mass continuity are also indicated in Fig. 9.10b.

Examination of Fig.9.10b reveals that the secondary circulation in the lower troposphere has the characteristics of a vertical deformation field. The advection of temperature by the low-level ageostrophic wind clearly will concentrate the temperature gradient at the surface to the south of the stream axis. Temperature advection by the upper-level secondary circulation, on the other hand, will tend to concentrate the temperature gradient north of the jet stream axis. As a result the frontal zone will slope northward with height. The differential vertical motion indicated in Fig. 9.10b, which has its maximum amplitude in the midtroposphere, will actually tend to weaken the front due to adiabatic temperature changes (adiabatic warming on the cold side of the front and adiabatic cooling on the warm side). Thus, differential vertical motion tends to weaken frontal zones in the middle troposphere. The most intense fronts are in the lower troposphere and near the tropopause.

9.4.2 Quasi-Geostrophic Frontogenesis

We can demonstrate that a synoptic scale deformation field as shown in Fig. 9.8b will indeed tend to concentrate a small preexisting temperature gradient. For this purpose we consider a level sufficiently close to the ground so that $\omega \simeq 0$ and we can write the thermodynamic energy equation (6.6) for adiabatic flow in the form

$$\frac{\partial T}{\partial t} + u\frac{\partial T}{\partial x} + v\frac{\partial T}{\partial y} = 0 \tag{9.40}$$

Thus, the temperature is changed solely by horizontal advection. In addition we assume that the temperature field depends only on y and has the initial functional dependence

$$T(y, 0) = -T_0(2/\pi)\tan^{-1}(y/L) \tag{9.41}$$

where T_0 and L are constants with dimensions of temperature and length, respectively. This temperature field is advected by a nondivergent velocity field whose geostrophic streamfunction has the form shown in Fig. 9.8b. Thus, $\psi = -Kxy$ and

$$v = \partial\psi/\partial x = -Ky \tag{9.42}$$

Combining (9.40) and (9.42) we have

$$\frac{\partial T}{\partial t} - Ky\frac{\partial T}{\partial y} = 0 \tag{9.43}$$

which shows that temperature is advected at the speed $-Ky$ in the y direction. Thus, it is obvious that the initial gradient will be concentrated toward $y = 0$.

Following the motion of an individual parcel, we have

$$\frac{dy}{dt} = -Ky \qquad \text{or} \qquad \frac{dy}{y} = -K\,dt$$

which may be integrated to give the position of a parcel as

$$y = y_0 e^{-Kt} \tag{9.44}$$

where y_0 is the initial position of the parcel, relative to the axis of dilation. Now since temperature is conserved following the motion, we can write an expression for the distribution of T at any time in terms of the initial distribution and the total parcel displacement:

$$T(y, t) = T(y_0, 0) = T(ye^{Kt}, 0) \tag{9.45}$$

This expression merely indicates that a parcel which is located at point y at time t was located at the point ye^{Kt} initially. Thus, the evolution of the temperature field is given by

$$T(y, t) = -T_0(2/\pi)\tan^{-1}(ye^{Kt}/L) \tag{9.46}$$

while the temperature gradient is

$$\frac{\partial T}{\partial y} = -\frac{T_0(2/\pi)Le^{Kt}}{L^2 + y^2e^{2Kt}} \tag{9.47}$$

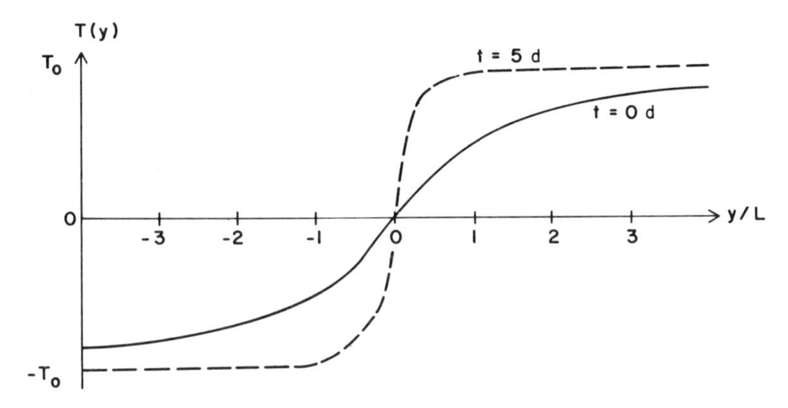

Fig. 9.11 Concentration of a small initial temperature gradient by horizontal deformation.

From (9.47) we see that as $t \rightarrow \infty$ the temperature gradient becomes infinite at $y = 0$ and vanishes everywhere else. Therefore a horizontal deformation field can lead to formation of a discontinuity in the temperature field in the absence of any compensating vertical motion.

Thus, even quasi-geostropic dynamics can lead to the formation of intense fronts near the ground where vertical motion is suppressed. However, the time scale for formation of fronts through quasi-geostrophic processes is unrealistically long. As an example we let $L = 1000$ km and $K = 5 \times 10^{-6}\ \mathrm{s}^{-1}$ (corresponding to a deformation velocity along the y axis of $\pm 5\ \mathrm{m\ s}^{-1}$ at $y = \pm 1000$ km). With this choice of K the lateral scale decreases by a factor of 10 in about 5 d. The resulting temperature profile given by (9.46) is shown in Fig. 9.11.

In reality frontogenesis occurs on a much shorter time scale because as the lateral scale decreases the frontogenetic role of the secondary circulation (i.e., vertical deformation) becomes increasingly important. The quasi-geostrophic equations, which neglect advection by the ageostrophic wind, thus are invalid beyond the initial stages of frontogenesis. Since the basic deformation field in this model was assumed to be geostrophic, no processes have been included here which are not present in the quasi-geostrophic model.

9.4.3 Steady-State Frontal Zones

One aspect of fronts which could not be incorporated in our two-level model is the characteristic observed slope of the zone of maximum temperature gradient with respect to height (see Fig. 6.4). That a stationary frontal zone must slope with height can be seen most easily by examining the limiting case of a frontal zone embedded in a purely geostrophic wind field

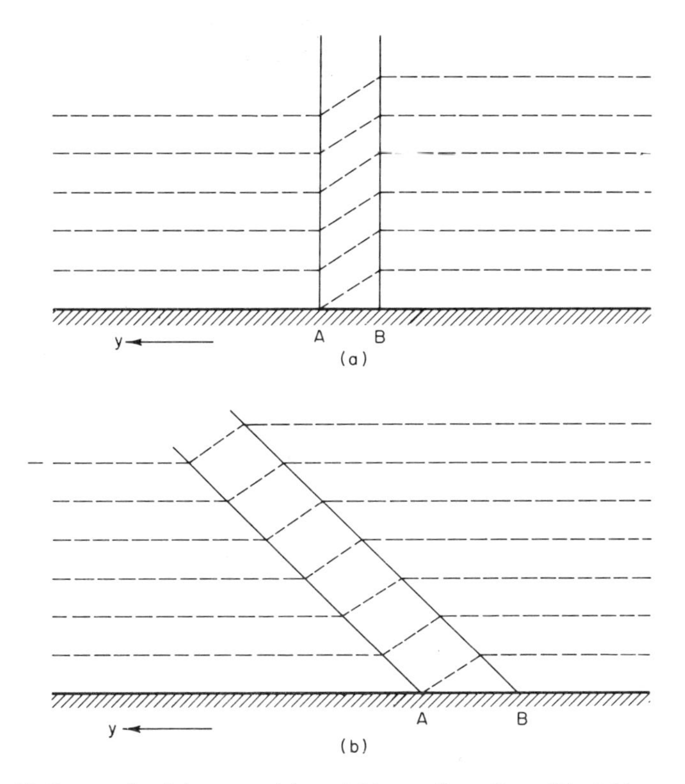

Fig. 9.12 Stationary frontal zones: (a) unstable configuration; (b) stable configuration. Dashed lines indicate isotherms.

which blows parallel to the isotherms. We assume that the entire temperature gradient is confined to a narrow zone of width δy with a horizontal temperature change of δT across the zone, and that outside this zone the temperature is constant in isobaric surfaces. In that case, the thermal wind equation (3.29) states approximately that

$$\frac{\delta U_g}{\delta \ln p} \simeq \frac{R}{f} \frac{\delta T}{\delta y}$$

Thus, if the frontal zone were vertically oriented as shown in Fig. 9.12a there would be discontinuities in U_g along the lines A and B since only in the zone between A and B would U_g have vertical shear. The resultant infinite lateral shear along A and B would be a highly unstable configuration and would soon be destroyed by the rapid growth of unstable perturbations along the shear line. Hence, a vertically oriented frontal surface is not a stable configuration. On the other hand, if the frontal zone slopes as shown in Fig. 9.12b, the thermal wind can provide a smooth transition between the constant geostrophic wind U_g in the cold section and the constant geostrophic wind $U_g + \delta U_g$ in the warm section. The frontal slope can then be estimated directly from the thermal wind relationship as

$$\frac{\delta \ln p}{\delta y} \simeq \frac{+f\,\delta U_g}{R\,\delta T}$$

or in terms of height coordinates

$$\frac{\delta z}{\delta y} \simeq \frac{-fH\,\delta U_g}{R\,\delta T} \tag{9.48}$$

where H is the mean scale height, $H = R\overline{T}/g$. Equation (9.48), although it results from a highly simplified model, does produce a reasonable estimate for the observed slopes of typical midlatitude fronts.

Problems

1. Show using Eq. (9.23) that the maximum growth rate for baroclinic instability when $\beta = 0$ occurs for

$$k^2 = 2\lambda^2(\sqrt{2} - 1)$$

How long does it take the most rapidly growing wave to amplify by a factor of e^1 if $\lambda = 2 \times 10^{-6}\ \text{m}^{-1}$ and $U_T = 20\ \text{m s}^{-1}$?

2. Solve for ψ'_3 and ω'_2 in terms of ψ'_1 for a baroclinic Rossby wave whose phase speed satisfies (9.21). Explain the phase relationship between ψ'_1, ψ'_3, and ω'_2 in terms of the quasi-geostrophic theory. (Note that $U_T = 0$ in this case).

3. For the case $U_1 = -U_3$ and $k^2 = \lambda^2$ solve for ψ'_3 and ω'_2 in terms of ψ'_1 for marginally stable waves [i.e., $\delta = 0$ in (9.19)].

4. For the case $\beta = 0$, $k^2 = \lambda^2$, and $U_m = U_T$ solve for ψ'_3 and ω'_2 in terms of ψ'_1. Explain the phase relationships between ω'_2, ψ'_1, and ψ'_3 in terms of the energetics of quasi-geostrophic waves for the amplifying wave.

5. Obtain a formula for the Rossby wave phase speed in a homogeneous barotropic fluid confined between rigid horizontal lids when friction is included in the form $\mathbf{Fr} = -\mu\mathbf{V}$ where μ is a constant drag coefficient, and $\mathbf{Fr}$ designates the horizontal friction force. Describe qualitatively the behavior of this sort of wave.

6. Suppose that a baroclinic fluid is confined between two rigid horizontal lids in a rotating tank in which $\beta = 0$ but frictional drag must be included. If the frictional force has the same form as given in Problem 5 (that is, friction is everywhere linearly proportional to the velocity) show that the two-level model perturbation vorticity equations in cartesian coordinates can be written as

$$\left(\frac{\partial}{\partial t} + U_1 \frac{\partial}{\partial x} + \mu\right)\frac{\partial^2 \psi'_1}{\partial x^2} - \frac{f}{\Delta p}\,\omega_2 = 0$$

$$\left(\frac{\partial}{\partial t} + U_3 \frac{\partial}{\partial x} + \mu\right)\frac{\partial^2 \psi'_3}{\partial x^2} + \frac{f}{\Delta p}\,\omega_2 = 0$$

where perturbations are assumed in the form given in (9.10). Assuming solutions of the form (9.14) show that the phase speed satisfies a relationship similar to (9.19) with β replaced everywhere by $i\mu k$, and that as a result the condition for baroclinic instability becomes

$$U_T > \frac{\mu}{(2\lambda^2 - k^2)^{1/2}}$$

***7.** For the case $\beta = 0$ determine the phase difference between the 250-mb and 750-mb geopotential fields for the most unstable baroclinic wave (see Problem 1). Show that the 500-mb geopotential and thickness fields are 90° out of phase.

***8.** For the conditions of Problem 7, given that the amplitude of ψ'_1 is $A = 10^7\ \mathrm{m^2\ s^{-1}}$ solve the system (9.15)–(9.17) to obtain B and C. Let $f_0 = 10^{-4}\ \mathrm{s^{-1}}$, $\Delta p = 50$ kPa, $\sigma = 2 \times 10^{-6}\ \mathrm{Pa^{-2}\ m^2\ s^{-2}}$, $\lambda = \sqrt{2} \times 10^{-6}\ \mathrm{m^{-1}}$, and $U_T = 15\ \mathrm{m\ s^{-1}}$.

***9.** For the situation of Problem 8 compute ω'_2 using the expression (9.26) and verify that the result agrees with the value found in Problem 8.

10. Compute the total potential energy per unit cross-sectional area for an atmosphere with an adiabatic lapse rate given that the temperature and pressure at the ground are $p = 10^5$ Pa and $T = 300$ K, respectively.

11. Consider two air masses at the uniform potential temperatures $\theta_1 = 320$ K and $\theta_2 = 340$ K which are separated by a vertical partition as shown in Fig. 9.5. Each air mass occupies a horizontal area of $10^4\ \mathrm{m^2}$ and extends from the surface ($p_0 = 10^5$ Pa) to the top of the atmosphere. What is the available potential energy for this sytem? What fraction of the total potential energy is available in this case?

***12.** For the unstable baroclinic wave which satisfies the conditions given in Problems 7 and 8 compute the energy conversion terms in (9.35) and (9.36) and hence obtain the instantaneous rates of change of the perturbation kinetic and available potential energies.

13. Compute the slope of a stationary geostrophic frontal zone at 43°N if the temperature and geostrophic wind change across the front by 10°C and 30 m s^{-1}, respectively. Assume a mean scale height of 8 km.

Suggested References

Phillips (1963) is a review paper on many aspects of geostrophic motions including a brief account of the principal results of baroclinic instability theory. This paper also contains an extensive bibliography of original literature on the subject.

Charney (1947) is the classical paper on baroclinic instability. The mathematical level is advanced, but Charney's paper contains an excellent qualitative summary of the main results which is very readable.

Wallace and Hobbs, *Atmospheric Science–An Introductory Survey*, contains a thorough synoptic case study of frontogenesis associated with a baroclinic disturbance.

Hoskins and Bretherton (1972) is a mathematically advanced paper which develops the modern theory of frontogenesis.

Chapter 10 The General Circulation

The *general circulation* of the atmosphere is usually considered to consist of the totality of motions which characterizes the global scale atmospheric flow. In particular, the study of the general circulation is concerned with the temporally and/or spatially averaged structures of winds, temperatures, and other climatic elements. For present purposes the general circulation will be regarded as consisting of the time averaged flow, where averages are taken over periods sufficiently long to remove the random variations associated with individual weather systems, but short enough to retain seasonal variations.

The time averaged circulation is highly longitudinally dependent due to longitudinally asymmetric forcing by orography and land–sea heating contrasts. The longitudinally dependent components of the general circulation consist both of *monsoonal circulations* (i.e., seasonally reversing circulations) and *stationary circulations* which vary little with time. A complete understanding of the physical basis for the general circulation thus requires a description involving the three spatial dimensions as well as time.

However, in attempting to develop a theory for the general circulation it has proven useful to isolate those processes which maintain the zonal mean flow, that is, the flow averaged around latitude circles. This approach is related to the perturbation method used in previous chapters, in which flow fields were split into zonal mean and perturbation components. Thus, we will here focus primarily on the zonally averaged general circulation, although we will include some qualitative discussion of important longitudinal dependencies.

Concentrating on the zonally averaged flow allows us to focus on those features which are not dependent on continentality, and should thus be common to all thermally driven rotating fluid systems. In particular, we will discuss the energy and momentum budgets of the zonally averaged flow. We will also show that the *mean meridional circulation* (i.e., the circulation consisting of the zonal mean vertical and meridional velocity components) is uniquely determined by a relationship analogous to the omega equation derived in Chapter 6 for midlatitude synoptic scale motions.

10.1 The Nature of the Problem

Theoretical speculation on the nature of the general circulation has quite a long history. Perhaps the most important early work on the subject was that of the eighteenth-century Englishman, George Hadley. Hadley, in seeking a cause for the trade wind circulation, realized that this circulation must be a form of thermal convection driven by the difference in solar heating between the equatorial and polar regions. He visualized the general circulation as the result of a zonally symmetric overturning in which the heated equatorial air rises and moves poleward where it cools, sinks, and moves equatorward again. This type of circulation is now called a *Hadley circulation* or Hadley cell.

Although the Hadley circulation is a mathematically possible circulation in the sense that it does not violate any of the laws of physics, it is not observed to be the primary mode of circulation for the earth's atmosphere. Evidence from a number of types of studies indicates that under the conditions existing in the earth's atmosphere, a symmetric *Hadley* circulation would be baroclinically unstable. The observed general circulation may then be thought of qualitatively as developing in the following way:

In the mean the net solar energy absorbed by the atmosphere and the earth must equal the infrared energy radiated back to space by the planet. However, solar heating is strongly dependent on latitude, with a maximum at the equator and a minimum at the poles. The outgoing infrared radiation, on the other hand, is only weakly latitude dependent. Thus, there is a radiation excess in the equatorial region and a deficit in the polar region. This differential heating creates a pole-to-equator temperature gradient, and hence produces a growing store of zonal mean available potential energy. However, at some point the zonal thermal wind (which must develop if the motion is to be geostrophically balanced in the presence of the pole-to-equator temperature gradient) becomes baroclinically unstable. As shown in the previous chapter, the resulting baroclinic waves transport heat poleward. These waves will intensify until their heat transport is sufficient to balance the radiation deficit in the polar regions so that the pole-to-equator temperature gradient ceases to grow. At the same time, these perturbations convert potential energy into kinetic energy, thereby maintaining the kinetic energy of the atmosphere against the effects of frictional dissipation.

From a thermodynamic point of view then, the atmosphere may be regarded as a "heat engine" which absorbs net heat at a high-temperature "reservoir" in the tropics and gives up heat to a low-temperature reservoir in the polar regions. In this manner radiation generates available potential energy which is in turn partially converted to kinetic energy which does work against friction. However, only a small fraction of the solar energy input actually gets converted to kinetic energy. Thus, from an engineer's viewpoint the atmosphere is a rather inefficient heat engine. However, if due account is taken of the many constraints operating on atmospheric motions, it appears that the atmosphere may in fact generate kinetic energy about as efficiently as dynamically possible.

The above qualitative discussion suggests that the gross features of the general circulation can be understood on the basis of quasi-geostrophic theory since, as we have previously seen, baroclinic instability processes are contained within the quasi-geostrophic framework. To keep the discussion as simple as possible, we will in this chapter concentrate on those aspects of the general circulation which are contained in the quasi-geostrophic model. Thus, we will focus primarily on the circulation of a dry atmosphere outside the tropics. The general circulation within the tropics cannot be understood without considering latent heat release in convective clouds. This subject will be postponed until Chapter 12.

It should be recognized that a quasi-geostrophic model cannot provide a complete theory of the general circulation because in the quasi-geostrophic model a number of assumptions are made about scales of motion which restrict in advance the possible types of solutions. Thus, most modern general circulation studies involve highly complex numerical models based on the primitive equations. The ultimate objective of these modeling efforts is to simulate the general circulation so faithfully that the climatological effects of any change in the external parameters can be accurately estimated. Present models are a long way from achieving this objective, but progress has been very rapid in recent years. The numerical simulation approach will be discussed in Section 10.6.

10.2 The Energy Cycle: A Quasi-Geostrophic Model

Perhaps the most important requirement for a theory of the general circulation is that it must provide a satisfactory explanation for the atmospheric energy cycle. It must, therefore, explain by what processes the atmosphere converts solar energy to kinetic energy in order to balance the kinetic energy losses by frictional dissipation. For a complete discussion of the energy cycle, it would be necessary to derive the energy equations starting with the primitive equations in spherical coordinates. However, in order to keep the discussion consistent with the general theme of this book we will here examine the energy cycle within the context of the quasi-geostrophic model. It turns out, in fact, that the basic energy cycle of the atmosphere can be described adequately by a properly formulated quasi-geostrophic model, at least in a qualitative sense. However, the quasi-geostrophic model in the form which we have previously used is not suitable for studying the atmospheric energy cycle because in that model we completely neglected the effects of friction and diabatic heating. For short-term forecasts, these omissions are perfectly justifiable since scaling arguments indicate that midlatitude synoptic scale motions are nearly adiabatic and frictionless. However, the flow must ultimately depend on the energy sources and sinks, so that in a general circulation model, friction and diabatic heating must be included in some manner.

In Chapter 5 we showed that for synoptic motions the primary effect of surface friction is to produce a secondary circulation in which the vertical velocity at the top of the boundary layer is proportional to the vorticity at that level. Further, we showed that the divergent motions associated with this secondary circulation always act to reduce the relative vorticity of the flow, that is, to "spin-down" the motion. In addition to the energy lost to the surface through this indirect spin-down effect there is also an energy loss due to internal frictional dissipation. However, for simplicity we will include only the Ekman layer friction in our model. This is most easily done by letting the quasi-geostrophic model apply only above the Ekman layer and setting as the lower boundary condition the vertical velocity at the top of the Ekman layer given in (5.31). In terms of ω this level terrain condition becomes

$$\omega(p_0) = -\left[\rho g\left(\frac{K}{2f_0}\right)^{1/2}\nabla^2\psi\right]_{p_0} \tag{10.1}$$

where p_0 is the pressure at the top of the Ekman layer.

For purposes of the present discussion, it is not necessary to specify the form of diabatic heating explicitly. It is merely necessary to note from (2.46) that when diabatic heating is included the first law of thermodynamics becomes

$$c_p \frac{d \ln \theta}{dt} = \frac{ds}{dt} \tag{10.2}$$

where s is the entropy per unit mass. In terms of the geostrophic streamfunction, $\psi = \Phi/f_0$, (10.2) can be written as

$$\frac{\partial}{\partial t}\left(\frac{\partial \psi}{\partial p}\right) = -\mathbf{V}_\psi \cdot \nabla\left(\frac{\partial \psi}{\partial p}\right) - \frac{\sigma}{f_0}\omega - \frac{\alpha}{f_0 c_p}\frac{ds}{dt} \tag{10.3}$$

where

$$\mathbf{V}_\psi = \mathbf{k} \times \nabla\psi$$

and where we have used all the usual approximations of quasi-geostrophic theory, except the neglect of diabatic heating [compare with Eq. (8.16)].

The equations of our model are thus the quasi-geostrophic vorticity equation (8.15) and the thermodynamic energy equation (10.3). Using the condition

$$\nabla \cdot \mathbf{V}_\psi = 0$$

to write the advection terms in the so-called *flux form* by letting

$$\mathbf{V}_\psi \cdot \nabla\chi = \nabla \cdot (\mathbf{V}_\psi \chi)$$

where χ here stands for any field variable, we can write these equations as

$$\frac{\partial}{\partial t}\nabla^2\psi = -\nabla \cdot [\mathbf{V}_\psi(\nabla^2\psi + f)] + f_0 \frac{\partial \omega}{\partial p} \tag{10.4}$$

$$\frac{\partial}{\partial t}\left(\frac{\partial \psi}{\partial p}\right) = -\nabla \cdot \left[\mathbf{V}_\psi \frac{\partial \psi}{\partial p}\right] - \frac{\sigma}{f_0}\omega - R \tag{10.5}$$

where for simplicity we have set

$$R \equiv \frac{\alpha}{f_0 c_p}\frac{ds}{dt}$$

We now assume that the atmosphere is confined to a zonal channel on the β plane with origin at 45° latitude and rigid walls at $y = \pm D$. This assumption may seem artificial, but it allows us to use cartesian coordinates and to avoid possible difficulties with the quasi-geostrophic model at the equator. The boundary conditions on the system (10.4) and (10.5) then become

$$\frac{\partial \psi}{\partial x} = 0 \quad \text{at} \quad y = \pm D \tag{10.6a}$$

Condition (10.6a) guarantees that the nondivergent part of the meridional velocity field vanishes at $y = \pm D$. However, in order to avoid nonzero flow across the boundaries in the Ekman layer we see from (5.21) that we must also require that at the top of the Ekman layer

$$[u_\psi]_{p_0} = -[\partial\psi/\partial y]_{p_0} = 0 \qquad \text{at} \qquad y = \pm D \tag{10.6b}$$

where the subscript p_0 again refers to the pressure at the top of the Ekman layer. Above the friction layer the condition (10.6b) can be replaced by the weaker restriction

$$\frac{\partial u_\psi}{\partial t} = -\frac{\partial^2\psi}{\partial t\,\partial y} = 0 \qquad \text{at} \qquad y = \pm D \tag{10.6c}$$

which is easily obtained by averaging the x component of the momentum equation zonally and using the fact that $v = 0$ at $y = \pm D$. For conditions at the horizontal boundaries we set $\omega = 0$ at $p = 0$ and apply the condition (10.1) at $p = p_0$.

In order to study the zonally averaged circulation with this model, we divide the nondivergent motion into a zonally averaged portion denoted by an overbar plus the deviation from a zonal average denoted by a prime:

$$\psi = \bar{\psi} + \psi', \qquad \mathbf{V}_\psi = \overline{\mathbf{V}}_\psi + \mathbf{V}'_\psi$$

In the following discussion we will often refer to the zonally averaged part of the flow as the *mean flow* (it must be kept in mind that a zonal mean and not a time mean is implied). In addition the deviation from the mean will be referred to as the *eddy* motion (we will generally reserve the term *perturbation* for small deviations from the mean).

The primary goal of this section is to derive energy equations for both the mean and eddy components of the motion and to discuss the interactions between the eddy and mean flow energy. However, it is instructive to consider first the energy cycle of the complete flow without separating the field into mean and eddy parts. In a manner analogous to the discussion in Section 9.3.2, we can obtain the kinetic energy equation by multiplying (10.4) through by $-\psi$ and integrating over the volume of the channel to get

$$\int_0^{p_0}\int_A -\psi\frac{\partial}{\partial t}\nabla^2\psi\,dA\,dp = \int_0^{p_0}\int_A \psi\,\nabla\cdot[\mathbf{V}_\psi(\nabla^2\psi + f)]\,dA\,dp$$
$$-\int_0^{p_0}\int_A f_0\psi\frac{\partial\omega}{\partial p}\,dA\,dp \tag{10.7}$$

The term on the left in (10.7) may be rewritten as

$$\int_A -\psi\frac{\partial}{\partial t}\nabla^2\psi\,dA = -\int_A \nabla\cdot\left(\psi\frac{\partial}{\partial t}\nabla\psi\right)dA + \int_A \nabla\psi\cdot\frac{\partial}{\partial t}\nabla\psi\,dA$$

Applying the divergence theorem to the first term on the right above yields

$$\int_A \nabla \cdot \left(\psi \frac{\partial}{\partial t} \nabla\psi\right) dA = \int_l \left(\psi \frac{\partial}{\partial t} \nabla\psi\right) \cdot \mathbf{n}\, dl$$

where the line integral on the right-hand side is evaluated along the boundaries at $y = \pm D$. Observing that

$$\nabla\psi \cdot \mathbf{n} = \pm \frac{\partial\psi}{\partial y} = \mp u_\psi \qquad \text{at} \quad y = \pm D$$

we see that from the boundary condition 10.6c) the line integral vanishes so that

$$\int_A -\psi \frac{\partial}{\partial t} \nabla^2\psi\, dA = \int_A \frac{\partial}{\partial t}\left(\frac{\nabla\psi \cdot \nabla\psi}{2}\right) dA$$

$$= \int_A \frac{\partial}{\partial t}\left(\frac{\mathbf{V}_\psi \cdot \mathbf{V}_\psi}{2}\right) dA \tag{10.8}$$

The first term on the right in (10.7) may also be shown to equal zero with the aid of the divergence theorem. Thus, we write

$$\int_A \psi\nabla \cdot [\mathbf{V}_\psi(\nabla^2\psi + f)]\, dA = \int_A \nabla \cdot [\psi\mathbf{V}_\psi(\nabla^2\psi + f)]\, dA$$

$$- \int_A (\nabla^2\psi + f)\mathbf{V}_\psi \cdot \nabla\psi\, dA \tag{10.9}$$

Applying the divergence theorem we find that the first term on the right in (10.9) is zero. The second term on the right is also zero because

$$\mathbf{V}_\psi \cdot \nabla\psi = (\mathbf{k} \times \nabla\psi) \cdot \nabla\psi \equiv 0$$

Therefore, there is no contribution to the change of average kinetic energy in a closed region due to horizontal advection of kinetic energy.

The final term in (10.7) can be expressed in a more revealing form if we integrate by parts in the vertical:

$$-f_0 \int_0^{p_0} \psi \frac{\partial\omega}{\partial p} dp = f_0 \int_0^{p_0} \omega \frac{\partial\psi}{\partial p} dp - f_0[\psi\omega]_{p_0} \tag{10.10}$$

where we have applied the boundary condition $\omega = 0$ at $p = 0$. Substituting from the boundary condition (10.1) into the last term of (10.10) and integrating over the area of the channel we obtain

$$-f_0 \int_A [\psi\omega]_{p_0}\, dA = \int_A \rho g \left(\frac{Kf_0}{2}\right)^{1/2} [\psi \nabla^2 \psi]_{p_0}\, dA$$

$$= - \int_A \left[\rho g \left(\frac{Kf_0}{2}\right)^{1/2} \frac{(\nabla\psi)^2}{2} \right]_{p_0} dA \equiv -\varepsilon g \qquad (10.11)$$

where we have again used the divergence theorem and the boundary condition (10.6b) in order to eliminate the term

$$\int_A \rho g \left(\frac{Kf_0}{2}\right)^{1/2} \nabla \cdot (\psi \nabla\psi) = \int_l \rho g \left(\frac{Kf_0}{2}\right)^{1/2} \psi \nabla\psi \cdot \mathbf{n}\, dl$$

in (10.11). The quantity ε defined in (10.11) is just the net rate of kinetic energy dissipation due to surface friction.

It is clear from the form of the integrand in (10.11) that ε will always be positive. Substituting from (10.8)–(10.11) back into (10.7) we obtain finally the quasi-geostrophic kinetic energy equation

$$\frac{d}{dt} \int_0^{p_0} \int_A \frac{(\nabla\psi)^2}{2}\, dA\, dp = \int_0^{p_0} \int_A f_0 \omega \frac{\partial\psi}{\partial p}\, dA\, dp - \varepsilon g \qquad (10.12)$$

To obtain an equation for the rate of change of available potential energy, we multiply (10.5) through by

$$\frac{f_0^2}{\sigma} \frac{\partial\psi}{\partial p}$$

and integrate over the entire channel. In this case, the first term on the right may be expressed as

$$-\frac{f_0^2}{\sigma}\left(\frac{\partial\psi}{\partial p}\right) \nabla \cdot \left[\mathbf{V}_\psi \frac{\partial\psi}{\partial p}\right] = -\frac{f_0^2}{2\sigma} \nabla \cdot \left[\mathbf{V}_\psi \left(\frac{\partial\psi}{\partial p}\right)^2\right] \qquad (10.13)$$

where we have used the relation $\nabla \cdot \mathbf{V}_\psi = 0$ to eliminate several terms. Clearly when we integrate over the area of the channel and apply the divergence theorem the expression in (10.13) will give zero contribution. Thus, the available potential energy equation becomes

$$\frac{d}{dt} \int_0^{p_0} \int_A \frac{f_0^2}{2\sigma} \left(\frac{\partial\psi}{\partial p}\right)^2 dA\, dp = - \int_0^{p_0} \int_A f_0 \omega \frac{\partial\psi}{\partial p}\, dA\, dp$$

$$- \int_0^{p_0} \int_A \frac{f_0^2}{\sigma} R \frac{\partial\psi}{\partial p}\, dA\, dp \qquad (10.14)$$

It is convenient at this point to introduce some new notation by defining the total kinetic energy K and the available potential energy P as

$$K \equiv \int_0^{p_0} \int_A \frac{(\nabla\psi)^2}{2} \, dA \, dp/g$$

$$P \equiv \int_0^{p_0} \int_A \frac{f_0^2}{2\sigma} \left(\frac{\partial\psi}{\partial p}\right)^2 dA \, dp/g$$

Further, we let

$$\{P \cdot K\} \equiv \int_0^{p_0} \int_A f_0 \omega \frac{\partial\psi}{\partial p} \, dA \, dp/g$$

$$G \equiv - \int_0^{p_0} \int_A \frac{f_0^2}{\sigma} R \frac{\partial\psi}{dp} \, dA \, dp/g$$

Then (10.12) and (10.14) can be written symbolically as

$$\frac{dK}{dt} = \{P \cdot K\} - \varepsilon \tag{10.15}$$

$$\frac{dP}{dt} = -\{P \cdot K) + G \tag{10.16}$$

From the form of (10.15) and (10.16) it is clear that $\{P \cdot K\}$ represents the conversion of available potential energy to kinetic energy due to the correlation of temperature and vertical motion, while G represents the generation of available potential energy due to the correlation of temperature and diabatic heating. Adding (10.15) and (10.16), we obtain the total energy equation for the quasi-geostrophic system

$$\frac{d}{dt}(P + K) = G - \varepsilon \tag{10.17}$$

In the long-term mean, $P + K$ must be constant so that in this model the generation of available potential energy by differential heating must just balance the kinetic energy loss due to surface friction.

In the above discussion we have illustrated the general technique of deriving energy equations for the quasi-geostrophic model. We now discuss the equations which arise when the energy is partitioned between its zonal mean and eddy portions.

Dividing the variables in (10.4) into their mean and eddy parts and taking the zonal mean we obtain the mean vorticity equation

$$\frac{\partial}{\partial t}\left(\frac{\partial^2 \bar{\psi}}{\partial y^2}\right) = \frac{\partial^2 M}{\partial y^2} + f_0 \frac{\partial \bar{\omega}}{\partial p} \tag{10.18}$$

Here, letting angle brackets as well as overbars denote zonally averaged quantities,

$$\begin{aligned}\frac{\partial^2 M}{\partial y^2} &= -\langle \nabla \cdot [(\overline{\mathbf{V}}_\psi + \mathbf{V}'_\psi)(\nabla^2 \overline{\psi} + \nabla^2 \psi' + f)] \rangle \\ &= -\frac{\partial}{\partial y}\left\langle \left[\frac{\partial \psi'}{\partial x} \nabla^2 \psi' \right] \right\rangle\end{aligned}$$

where we have used the facts that $\partial \langle [\ \ ] \rangle / \partial x = 0$ for any function and $\overline{\mathbf{V}'_\psi} = 0$. The quantity M may be given a simple physical interpretation as follows:

$$\begin{aligned}\frac{\partial M}{\partial y} &= -\left\langle \frac{\partial \psi'}{\partial x} \nabla^2 \psi' \right\rangle = -\left\langle \frac{\partial \psi'}{\partial x}\frac{\partial^2 \psi'}{\partial x^2} + \frac{\partial \psi'}{\partial x}\frac{\partial^2 \psi'}{\partial y^2} \right\rangle \\ &= -\frac{\partial}{\partial y}\left\langle \left(\frac{\partial \psi'}{\partial x}\frac{\partial \psi'}{\partial y} \right) \right\rangle = \frac{\partial}{\partial y}\langle u'v' \rangle \qquad (10.19)\end{aligned}$$

where we have again used $\partial \langle [\ \ ] \rangle / \partial x = 0$. Thus, $M = \langle u'v' \rangle$ is just the correlation between the eddy zonal and meridional velocity components averaged around a latitude circle.

Subtracting (10.18) from (10.4) we obtain the eddy vorticity equation

$$\frac{\partial}{\partial t}\nabla^2 \psi' = -\nabla \cdot [(\overline{\mathbf{V}}_\psi + \mathbf{V}'_\psi)(\nabla^2 \overline{\psi} + \nabla^2 \psi' + f)] + f_0 \frac{\partial \omega'}{\partial p} - \frac{\partial^2 M}{\partial y^2} \qquad (10.20)$$

We obtain the mean kinetic energy equation by multiplying (10.18) by $-\overline{\psi}$ and integrating over the entire volume. After some manipulation we obtain

$$\frac{d\overline{K}}{dt} = \{K' \cdot \overline{K}\} + \{\overline{P} \cdot \overline{K}\} - \bar{\varepsilon} \qquad (10.21)$$

where

$$\overline{K} \equiv \int_0^{p_0} \int_A \frac{(\nabla \overline{\psi})^2}{2} \, dA \, dp/g$$

is the mean kinetic energy,

$$\{K' \cdot \overline{K}\} \equiv \int_0^{p_0} \int_A \overline{\psi} \frac{\partial^2 M}{\partial y^2} \, dA \, dp/g$$

is the conversion of eddy kinetic energy to mean kinetic energy,

$$\{\overline{P} \cdot \overline{K}\} \equiv \int_0^{p_0} \int_A f_0 \, \overline{\omega} \frac{\partial \overline{\psi}}{\partial p} \, dA \, dp/g$$

is the conversion of mean available potential energy to mean kinetic energy, and

$$\bar{\varepsilon} \equiv \int_A \left[\rho\left(\frac{Kf_0}{2}\right)^{1/2} \frac{(\nabla\bar{\psi})^2}{2}\right]_{p_0} dA$$

is the surface frictional dissipation for the mean wind. Similarly, multiplying through (10.20) by ψ' and integrating over the entire volume, we obtain

$$\frac{dK'}{dt} = -\{K' \cdot \bar{K}\} + \{P' \cdot K'\} - \varepsilon' \tag{10.22}$$

where

$$K' \equiv \int_0^{p_0} \int_A \frac{(\nabla\psi')^2}{2} dA\, dp/g$$

is the eddy kinetic energy,

$$\{P' \cdot K'\} = \int_0^{p_0} \int_A f_0 \omega' \frac{\partial \psi'}{\partial p} dA\, dp/g$$

is the conversion of eddy available potential energy to eddy kinetic energy, and

$$\varepsilon' \equiv \int_A \left[\rho\left(\frac{Kf_0}{2}\right)^{1/2} \frac{(\nabla\psi')^2}{2}\right]_{p_0} dA$$

is the eddy surface frictional dissipation.

We now turn to the thermodynamic energy equation. Again dividing the variables into their mean and eddy parts, we obtain from (10.5) the mean equation

$$\frac{\partial}{\partial t}\left(\frac{\partial \bar{\psi}}{\partial p}\right) = +\frac{\partial B}{\partial y} - \frac{\sigma}{f_0}\bar{\omega} - \bar{R} \tag{10.23}$$

where

$$B \equiv -\left\langle \frac{\partial \psi'}{\partial x} \frac{\partial \psi'}{\partial p} \right\rangle$$

is proportional to the correlation between temperature and meridional velocity averaged around a latitude circle.

Subtracting (10.23) from (10.5) we obtain the eddy thermodynamic energy equation

$$\frac{\partial}{\partial t}\left(\frac{\partial \psi'}{\partial p}\right) = -\nabla \cdot \left[(\bar{\mathbf{V}}_\psi + \mathbf{V}'_\psi) \cdot \nabla\left(\frac{\partial \bar{\psi}}{\partial p} + \frac{\partial \psi'}{\partial p}\right)\right] - \frac{\sigma}{f_0}\omega' - R' - \frac{\partial B}{\partial y} \tag{10.24}$$

Multiplying (10.23) and (10.24) through by

$$\frac{f_0^2}{\sigma}\frac{\partial\bar{\psi}}{\partial p} \quad \text{and} \quad \frac{f_0^2}{\sigma}\frac{\partial\psi'}{\partial p}$$

respectively, we obtain the mean available potential energy equation

$$\frac{d\bar{P}}{dt} = -\{\bar{P}\cdot P'\} - \{\bar{P}\cdot\bar{K}\} + \bar{G} \tag{10.25}$$

and the eddy available potential energy equation

$$\frac{dP'}{dt} = +\{\bar{P}\cdot P'\} - \{P'\cdot K'\} + G' \tag{10.26}$$

where

$$\bar{P} \equiv \int_0^{p_0}\int_A \frac{f_0^2}{2\sigma}\left(\frac{\partial\bar{\psi}}{\partial p}\right)^2 dA\, dp/g$$

is the mean available potential energy,

$$P' = \int_0^{p_0}\int_A \frac{f_0^2}{2\sigma}\left(\frac{\partial\psi'}{\partial p}\right)^2 dA\, dp/g$$

is the eddy available potential energy,

$$\{\bar{P}\cdot P'\} \equiv -\int_0^{p_0}\int_A \frac{f_0^2}{\sigma}\frac{\partial\bar{\psi}}{\partial p}\frac{\partial B}{\partial y}\, dA\, dp/g$$

is the conversion of mean available potential energy to eddy available potential energy,

$$\bar{G} \equiv -\int_0^{p_0}\int_A \frac{f_0^2}{\sigma}\bar{R}\frac{\partial\bar{\psi}}{\partial p}\, dA\, dp/g$$

is the generation of mean available potential energy by the mean heating, and

$$G' \equiv -\int_0^{p_0}\int_A \frac{f_0^2}{\sigma}R'\frac{\partial\psi'}{\partial p}\, dA\, dp/g$$

is the generation of eddy available potential energy by the eddy diabatic heating.

Combining (10.21), (10.22), (10.25), and (10.26), we obtain the total energy equation

$$\frac{d}{dt}(\bar{K} + K' + \bar{P} + P') = \bar{G} + G' - \bar{\varepsilon} - \varepsilon' \tag{10.27}$$

which should be compared with (10.17). Thus, in the long-term time mean

$$\bar{G} + G' = +\bar{\varepsilon} + \varepsilon' \tag{10.28}$$

which states that in the long-term mean the generation of available potential energy by diabatic processes must just balance the frictional dissipation of kinetic energy. Since $\bar{\varepsilon}$ and ε' are invariably positive, the sum of $\bar{G}$ and G' must be positive in the long-term mean. Now, since solar radiative heating is a maximum in the tropics, where the temperatures are high, it is clear that $\bar{G}$, the generation of zonal mean potential energy by the zonal mean heating, will be positive. For a *dry* atmosphere in which R' is due entirely to radiation and diffusion, G' should be negative because the infrared radiation of the atmosphere to space increases with increasing temperature. However, for the earth's atmosphere the presence of clouds and precipitation greatly alters the distribution of G'. Present estimates (see Fig. 10.1) suggest that in the Northern Hemisphere G' is positive and nearly half as large as $\bar{G}$. Thus, diabatic heating generates both zonal mean and eddy available potential energy.

The equations (10.21), (10.22), (10.25), and (10.26) together provide a complete description of the atmospheric energy cycle, within the limits of the quasi-geostrophic theory. The content of these equations is summarized by means of the energy cycle diagram shown in Fig. 10.1. In this diagram the squares represent reservoirs of energy and the arrows indicate sources, sinks, and conversions of energy. The observed direction of the conversion terms in the troposphere for the Northern Hemisphere annual mean is indicated by

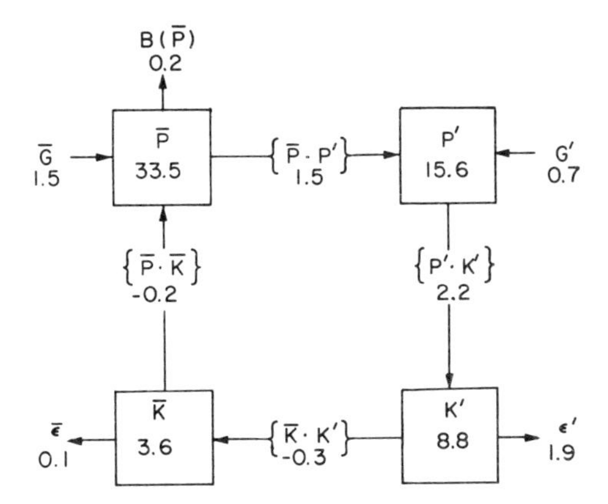

Fig. 10.1 The observed annual mean energy cycle for the Northern Hemisphere. Numbers in the squares are energy amounts in units of 10^5 J m^{-2}. Numbers next to the arrows are energy transformation rates in units of W m^{-2}. $B(\bar{P})$ represents a net energy flux into the Southern Hemisphere. Other symbols are defined in the text. (Adapted from Oort and Peixoto, 1974.)

the arrows. It should be emphasized that the direction of the various conversions cannot be theoretically deduced by reference to the energy equations alone. However, the observed energy cycle as summarized in Fig. 10.1 suggests the following qualitative picture:

1. The zonal mean radiative heating generates mean zonal available potential energy through a net heating of the tropics and cooling of the polar regions.
2. Baroclinic eddies transport warm air northward, cold air southward, and transform the mean available potential energy to eddy available potential energy.
3. At the same time eddy available potential energy is transformed into eddy kinetic energy by the vertical motions in the eddies.
4. The zonal kinetic energy is maintained primarily by the conversions from eddy kinetic energy due to the correlation $\langle u'v' \rangle$. This will be discussed further in the next section.
5. The energy is dissipated by surface and internal friction in the eddies and mean flow. The eddies tend to have higher vorticity than the mean flow at the top of the Ekman layer, hence much more of the surface dissipation is in the eddies than in the mean flow despite the fact that the mean flow kinetic energy is larger than the eddy kinetic energy.

In summary, the observed atmospheric energy cycle is consistent with the notion that eddies which result from the baroclinic instability of the mean flow are to a large extent responsible for the energy exchange in the atmosphere. It is through the eddy motions that the kinetic energy lost through friction is replaced, and it is the eddies which are primarily responsible for the poleward heat transport to balance the radiation deficit in the polar regions. In addition to the transient baroclinic eddies, forced stationary orographic waves and free Rossby waves may also contribute substantially to the poleward heat flux. The direct conversion of mean available potential energy to mean kinetic energy by symmetric overturning is, on the other hand, small and negative in middle latitudes, but positive in the tropics where it plays an important role in the maintenance of the mean Hadley circulation.

10.3 The Momentum Budget

In the previous section we saw that the observed zonal mean kinetic energy was maintained primarily by the eddies through the conversion $\{K' \cdot \bar{K}\}$. This process can alternatively be interpreted in terms of the atmospheric momentum budget. The angular momentum of the earth and atmosphere combined must remain constant in time except for the small effects of tidal friction. Since the average rotation rate of the earth is itself observed to be very close to constant, the atmosphere must also on the average conserve its angular momentum. Since the atmosphere gains angular momentum from the earth in the tropics where the surface winds are easterly and gives up angular momentum to the earth in middle latitudes where the surface winds are westerly, there must be a net poleward transport of angular momentum within the atmosphere, otherwise the torque due to surface friction would decelerate both the easterlies and westerlies. Furthermore, the angular momentum given by the earth to the atmosphere in the belt of easterlies must just balance the angular momentum given up to the earth in the westerlies if the angular momentum of the atmosphere is to remain constant.

In the equatorial regions the poleward momentum transport is divided between the advection of absolute angular momentum by the poleward flow in the axially symmetric Hadley circulation and transport by eddies. However, in the middle latitudes the zonal mean meridional velocity $\bar{v}$ is much too small to account for a significant fraction of the required transport. Thus, it must be primarily the eddy motions which transport momentum poleward in the middle latitudes and the angular momentum budget of the atmosphere must be qualitatively as shown in Fig. 10.2.

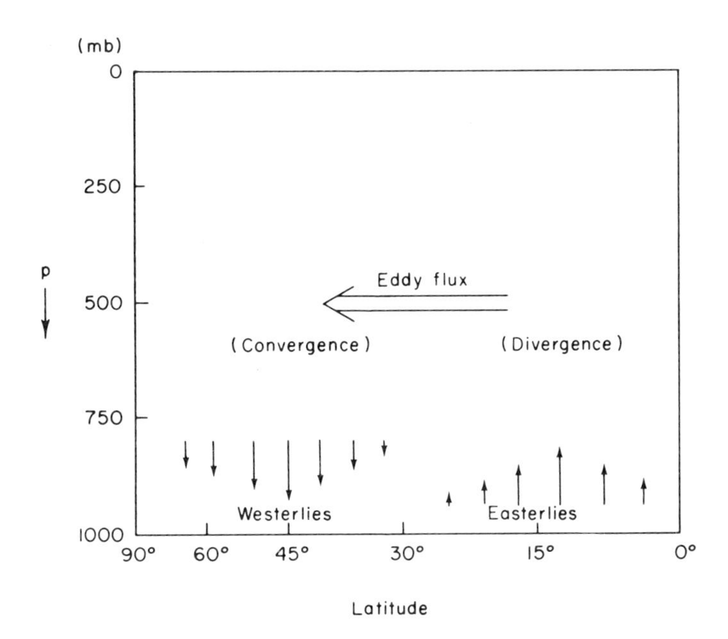

Fig. 10.2 Schematic diagram indicating the annual mean angular momentum budget in the atmosphere.

Observations in the Northern Hemisphere indicate a maximum poleward flux of angular momentum at about 30°N and a maximum horizontal flux convergence at about 45°N. This maximum in the flux convergence is a reflection of the strong $\{K' \cdot \bar{K}\}$ energy conversion in the upper-level westerlies and is the mechanism whereby the atmosphere can maintain a positive zonal wind in the middle latitudes despite the momentum lost to the surface.

Since it is the *absolute* angular momentum which is conserved for an air parcel in the absence of frictional or pressure torques, it is convenient to analyze the momentum budget in terms of absolute angular momentum. The absolute angular momentum per unit mass of atmosphere is

$$\mu = (\Omega a \cos \phi + u) a \cos \phi$$

The absolute angular momentum of an individual air parcel can be changed only by the torques, due to the zonal pressure gradient and the viscous stresses. Thus, from Newton's second law in its angular momentum form

$$\frac{d\mu}{dt} = -\frac{a \cos \phi}{\rho} \frac{\partial p}{\partial x} + \frac{a \cos \phi}{\rho} \frac{\partial \tau_x}{\partial z}$$

where it is assumed that horizontal viscous stresses are negligible compared to the vertical stress.

In order to analyze the angular momentum budget for a zonal ring of air it is useful to transform the above equation to the sigma coordinate system discussed in Section 8.7.1. Using the hydrostatic approximation to transform the viscous stress term we obtain

$$\left(\frac{\partial}{\partial t} + \mathbf{V} \cdot \nabla + \dot{\sigma} \frac{\partial}{\partial \sigma}\right) \mu = -a \cos \phi \left(\frac{\partial \Phi}{\partial x} + \frac{RT}{p_s} \frac{\partial p_s}{\partial x}\right) - \frac{g a \cos \phi}{p_s} \frac{\partial \tau_x}{\partial \sigma} \tag{10.29}$$

Multiplying the continuity equation

$$\frac{\partial p_s}{\partial t} + \nabla \cdot (p_s \mathbf{V}) + p_s \frac{\partial \dot{\sigma}}{\partial \sigma} = 0 \tag{10.30}$$

by μ and adding the result to (10.29) multiplied by p_s, we obtain the flux form of the angular momentum equation:

$$\frac{\partial}{\partial t}(p_s \mu) = -\nabla \cdot (p_s \mu \mathbf{V}) - \frac{\partial}{\partial \sigma}(p_s \mu \dot{\sigma})$$

$$- a \cos \phi \left[p_s \frac{\partial \Phi}{\partial x} + RT \frac{\partial p_s}{\partial x}\right] - g a \cos \phi \frac{\partial \tau_x}{\partial \sigma} \tag{10.31}$$

To obtain the zonal mean momentum budget we must average (10.31) in longitude. Using the spherical coordinate expansion for the horizontal divergence as given in Appendix C, we have

$$\nabla \cdot (p_s \mu \mathbf{V}) = \frac{1}{a \cos\phi} \frac{\partial}{\partial \lambda} (p_s \mu u) + \frac{1}{a \cos\phi} \frac{\partial}{\partial \phi} (p_s \mu v \cos\phi) \tag{10.32}$$

We also observe that the bracketed term on the right in (10.31) can be rewritten as

$$\left[p_s \frac{\partial}{\partial x} (\Phi - RT) + \frac{\partial}{\partial x} (p_s RT) \right] \tag{10.33}$$

But with the aid of the hydrostatic equation (8.63) we can write

$$(\Phi - RT) = \Phi + \sigma \frac{\partial \Phi}{\partial \sigma} = \frac{\partial}{\partial \sigma} (\Phi \sigma)$$

Thus, recalling that p_s does not depend on σ, (10.33) becomes

$$\left[\frac{\partial}{\partial \sigma} \left(p_s \sigma \frac{\partial \Phi}{\partial x} \right) + \frac{\partial}{\partial x} (p_s RT) \right] \tag{10.34}$$

Taking the zonal average of (10.31) and letting $\langle \quad \rangle \equiv (1/2\pi) \int_0^{2\pi} (\quad)\, d\lambda$, we obtain with the aid of (10.32) and (10.34)

$$\begin{aligned} \frac{\partial}{\partial t} \langle p_s \mu \rangle = & - \frac{1}{\cos\phi} \frac{\partial}{\partial y} (\langle p_s \mu v \rangle \cos\phi) \\ & - \frac{\partial}{\partial \sigma} \left(\langle p_s \mu \dot{\sigma} \rangle + g a \cos\phi \langle \tau_x \rangle + (a \cos\phi) \left\langle p \frac{\partial \Phi}{\partial x} \right\rangle \right) \end{aligned} \tag{10.35}$$

The terms on the right in (10.35) represent the convergence of the horizontal flux of angular momentum and the convergence of the vertical flux of angular momentum, respectively.

Integrating (10.35) vertically from the surface of the earth ($\sigma = 1$) to the top of the atmosphere ($\sigma = 0$), and recalling that $\dot{\sigma} = 0$ for $\sigma = 0, 1$ we have

$$\begin{aligned} \int_0^1 \frac{1}{g} \frac{\partial}{\partial t} \langle p_s \mu \rangle \, d\sigma = & \frac{-1}{g \cos\phi} \int_0^1 \frac{\partial}{\partial y} (\langle p_s \mu v \rangle \cos\phi) \, d\sigma \\ & - a \cos\phi \left[\langle \tau_x \rangle_{\sigma=1} + \left\langle p_s \frac{\partial h}{\partial x} \right\rangle \right] \end{aligned} \tag{10.36}$$

where $h(x, y)$ is the height of the lower boundary ($\sigma = 1$), and we have assumed that $\tau_x = 0$ at $\sigma = 0$.

Equation (10.36) expresses the angular momentum budget for a zonal ring of air of unit meridional width, extending from the ground to the top of the atmosphere. In the long-term mean the three terms on the right, representing the convergence of the meridional angular momentum flux, the surface viscous torque, and surface pressure torque, respectively, must balance. In the sigma coordinate system the surface pressure torque takes the particularly simple form $-\langle p_s\, \partial h/\partial x\rangle$. Thus, the pressure torque will act as an angular momentum sink for the atmosphere provided that surface pressure and the slope of the ground ($\partial h/\partial x$) are positively correlated. Observations indicate that this is indeed the case in middle latitudes because there is a tendency for the surface pressure to be higher on the western sides of mountains than on the eastern sides (see Fig. 4.9). In the Northern Hemisphere the surface pressure torque in middle latitudes is nearly as large a momentum sink as the surface frictional torque (Fig. 10.4.).

The role of eddy motions in providing the meridional angular momentum transport necessary to balance the surface angular momentum sinks can be elucidated best if we divide the flow into zonal mean and eddy components by letting

$$\mu = \langle \mu \rangle + \mu' = (\Omega a \cos\phi + \langle u \rangle + u')a \cos\phi$$
$$p_s v = \langle p_s v \rangle + (p_s v)'$$

where primes indicate deviations from the zonal mean. Thus, the meridional flux becomes

$$\langle p_s \mu v \rangle = (\Omega a \cos\phi\, \langle p_s v \rangle + \langle u \rangle \langle p_s v \rangle + \langle u'(p_s v)' \rangle)a \cos\phi \quad (10.37)$$

The three terms on the right in this expression are called the Ω-momentum flux, the drift, and the eddy momentum flux, respectively. The drift term is important in the tropics, but in midlatitudes it is small compared to the eddy flux and can be neglected in an approximate treatment. Furthermore, we can show that the Ω-momentum flux does not contribute to the vertically integrated flux. Averaging the continuity equation zonally and integrating vertically we obtain

$$\frac{1}{g}\int_0^1 \frac{\partial}{\partial t}\langle p_s \rangle\, d\sigma = -\frac{1}{g\cos\phi}\frac{\partial}{\partial y}\int_0^1 \langle p_s v \rangle\, d\sigma$$

Thus, in the long-term mean $\int_0^1 \langle p_s v \rangle\, d\sigma$ must vanish (i.e., there is no net mass flow across latitude circles). The vertically integrated angular momentum flux is therefore given approximately by

$$\int_0^1 \langle p_s \mu v \rangle\, d\sigma = \int_0^1 \langle u'(p_s v)' \rangle a \cos\phi\, d\sigma \simeq \int_0^1 a \cos\phi\, \langle p_s \rangle M\, d\sigma \quad (10.38)$$

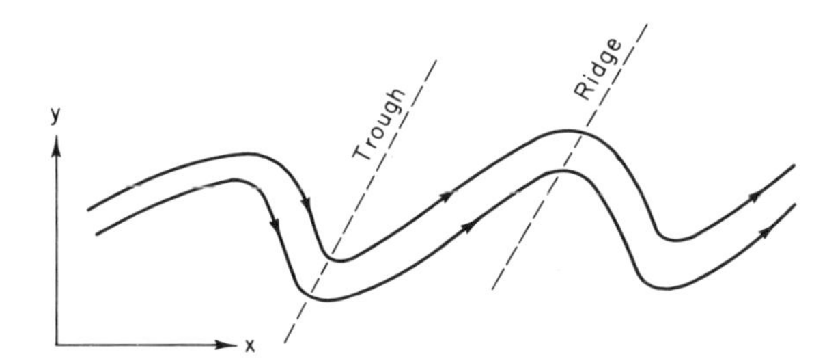

Fig. 10.3 Schematic streamlines for a positive eddy momentum flux.

where $M \equiv \langle u'v' \rangle$ and we have used the fact that the fractional change in p_s is small compared to the change in v' so that $(p_s v)' \simeq \langle p_s \rangle v'$. M is the correlation between the eddy zonal and meridional velocity components averaged around a latitude circle [see (10.19)].

According to the scheme shown in Fig. 10.2, M should be positive and decreasing with latitude in the belt of westerlies. If M is to be positive, the eddies must be asymmetric in the horizontal plane with the trough and ridge axes tilting as indicated in Fig. 10.3. When the troughs and ridges on the average have a southwest to northeast tilt the zonal flow will be larger than average ($u' > 0$) when the meridional flow is poleward ($v' > 0$) and the zonal flow will be less than average ($u' < 0$) for equatorward flow. Thus, $\langle u'v' \rangle > 0$ and the eddies will systematically transport positive zonal momentum poleward.

As shown in (10.35) the total vertical momentum flux consists of the flux due to large-scale motions $\langle p_s \mu \dot{\sigma} \rangle$, the flux due to the pressure torque $a \cos \phi \langle p \partial \Phi / \partial x \rangle$, and the flux due to small scale turbulent stresses $ga \cos \phi \langle \tau_x \rangle$. As was previously mentioned, the last two are responsible for the transfer of momentum from the earth to the atmosphere in the tropics and from the atmosphere to the earth in midlatitudes. Outside the planetary boundary layer, however, the vertical momentum transport is primarily due to the Ω-momentum flux $\Omega a \cos \phi \langle p_s \dot{\sigma} \rangle$, although the eddy momentum flux $\langle u'(p_s \dot{\sigma})' \rangle$ plays an essential role in the tropical stratosphere as we shall see in Chapter 11.

Estimates of the torques due to surface friction and the pressure torque have been attempted by several investigators despite the lack of adequate global data coverage and the difficulty in estimating the surface stress. One such estimate of the latitudinal variation of the eastward torque exerted by the earth on the atmosphere is shown in Fig. 10.4a. The mountain torque has been included implicitly in this estimate by raising the measured midlatitude surface torque by 40% so that the net latitudinally integrated torque balances to zero. The required northward flux of angular momentum to balance the estimated surface torque is indicated in Fig. 10.4b. This flux can also be directly estimated from wind data using (10.38). In Fig. 10.4b an estimate based on the observed winds is compared with the northward angular momentum flux required to balance the torques shown in Fig. 10.4a. Considering the many uncertainties in these measurements, the agreement is remarkable. It should also be mentioned here that except for the belt within 10° of the equator, almost all of the northward flux is due to the eddy flux term $\langle u'v' \rangle$. Thus, the momentum budget and the energy cycle both depend critically on the transports by the eddies.

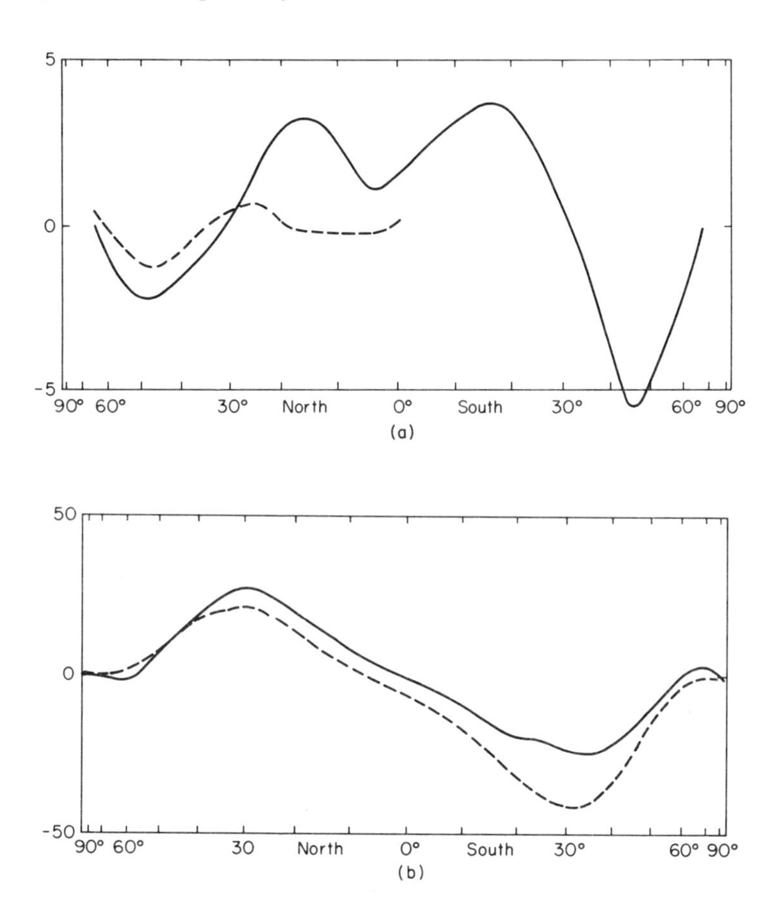

Fig. 10.4 (a) Average eastward torque per unit horizontal area exerted on the atmosphere by surface friction (solid curve) and by mountains in the Northern Hemisphere (dashed curve) in units of 10^5 kg s^{-2}. (b) The observed transport of angular momentum in units of 10^{18} kg m^2 s^{-2} (solid curve) and the required transport (dashed curve) as given by the observed surface torques. (After Lorenz, 1967.)

10.4 The Dynamics of Zonally Symmetric Circulations

In order to study the distribution of the mean zonal wind quantitatively, it is necessary to derive an equation which relates the zonal mean momentum to the sources, sinks, and transports of momentum. In this section we will show that zonally symmetric motions may be understood as a special case of the quasi-geostrophic theory, and that the vertical flow associated with an axially symmetric vortex is dynamically analogous to the vertical circulation in a baroclinic wave.

In the quasi-geostrophic model which we used to discuss the energy cycle in Section 10.2 internal friction and radiative dissipation were neglected in the equations for the mean zonal motion. Actually, since the mean zonal circulation changes rather slowly in time, internal dissipation should not be neglected. Internal damping is essential as a mechanism to balance the eddy momentum and heat sources in the long-term mean. A qualitative understanding of the role of internal dissipation may be obtained by representing this process very simply as a linear drag (Rayleigh friction) in the momentum equation and a linear thermal damping (Newtonian cooling) in the thermodynamic energy equation. For symmetry, we assume that the rate coefficient d is the same for both of these processes. The zonal vorticity equation (10.18) and zonal mean thermodynamic energy equation (10.23) then become

$$\left(\frac{\partial}{\partial t} + d\right)\left(\frac{\partial^2 \bar{\psi}}{\partial y^2}\right) = \frac{\partial^2 M}{\partial y^2} + f_0 \frac{\partial \bar{\omega}}{\partial p} \tag{10.39}$$

$$\left(\frac{\partial}{\partial t} + d\right)\left(\frac{\partial \bar{\psi}}{\partial p}\right) = \frac{\partial B}{\partial y} - \frac{\sigma}{f_0}\bar{\omega} - \bar{R} \tag{10.40}$$

This is a closed set in the variables $\bar{\psi}$ and $\bar{\omega}$ provided that the momentum flux M, the heat flux B, and the diabatic heating rate $\bar{R}$ are known,

If we eliminate $\bar{\omega}$ between (10.39) and (10.40) we obtain a *potential vorticity* equation for the mean flow:

$$\left(\frac{\partial}{\partial t} + d\right)\bar{q} = \frac{\partial^2 M}{\partial y^2} + \frac{f_0^2}{\sigma}\frac{\partial^2 B}{\partial y\, \partial p} - \frac{f_0^2}{\sigma}\frac{\partial \bar{R}}{\partial p} \tag{10.41}$$

where

$$\bar{q} = \frac{\partial^2 \bar{\psi}}{\partial y^2} + \frac{f_0^2}{\sigma}\frac{\partial^2 \bar{\psi}}{\partial p^2}$$

Now from the definition of B we can write, again using angle brackets instead of overbars to indicate averaged quantities,

$$\frac{\partial B}{\partial p} = -\frac{\partial}{\partial p}\left\langle\left(\frac{\partial\psi'}{\partial x}\frac{\partial\psi'}{\partial p}\right)\right\rangle = -\left\langle\frac{\partial\psi'}{\partial x}\frac{\partial^2\psi'}{\partial p^2}\right\rangle$$

where we have used the fact that

$$\left\langle\left(\frac{\partial\psi'}{\partial p}\right)\frac{\partial}{\partial x}\left(\frac{\partial\psi'}{\partial p}\right)\right\rangle = \frac{1}{2}\left\langle\frac{\partial}{\partial x}\left(\frac{\partial\psi'}{\partial p}\right)^2\right\rangle = 0$$

Recalling that

$$-\frac{\partial M}{\partial y} = \left\langle\left(\frac{\partial\psi'}{\partial x}\nabla^2\psi'\right)\right\rangle$$

we can now combine the terms in (10.41) involving B and M according to

$$\frac{\partial^2 M}{\partial y^2} + \frac{f_0^2}{\sigma}\frac{\partial}{\partial y}\left(\frac{\partial B}{\partial p}\right) = -\frac{\partial}{\partial y}\langle(q'v')\rangle \tag{10.42}$$

where

$$v' = \frac{\partial\psi'}{\partial x}, \qquad \text{and} \qquad q' \equiv \nabla^2\psi' + \frac{f_0^2}{\sigma}\frac{\partial^2\psi'}{\partial p^2}$$

is the eddy potential vorticity.

Therefore, in the absence of diabatic heating and internal damping (10.41) can be written simply as

$$\frac{\partial\bar{q}}{\partial t} = -\frac{\partial}{\partial y}\langle(v'q')\rangle \tag{10.43}$$

which states that *the zonal mean potential vorticity is changed only if there is potential vorticity transport by the eddies.* Thus, the net forcing of the mean flow by the eddies cannot be deduced from considering the heat or momentum fluxes separately. As can be seen from examination of (10.42), both M and B can be nonzero and the potential vorticity flux may still vanish, provided that

$$\frac{\partial M}{\partial y} = -\frac{f_0^2}{\sigma}\frac{\partial B}{\partial p}$$

Some insight into the conditions under which the potential vorticity flux vanishes can be obtained by considering eddy motions governed by the linearized perturbation potential vorticity equation

$$\frac{\partial q'}{\partial t} + \bar{u}\frac{\partial q'}{\partial x} + v'\frac{\partial\bar{q}}{\partial y} = 0 \tag{10.44}$$

For perturbations in the form of steady, zonally harmonic waves the x and time dependence of the perturbations is of the form $\exp[ik(x - ct)]$ where k is the zonal wave number and c the zonal phase speed. For such disturbances $\partial q'/\partial t = -c\,\partial q'/\partial x$ so that using the fact that $v' = \partial\psi'/\partial x$ we can rewrite (10.44) as

$$(c - \bar{u})\frac{\partial q'}{\partial x} = \frac{\partial\psi'}{\partial x}\frac{\partial\bar{q}}{\partial y} \tag{10.45}$$

Multiplying (10.45) through by ψ' and averaging zonally we obtain

$$(c - \bar{u})\left\langle \psi'\frac{\partial q'}{\partial x}\right\rangle = (\bar{u} - c)\left\langle q'\frac{\partial\psi'}{\partial x}\right\rangle = 0$$

or

$$(\bar{u} - c)\langle q'v'\rangle = 0 \tag{10.46}$$

Thus, provided that $\bar{u} - c \neq 0$ there is no potential vorticity transport by the eddies. We can therefore conclude that in the absence of *critical lines*, where $\bar{u} = c$, the forcing of the mean flow by the eddies vanishes for steady, nondissipative waves. The role of the eddies in driving the zonal mean flow is thus dependent on wave transience (growth or decay in time) and damping processes.

For physical interpretation of the processes which maintain the zonal mean circulation, it is easier to work with the zonal momentum equation rather than the potential vorticity equation. Using the zonally averaged continuity equation

$$\frac{\partial\bar{v}}{\partial y} = -\frac{\partial\bar{\omega}}{\partial p}$$

where $\bar{v}$ is the zonally averaged meridional velocity, we may eliminate $\partial\bar{\omega}/\partial p$ from (10.39) and immediately integrate once with respect to y to obtain

$$\left(\frac{\partial}{\partial t} + d\right)\left(-\frac{\partial\bar{\psi}}{\partial y}\right) = -\frac{\partial M}{\partial y} + f_0\,\bar{v} \tag{10.47}$$

We next define a stream function X for the mean meridional circulation by letting

$$\bar{\omega} = \frac{\partial X}{\partial y} \qquad \text{and} \qquad \bar{v} = -\frac{\partial X}{\partial p}$$

The continuity equation is then identically satisfied. Differentiating (10.40) with respect to y, (10.47) with respect to p, and adding the resulting equations, we obtain after representing $\bar{\omega}$ and $\bar{v}$ in terms of X

$$\left(\frac{\partial^2}{\partial y^2} + \frac{f_0^2}{\sigma}\frac{\partial^2}{\partial p^2}\right)X = -\frac{f_0}{\sigma}\left[\frac{\partial}{\partial p}\left(\frac{\partial M}{\partial y}\right) - \frac{\partial^2 B}{\partial y^2} + \frac{\partial \bar{R}}{\partial y}\right] \tag{10.48}$$

Equation (10.48) is a diagnostic equation for the mean meridional circulation which is exactly analogous to the omega equation (6.21) for quasi-geostrophic disturbances. The terms in (10.48) thus have physical interpretations similar to those of the analogous terms in the omega equation. The term on the left, for example, is identical to the term on the left in the omega equation but with ω replaced by X. By the arguments presented in Section 6.2.4 we can thus write

$$\left(\frac{\partial^2}{\partial y^2} + \frac{f_0^2}{\sigma}\frac{\partial^2}{\partial p^2}\right)X \propto -X$$

Assuming that there is no circulation across the equator, X must be zero at the equator and at the pole; otherwise, the average value of $\bar{\omega}$ integrated over the hemisphere would not be zero since

$$\bar{\omega}\, dy = dX$$

Thus, as indicated in Fig. 10.5, in the Northern Hemisphere a negative X implies a *direct* meridional circulation with rising motion to the south, poleward motion above, sinking motion to the north, and equatorward motion below the circulation center. Conversely, positive X implies an indirect meridional cell. Thus, by evaluating the forcing terms on the right in

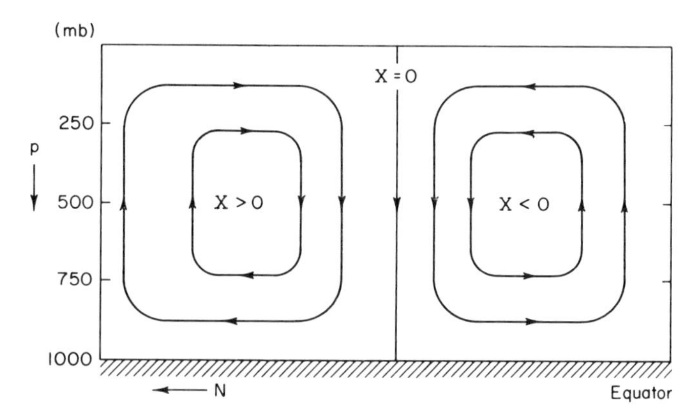

Fig. 10.5 Relation of the meridional streamfunction X to the vertical and meridional velocities.

(10.48) we can immediately deduce the sense of the mean meridional circulation since

$$X \propto \frac{\partial}{\partial p}\left(\frac{\partial M}{\partial y}\right) - \frac{\partial^2 B}{\partial y^2} + \frac{\partial \bar{R}}{\partial y} \tag{10.49}$$

The first term on the right in (10.49) is proportional to the vertical gradient of the horizontal eddy momentum flux convergence. This momentum term plays the same role in forcing a mean meridional circulation as the differential vorticity advection term of the omega equation plays in forcing vertical motions in a baroclinic disturbance. To interpret the momentum forcing physically, we suppose as shown in Fig. 10.6 that the momentum flux convergence is positive and increasing with height. Then

$$\frac{\partial}{\partial p}\left(\frac{\partial M}{\partial y}\right) > 0$$

which implies that X is also greater than zero. But from (10.47) we see that unless this term is balanced everywhere by either the linear drag or the Coriolis torque, the zonal mean thermal wind must be increasing in time, that is,

$$\frac{\partial}{\partial t}\left(-\frac{\partial \bar{u}}{\partial p}\right) > 0$$

And in order to maintain geostrophic and hydrostatic balance the latitudinal temperature gradient must be simultaneously increasing. However, in the absence of eddy heat flux and diabatic heating the increased temperature gradient can only be provided by an indirect meridional circulation as

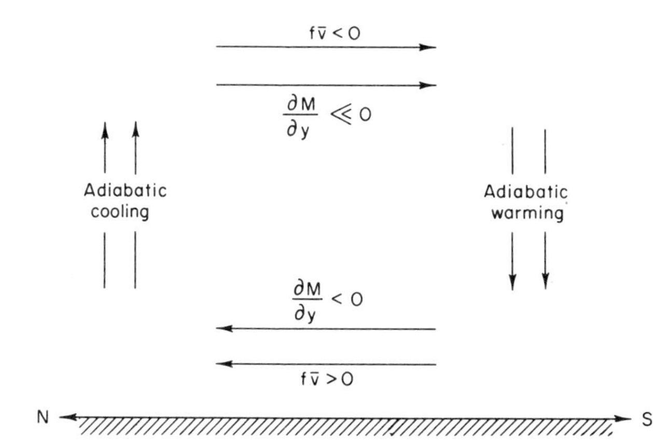

Fig. 10.6 Meridional circulation driven by a vertical gradient in the eddy momentum flux convergence.

shown in Fig. 10.5 with rising and adiabatic cooling north of the momentum source region and sinking and adiabatic warming south of the momentum source. Thus the circulation implied by (10.49) is consistent with the dynamical constraints of geostrophic and hydrostatic balance. In this case the meridional velocities required by continuity create Coriolis torques which tend to offset partially the effects of the momentum source by contributing easterly acceleration above and a westerly acceleration below. Thus, the meridional circulation here plays a role analogous to the divergent motions in baroclinic waves which, as we found in Section 6.2.4, tend to oppose the effects of differential vorticity advection.

Turning now to the second term on the right in (10.49) we first recall that

$$B = -\left\langle \frac{\partial \psi'}{\partial x} \frac{\partial \psi'}{\partial p} \right\rangle \propto \langle v'T' \rangle$$

is a measure of the northward eddy heat flux. Thus, $\partial^2 B/\partial y^2$ in (10.49) is analogous to the term involving the Laplacian of the temperature advection in the omega equation. At the latitude of maximum eddy heat flux B will be a maximum and $\partial^2 B/\partial y^2 < 0$, so that from (10.49) $X > 0$ and there will be an indirect meridional circulation centered about that latitude. The physical requirement for such an indirect meridional circulation can again be understood in terms of the need to maintain geostrophic and hydrostatic balance. North of the latitude where B is a maximum there is a convergence of eddy heat flux, while south of that latitude there is a divergence. Thus, the eddy heat transport will tend to reduce the pole-to-equator mean temperature gradient. If the mean zonal flow is to remain geostrophic, the thermal wind must then also decrease. In the absence of eddy momentum transport, this decrease in the thermal wind can only be produced by the Coriolis torques due to a mean meridional circulation as shown in Fig. 10.7. At the same time, it is not surprising to find that the vertical mean motions required by continuity oppose the temperature changes due to the eddy heat flux through adiabatic warming in the region of eddy heat flux divergence, and adiabatic cooling in the region of eddy heat flux convergence.

The final term in (10.49) is simply the zonal mean differential heating. In the usual case for the troposphere there is net diabatic heating in the tropics and cooling in the polar regions. Thus, $\partial \bar{R}/\partial y < 0$ so that $X < 0$ and the diabatic heating drives a direct meridional circulation. Again this circulation may be understood in terms of the hydrostatic and geostrophic constraints. The differential heating increases the pole-to-equator temperature gradient, which requires an increase in the zonal thermal wind. In the absence of eddy momentum sources the increased thermal wind can only be produced by

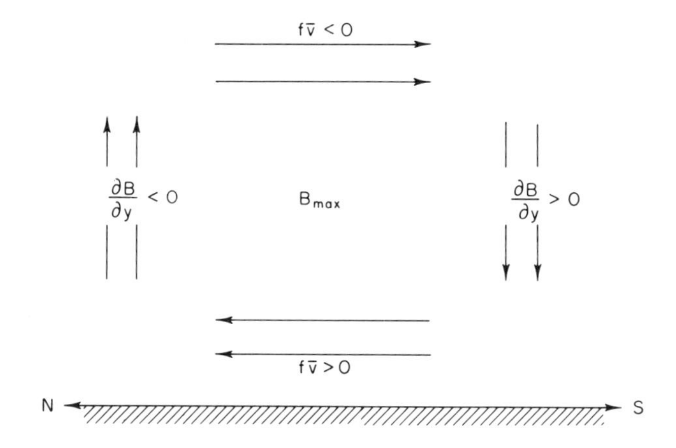

Fig. 10.7 Meridional circulation driven by an eddy heat flux.

the Coriolis torques of a direct circulation with poleward motion aloft and equatorward motion near the surface. Again, the adiabatic temperature changes due to the vertical motions associated with the direct cell act to oppose the diabatic heating.

The direct circulation which we have just described is, of course, just the well-known Hadley cell. From (10.49) we now see that in the absence of eddy transports of heat and momentum the circulation forced by solar differential heating will be the Hadley cell circulation. In the equatorial area where the eddies are very weak and the heating is strong the observed meridional circulation is indeed that of a direct Hadley cell. However, in the middle-latitude troposphere where both eddy heat flux and eddy momentum fluxes are strong, the eddy terms will tend to dominate in (10.49) so that the actual meridional circulation is a thermally indirect cell usually referred to as the *Ferrel cell.* For this reason, the mean meridional motion is difficult to measure directly from zonal averaging of the observed data. However, if the eddy flux terms and the differential heating can be determined, it is possible to deduce the mean meridional motion by solving (10.48) just as the vertical motions in a baroclinic wave can be deduced indirectly from the omega equation. Once $\bar{v}$ has been deduced in this manner, (10.47) can be used to determine the steady-state mean zonal wind distribution implied by the simple linear drag friction law. Present evidence, both from direct measurements of $\bar{v}$ and indirect deduction from (10.48), indicates that the mean meridional circulation is indirect in the middle latitude troposphere, but that it is so weak that it must play only a secondary role in the maintenance of the middle-latitude zonal mean circulation.

10.5 Laboratory Simulation of the General Circulation

In the previous sections we have seen that the gross features of the general circulation can be understood within the framework of the quasi-geostrophic model. In particular, we found that the observed zonal mean circulation and the atmospheric energy cycle are both rather well described by quasi-geostrophic theory. The fact that the quasi-geostrophic model, in which the spherical earth is replaced by the β plane, can successfully model many of the essential features of the general circulation suggests that the fundamental properties of the general circulation are not dependent on the spherical geometry or other parameters unique to the earth, but may be common to all rotating differentially heated fluids. That this conjecture is in fact correct can be demonstrated in the laboratory with rather simple apparatus.

Laboratory simulation experiments are a valuable method of testing hypotheses concerning the nature of the general circulation because parameters in experiments are subject to external control. For example, the rotation rate or the intensity of differential heating may be varied in an experiment, whereas in the atmosphere we can only accept the conditions given by nature.

In the past several years there have been a number of experimental studies directed toward an understanding of rotating differentially heated fluid systems. In most of these experiments the working fluid has been water rather than air, primarily because it is easier to follow motions by various sorts of tracers in a liquid than in a gas. In one group of experiments the apparatus consists of a cylindrical vessel which is rotated about its vertical axis. The vessel is heated at its rim and cooled at the center. These experiments are often referred to as "dishpan" experiments since some of the first experiments of this type utilized an ordinary dishpan. The fluid in the dishpan experiments crudely represents one hemisphere in the atmosphere with the rim of the dishpan corresponding to the equator and the center to the pole. However, because the geometry is cylindrical rather than spherical, the β effect is not modeled for baroclinic motions.[1] Thus the dishpan experiments omit the dynamical effect of the earth's vorticity gradient as well as the geometrical curvature terms which were neglected in the quasi-geostrophic model.

Qualitatively the experiments indicate that for certain combinations of rotation and heating rates, the flow in the cylinder appears to be axially symmetric with a steady zonal flow which is in thermal-wind equilibrium with the radial temperature gradient, and a superposed direct meridional circula-

[1] For barotropic motions the radial height gradient of the rotating fluid in the cylinder creates an "equivalent" β effect (see Section 4.3).

tion with rising motion near the rim and sinking near the center. This symmetric flow is usually called the *Hadley regime* since the flow is essentially that of a Hadley cell.

However, for other combinations of rotation and heating rates the observed flow is not symmetric but consists, rather, of irregular wavelike fluctuations and meandering zonal jets. In such experiments velocity tracers on the surface of the fluid reveal patterns very similar to those on middle-latitude upper-air weather charts. This type of flow is usually called the *Rossby regime*, although it should be noted that the observed waves are not Rossby waves since there is no β effect in the tank.

In addition to the dishpan experiments a number of similar experiments have been carried out in which the fluid is contained in the annular region between two coaxial cylinders of different radii. In these experiments the walls of the inner and outer cylinders are held at constant temperatures so that a precisely controlled temperature difference can be maintained across the annular region. In these "annulus" experiments very regular wave patterns can be obtained for certain combinations of rotation and heating. In some cases the wave patterns are steady, while in other cases they undergo regular periodic fluctuations called "vacillation" cycles (see Fig. 10.8). Although these flow patterns do not have the striking resemblance to atmospheric flow patterns that can be seen in the dishpan experiments, the annulus experiments have had a greater scientific impact than the dishpan experiments because the regularity of the waves in the annulus and the abrupt transitions between flow regimes as external parameters are changed make the annulus experiments far more amenable to theoretical analysis. For this reason, the annulus experiments have led to important advances in our understanding of thermal convection in rotating fluid systems.

In order to determine whether the laboratory experiments are really valid analogs of the atmospheric circulation or merely bear an accidental resemblance to atmospheric flow, it is necessary to analyze the experiments quantitatively. The mathematical analysis of the experiments can proceed from essentially the same equations which are applied to the atmosphere except that cylindrical geometry replaces spherical geometry and temperature replaces potential temperature in the heat equation. (Since water is nearly incompressible, *adiabatic* temperature changes are negligible following the motion.) In addition the equation of state must be replaced by an appropriate measure of the relationship between temperature and density:

$$\rho = \rho_0[1 - \varepsilon(T - T_0)] \tag{10.50}$$

where ε is the thermal expansion coefficient ($\varepsilon \simeq 2 \times 10^{-4}\,^\circ\mathrm{C}^{-1}$ for water) and ρ_0 is the density at the mean temperature T_0.

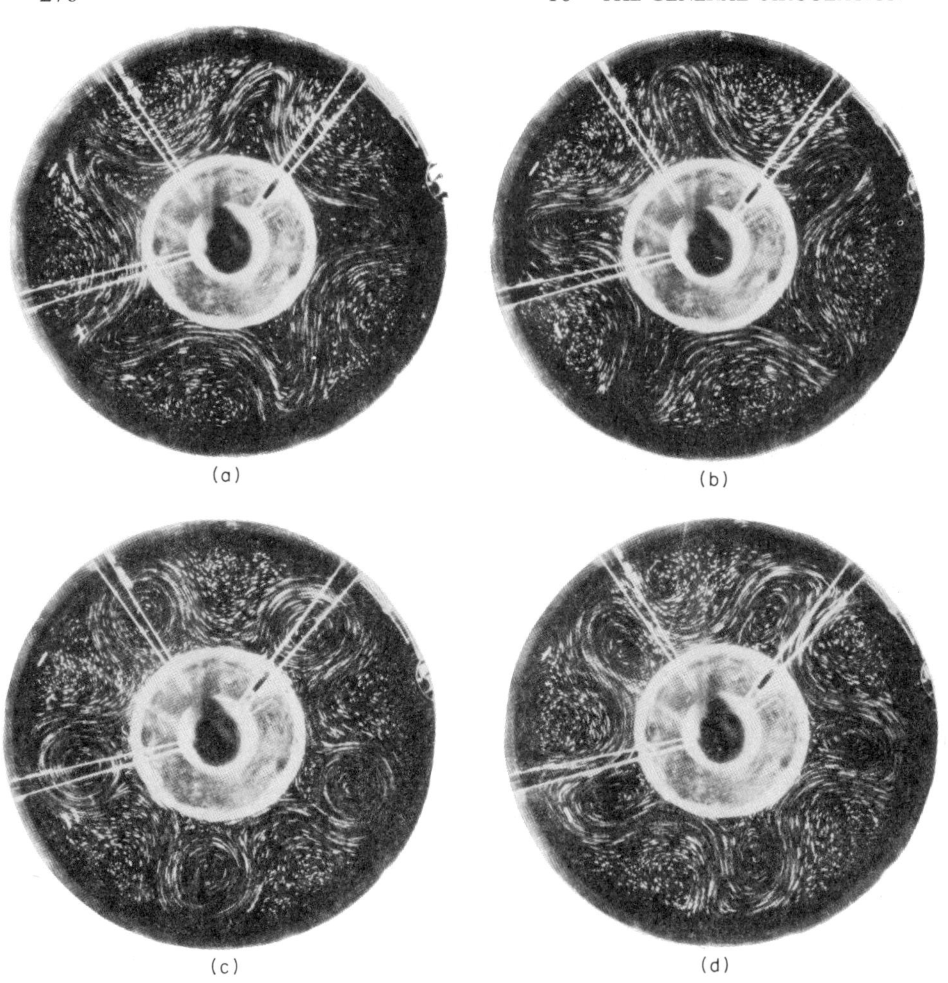

Fig. 10.8 Time exposures showing the motion of surface tracer particles in a rotating annulus. The four photographs illustrate various stages of a five-wave tilted trough vacillation cycle. The period of the vacillation cycle is $16\frac{1}{4}$ revolutions and the photographs are at intervals of 4 revolutions. (Photographs by Dave Fultz.)

As we saw already in Chapter 1, the character of the motion in a fluid is crucially dependent on certain characteristic scales of parameters such as the velocity, pressure fluctuations, length, time, etc. In the laboratory the scales of these parameters are generally many orders of magnitude different from their scales in nature. However, it is still possible to produce quasi-geostrophic motions in the laboratory provided that the motions are slow enough so that the flow is approximately hydrostatic and geostrophic. As we saw in Section 2.4.2, the geostrophy of a flow does not depend on the absolute value of the scaling parameters, but rather on a nondimensional ratio of these parameters called the Rossby number. For the annulus experiments, the maximum horizontal scale is set by the dimensions of the tank so that it is convenient to define the Rossby number as

$$\mathrm{R}o \equiv \frac{U}{\Omega(b-a)}$$

where Ω is the angular velocity of the tank, U is a typical relative velocity of the fluid, and $b - a$ is the difference between the radius b of the outer wall and the radius a of the inner wall of the annular region.

Using the hydrostatic approximation

$$\frac{\partial p}{\partial z} = -\rho g \tag{10.51}$$

Together with the geostrophic relationship

$$2\Omega \mathbf{V}_g = \frac{\mathbf{k} \times \nabla p}{\rho_0} \tag{10.52}$$

and the equation or state (10.50) we can obtain a "thermal wind" relationship in the form

$$\frac{\partial \mathbf{V}_g}{\partial z} = \frac{\varepsilon g}{2\Omega} \mathbf{k} \times \nabla T \tag{10.53}$$

Letting U denote the scale of the geostrophic velocity, H the mean depth of fluid, and ΔT the radial temperature difference across the width of the annulus, we obtain from (10.53)

$$U \sim \frac{\varepsilon g H\, \Delta T}{2\Omega(b-a)} \tag{10.54}$$

Substituting the value of U in (10.54) into the formula for the Rossby number yields the *thermal Rossby number*

$$\mathrm{R}o_{\mathrm{T}} = \frac{\varepsilon g H\, \Delta T}{2\Omega^2(b-a)^2} \tag{10.55}$$

This nondimensional number is the best measure of the range of validity of quasi-geostrophic dynamics in the annulus experiments. Provided $\mathrm{R}o_{\mathrm{T}} \ll 1$ the quasi-geostrophic theory should be valid for motions in the annulus. For example, in a typical experiment

$$H \simeq b - a \simeq 10 \quad \text{cm}, \qquad \Delta T \simeq 10°\text{C}, \qquad \text{and} \qquad \Omega \simeq 1\ \text{s}^{-1}$$

so that

$$\mathrm{R}o_{\mathrm{T}} \simeq 10^{-1}$$

There is, however, one difficulty with the thermal Rossby number as derived here. The difficulty arises because near the vertical boundaries of the annulus there are *conduction boundary layers* in which the temperature changes rather rapidly away from the boundaries. The thermal wind relationship is not valid in these boundary layers; thus the ΔT in (10.55) really represents only the temperature gradient across the interior region of fluid away from the sidewall boundary layers. However, this ΔT is not an external parameter, but is in itself dependent on the flow. Therefore, in order to describe the experimental results entirely in terms of external parameters, it is usual to define an *imposed* thermal Rossby number

$$Ro_T^* \equiv \frac{\varepsilon g H}{2\Omega^2(b-a)^2}(T_b - T_a) \tag{10.56}$$

where T_b and T_a are the imposed temperatures of the outer and inner walls, respectively. From the discussion above it is clear that $Ro_T^* > Ro_T$, so that $Ro_T^* \ll 1$ is certainly a sufficient condition for quasi-geostrophic theory to be valid, but is not clearly a necessary condition.

Annulus experiments carried out over a wide range of rotation rates and temperature contrasts produce results which can be classified consistently on a log–log plot of the imposed thermal Rossby number versus a nondimensional measure of the rotation rate $(G^*)^{-1} \equiv (b-a)\Omega^2/g$. The results for Fultz's experiments are summarized in Fig. 10.9. The heavy solid line separates the axially symmetric Hadley regime from the wavy Rossby regime. These results can best be understood qualitatively by considering an experiment in which the thermal Rossby number is slowly increased from zero by gradually imposing a temperature difference $T_b - T_a$. For very weak differential heating ($T_b - T_a$ very small) the motion is of the Hadley type with weak horizontal and vertical temperature gradients in the fluid. In the absence of viscosity this flow would probably be baroclinically unstable. Viscous boundary layer dissipation removes the possibility of unstable wave growth. However, as the horizontal temperature contrast is increased, the mean thermal wind must also increase until at some critical value of Ro_T^* the flow becomes baroclinically unstable. According to the theory presented in Chapter 9, the wavelength of maximum instability is proportional to the ratio of the static stability to the square of the rotation rate. Thus, as can be seen in Fig. 10.9, the wave number observed when the flow becomes unstable decreases (that is, wavelength increases) as the rotation rate is reduced. Furthermore, since baroclinic waves transport heat vertically as well as laterally, they will tend to increase the static stability of the fluid. Therefore, as the thermal Rossby number is increased within the Rossby

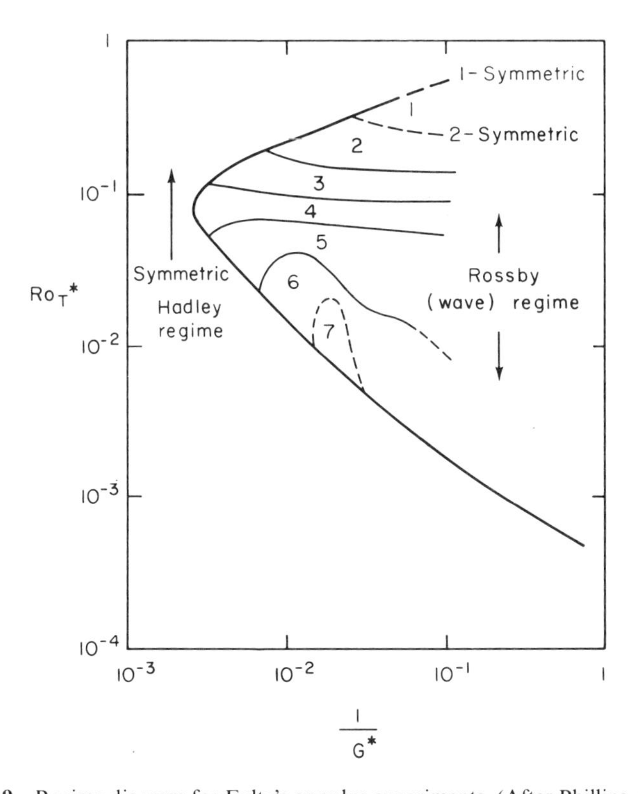

Fig. 10.9 Regime diagram for Fultz's annulus experiments. (After Phillips, 1963.)

regime the increased heat transport by the waves will raise the static stability, and hence increase the wavelength of the wave of maximum instability. Thus, as Ro_T^* is increased the flow will undergo transitions in which the observed wave number decreases until finally the static stability becomes so large that the flow is stable to even the largest wave that can fit the tank. The flow then returns to a symmetric Hadley circulation stabilized by a high static stability which is maintained by a vigorous direct meridional circulation. This regime is usually called the *upper* symmetric regime to distinguish it from the *lower* symmetric regime which occurs for very weak heating.

In the annulus where the flow is highly constrained by the geometry of the system the transitions between wave numbers in the Rossby regime as indicated in Fig. 10.9 are quite abrupt and regular, at least for moderately small values of $(G^*)^{-1}$. In the dishpan experiments, the waves are not so regular and transitions are more continuous. However, the basic features of the flow do remain the same.

So far we have discussed only the qualitative similarity between the Rossby regime and the general circulation of the atmosphere. Detailed temperature and velocity measurements have revealed that the energy cycle of the Rossby regime flow in the annulus and dishpan experiments is the same as that observed for the troposphere as shown in Fig. 10.1. Furthermore, the waves in the experiments under some conditions have an eddy momentum transport $\langle u'v' \rangle$ similar to that observed in the atmosphere, with eddy momentum convergence in the region of strong zonal jets. In fact, it is this interchange of momentum between the eddies and the zonal flow which accounts for the vacillation cycle in the annulus shown in Fig. 10.8. It is clear from Fig. 10.8a that during this stage of the vacillation cycle the troughs and ridges have a strong southwest to northeast tilt (assuming that the axis of rotation corresponds to the North Pole) and that as a consequence there is a strong northward eddy momentum transport. As a result of this eddy momentum transport the zonal flow is concentrated in a meandering jet stream. Temperature measurements in this type of experiment have revealed that the horizontal temperature gradient is concentrated in a "frontal" zone associated with the meandering jet, as would be required for the jet to satisfy the thermal-wind equation.

In summary, the laboratory studies despite their many idealizations can model the most important of those features of the general circulation which are not dependent on the earth's topography or continent–ocean heating contrasts. Specifically, we find, perhaps surprisingly, that the β effect is not essential for the development of circulations which look very much like tropospheric synoptic systems. Apparently the role of the β effect is primarily a modifying influence which, although it certainly changes the zonal phase speed of atmospheric waves, is not essential to modeling the primary features of the middle-latitude circulation. Thus, the observed middle latitude waves should be regarded essentially as *baroclinic waves* modified by the β effect, not as Rossby waves in a baroclinic current.

In addition to demonstrating the primacy of baroclinic instability, the experiments also confirm that condensation heating is not an essential mechanism for simulation of midlatitude circulations. The laboratory experiments thus enable us to separate the essential mechanisms from second-order effects in a manner not easily accomplished by observation of the atmosphere itself. Also, since laboratory simulation experiments have typical rotation rates of 10 rpm it is possible to simulate many years of atmospheric flow in a short time. Thus, the model experiments provide a unique opportunity for collecting "general circulation" statistics. In addition, since temperatures and velocities can be measured at uniform intervals in the experiments, the experiments can provide excellent sets of data for testing long-term numerical weather prediction models.

10.6 Numerical Simulation of the General Circulation

In the previous section we have discussed the role of laboratory experiments in contributing toward a qualitative understanding of the general circulation of the atmosphere. Although laboratory experiments can elucidate most of the gross features of the general circulation, there are many details which cannot possibly be duplicated in the laboratory. For example, the possible long-term climatic effects of aerosols and trace gases which are added to the atmosphere as a result of man's activities could not possibly be predicted on the basis of laboratory experiments. As another example, the influence of an ice-free Arctic Ocean on the general circulation would also be extremely difficult to simulate in the laboratory. Since all the conditions of the atmosphere cannot be duplicated in the laboratory, it would appear that the only practical manner in which to predict possible climate modifications resulting from intentional or unintentional human intervention is by numerical simulation on high-speed computers.

Progress in numerical modeling of the general circulation has been to some degree dictated by the rate of development in the field of computer technology. However, our limited ability to parameterize the effects of small-scale processes in terms of the large-scale motions has been an equally important limiting factor. Essentially, the problem of numerical modeling of the general circulation is simply that of producing a very long-range numerical weather forecast. However, the equations used in general circulation models must be more sophisticated than those in numerical weather prediction models because as the length of the "forecast" is extended, physical processes which are unimportant for the short-term evolution may become crucial. For example, the interaction between the atmosphere and the upper layers of the ocean is probably important for understanding variations in the general circulation with periods greater than a month or so. On the other hand, since a forecast of the flow evolution from a specific initial state is not needed, the problems associated with specifying initial data in primitive equation forecast models do not arise in general circulation models. The object of most general circulation models is to simulate the climate. Since climate (here defined as the time-averaged circulation) should be independent of the initial conditions, it is customary to begin most general circulation model integrations with a resting atmosphere as the initial state.

Because the values of the field variables in a numerical model are known on a regular array of grid points at regular time intervals, it is very simple to compute the statistics needed to completely diagnose the energy cycle and momentum budget for a model general circulation. Thus, a general circulation model can be used like the dishpan experiments to test the quasi-geostrophic energy equations discussed in Section 10.2. In the first subsection below we will discuss a very simple quasi-geostrophic general circulation model suitable for this purpose, while in the second subsection we will describe briefly some results from a highly sophisticated primitive equation general circulation model.

10.6.1 A Quasi-Geostrophic Model

The success of the quasi-geostrophic model in short-range weather prediction suggests that for simulating the gross features of the general circulation such a model might be adequate provided that diabatic heating and frictional dissipation were included in a suitable manner. In fact, we have already shown in Section 10.2 that the quasi-geostrophic model is capable of at least qualitatively describing the observed energy cycle and momentum budget in the midlatitude troposphere.

Phillips (1956) made the first attempt to model the general circulation numerically. His classic experiment employed a two-level quasi-geostrophic forecast model similar to the model discussed in Section 8.5. However, the model was modified so that it included friction, primarily in the form of an Ekman layer dissipation. Phillips realized that to represent accurately the diabatic heating would have required extensive radiative transfer computations as well as description of the water vapor budget. In order to avoid such complexities, he simply specified a net heating rate, based on observations, which was a linear function of latitude only and had a zero horizontal mean. The domain of integration was a zonal channel on the β plane with a latitudinal width of 10,000 km and an assumed zonal periodicity with 6000-km wavelength. The heating rates chosen by Phillips were based on estimates of the *net* diabatic heating rates necessary to balance the poleward heat transport at 45°N computed from observational data. Despite the severe limitations of the computer available, and the eventual breakdown of the integration due to a nonlinear computational instability, the computations resulted in a model circulation which in many respects resembled the observed circulation. Perhaps this does not seem surprising since we know that the quasi-geostrophic model can produce reasonable short-range forecasts, and scaling considerations indicate that the quasi-geostrophic model contains the essential physical mechanisms for the development of synoptic scale eddies. However, it should be kept in mind that Phillips' experiment was quite a bold departure from the previous applications of the quasi-geostrophic model to short-term forecasts. In the forecast applications, the model equations were integrated from an observed initial state already containing fully developed systems. In Phillips' model the atmosphere was started from an initial state of rest and the motion was allowed to develop as a result of the differential diabatic heating.

The motion field which initially developed was a pure Hadley cell circulation with rising motion south of the center of the channel and sinking motion north of the center. As the diabatic heating increased the latitudinal temperature gradient, the zonal thermal wind also increased until the flow became baroclinically unstable. At that point, small random perturbations introduced into the calculations caused the development of baroclinic waves. These waves grew rapidly until the rate of energy conversion from the mean flow was approximately balanced by frictional dissipation in the waves. Although the accumulation of truncation errors in Phillips' experiment prevented the flow from reaching a statistically steady state, the circulation for about 20 model days after the initiation of baroclinic instability did remarkably resemble that of a midlatitude weather chart as Fig. 10.10 reproduced from Phillips' paper indicates.

Phillips computed the energy transformations in his model and found that the energy cycle of the model atmosphere agreed remarkably with that shown in Fig. 10.1. Examination of the momentum budget indicated further that even the transfer of eddy momentum to mean zonal momentum was modeled in this simple experiment. Therefore, Phillips' numerical experiment provides additional evidence that the quasi-geostrophic model is capable of describing the essential characteristics of the general circulation outside the tropics.

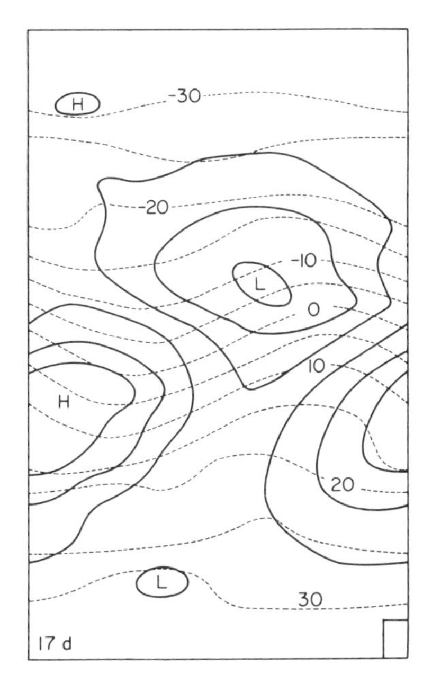

Fig. 10.10 Distribution of 1000-mb contour height at 200-ft intervals (solid lines) and 500-mb temperature at 5°C intervals (dashed lines) at 17 d. The small rectangle in the lower right corner shows the size of the finite difference grid intervals Δx and Δy. (After Phillips, 1956.)

10.6.2 Primitive Equation Models

Phillips' experiment, although it was an extremely important advance in dynamic meteorology, suffered from a number of shortcomings as a general circulation model. Perhaps the gravest shortcoming in his model was the specification of diabatic heating as a fixed function of latitude only. In reality the atmosphere must to some degree determine the distribution of its own heat sources. This is true not only for condensation heating, which obviously depends on the distribution of vertical motion and water vapor, but also holds for radiative heating as well. Both the net solar heating and net infrared heating are sensitive to the distribution of clouds, and infrared heating depends on the atmospheric temperature as well.

Another important limitation of Phillips' two-level quasi-geostrophic model was that static stability could not be predicted (since temperature was computed at only a single level) but had to be specified as an external parameter. This limitation is a serious one because the static stability of the atmosphere is obviously controlled by the motions. Indeed, in the discussion of the annulus experiments in Section 10.5, we pointed out that the existence of the upper symmetric regime can be attributed to the stabilization of the flow due to the very high static stability established by the strong vertical heat transport of the meridional circulation. In the atmosphere the vertical temperature profile, and hence the static stability, is determined by the combined effects of diabatic heating and the vertical heat flux due to large-scale eddies and small-scale convection.

It would be possible to design a quasi-geostrophic model in which the diabatic heating and static stability were motion dependent. Indeed such models have been used to some extent, especially in theoretical studies of the annulus experiments. However, to model the global circulation clearly requires a dynamical framework which is valid in the equatorial zone. Thus, it is highly desirable to base a general circulation model on the primitive equations. Because, as was discussed in Chapter 8, gravity waves are not filtered by the primitive equations, time increments in a primitive equation model must be much smaller than for a quasi-geostrophic model. Therefore, an enormous amount of computation is necessary to simulate even a single season of the global circulation. However, the task is well within the capabilities of modern computers. Because of its enormous complexity and many important applications, general circulation modeling has become a highly specialized activity which cannot possibly be adequately covered in a short space. To keep the discussion within the scope of this text, we will here discuss only one of the primitive equation models, the model developed at the Geophysical Fluid Dynamics Laboratory (GFDL) of the National Oceanic and Atmospheric Administration (NOAA).

The GFDL model is based on the primitive equations in the σ-coordinate system. The eleven prediction levels are arranged at unequal vertical intervals designed to give an adequate resolution both in the planetary boundary layer near the ground and in the lower stratosphere. In the simulation described here, the flux of solar radiation at the top of the atmosphere varies with latitude and season (the diurnal variation is eliminated). The radiative heating calculation includes the effects of cloudiness, carbon dioxide, ozone, and water vapor. However, the radiative heating calculation in this model uses a zonally symmetric annual mean observed cloudiness distribution, rather than model computed cloudiness. In addition, the ozone mixing ratio is a prescribed function of latitude, height, and season, rather than a prognostic variable. Water vapor mixing ratio is, however, computed as one of the prognostic variables of the model.

The horizontal grid resolution of approximately 265 km is adequate to incorporate the condensation heating due to synoptic scale uplift. However, it is obviously not adequate to incorporate explicitly cumulus convection. The latter is implicitly included in the model through a *convective adjustment* process which operates as follows: Any time the lapse rate at some grid point exceeds moist adiabatic and the relative humidity is 100%, the lapse rate is adjusted back to the moist adiabatic value using a scheme which conserves the potential energy for the entire vertical column. This process at least crudely simulates the stabilizing effect of cumulus convection. In addition, any time that the lapse rate exceeds the dry adiabatic rate, it is similarly adjusted back to the dry adiabatic value.

The model domain is global with a realistic distribution of continents and oceans, and a smoothed representation of the topography. The land is assumed to have zero heat capacity and the ocean is assumed to have infinite heat capacity. The ocean surface temperature is a prescribed function of position and season, based on observations.

The initial state for the integration was taken to be the long-term mean from a previous study in which the solar heating and sea surface temperatures were fixed at their January values. The simulation was carried out for about 3.5 model years, and seasonal mean statistics were computed for the last model year.

Meridional cross sections of the computed seasonally averaged zonal mean temperature fields for both summer and winter are shown together with the observed fields in Fig. 10.11. The model reproduces most features of the observed temperature structure quite well including the high cold tropical tropopause and the low polar tropopause. Thus, it is certain that this model contains the basic physical processes necessary for the formation and maintenance of the tropopause. The major deficiency in the simulated temperature structure is in the winter polar stratosphere where the temperatures are more

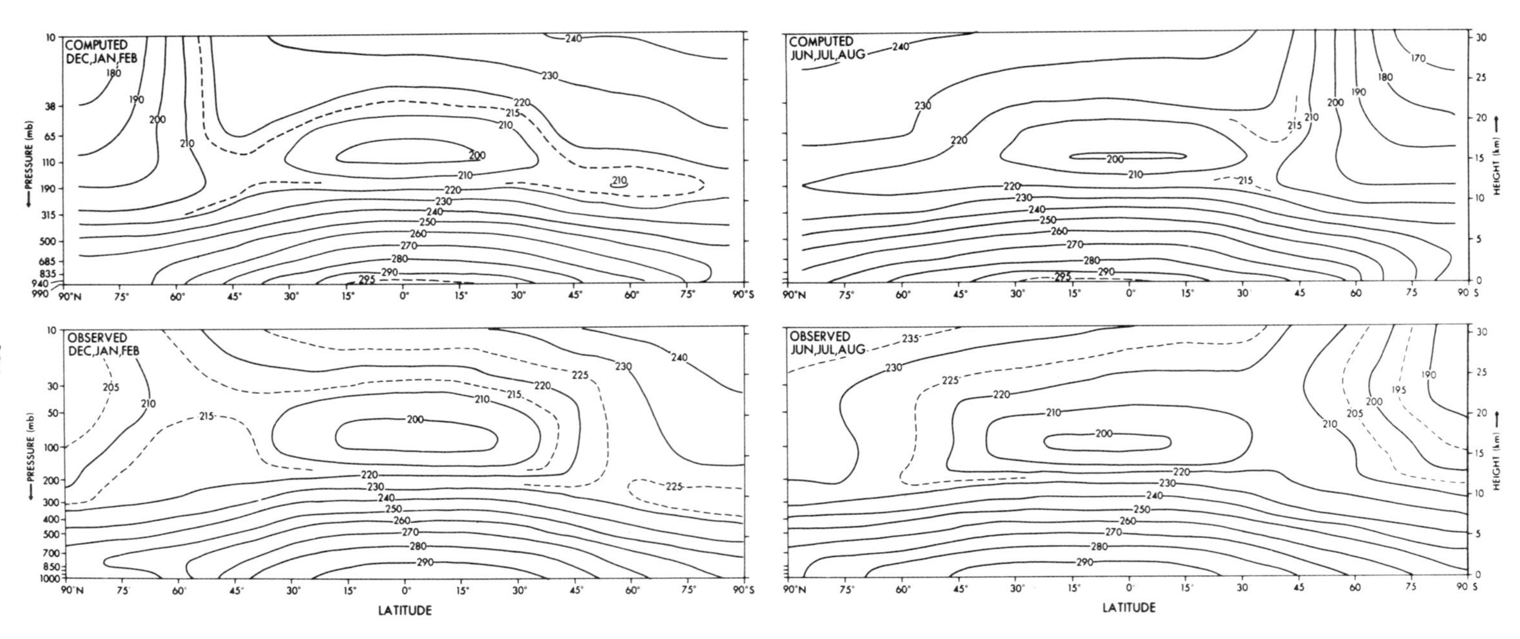

Fig. 10.11 Latitude–height distributions of zonal mean temperature (K). Top, computed distribution: bottom, observed distribution. (After Manabe and Mahlman, 1976. Reproduced with permission of the American Meteorological Society.)

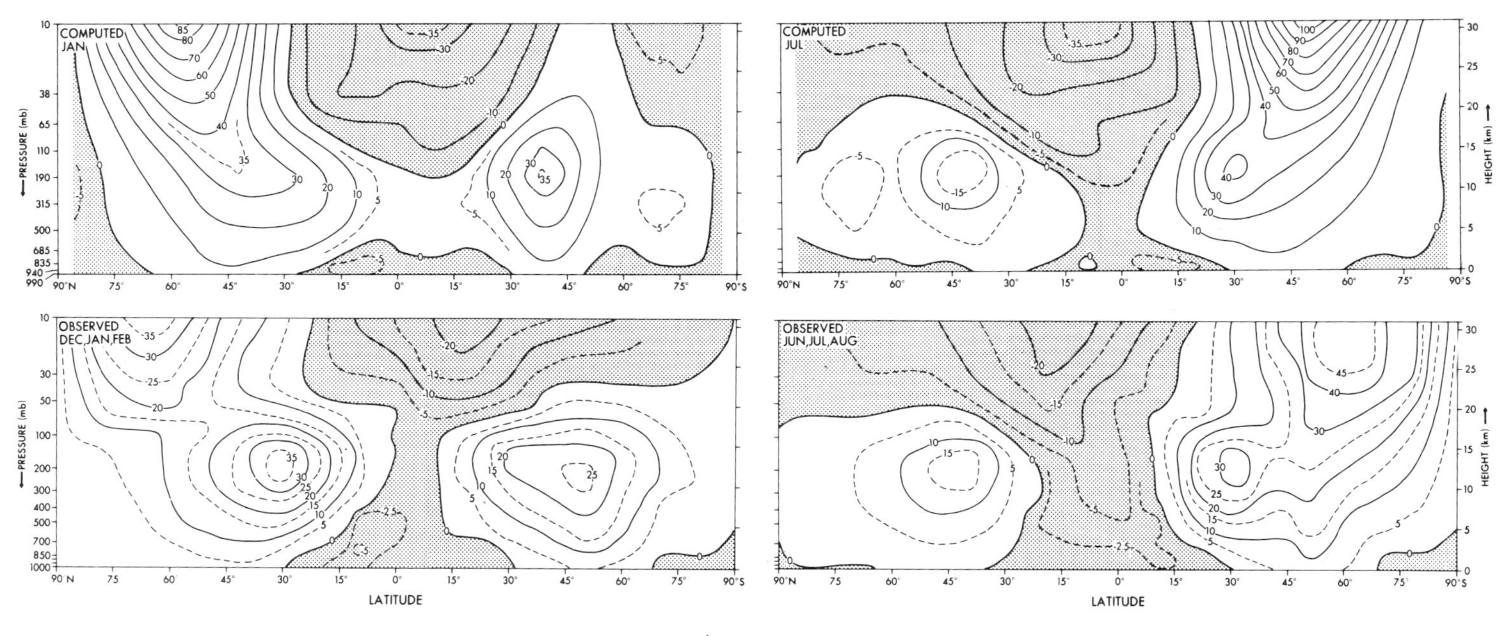

Fig. 10.12 Lattitude–height distributions of mean zonal wind ($m\ s^{-1}$). Top, computed distributions; bottom, observed distributions. (After Manabe and Mahlman, 1976. Reproduced with permission of the American Meteorological Society.)

than 20 K lower than observed. This difficulty may be related to the poor vertical resolution at the upper levels and the "rigid lid" upper boundary condition (i.e., $\dot{\sigma} = 0$ at the top boundary).

The mean zonal wind distributions corresponding to the temperature fields of Fig. 10.11 are shown in Fig. 10.12. The model produces tropospheric jet streams in fair accord with observations. The stratospheric summer easterlies are also reasonably well simulated; however, the winter stratospheric westerlies are too strong by a factor of two due to the establishment of a thermal wind balance with the unrealistically intense meridional temperature gradients. An additional difference between the observed and computed fields is that the equatorial tropospheric easterlies are too weak in the model (in fact westerlies extend right across the equator in the computed January).

The energy cycle computed for the simulation agrees quite well with the observed cycle indicated in Fig. 10.1. The angular momentum budget for the model is also in general accord with the observed budget. In particular the horizontal eddy momentum flux is a maximum in midlatitudes while the flux by the mean meridional circulation is significant only in the tropics.

A comparison of the results from this model with those of a similar run without a hydrological cycle indicate that in middle latitudes there is little difference between the general circulation of a dry atmosphere and that of a wet atmosphere. However, in the tropics the circulation for a dry atmosphere is much weaker than that of the moist model. Thus, it appears that latent heat release is an essential driving force for the tropical atmosphere, but not for the extratropical regions. This comparison illustrates one of the most powerful features of the numerical models, namely that various physical processes can be included or excluded at will by modifying the equations; and by comparing the output from runs with and without a given process, the effect of that process on the circulation can be easily evaluated. Using this sort of procedure, it should eventually be possible to make very good predictions of the climatological consequences of various man-caused environmental changes.

10.7 Longitudinally Varying Features of the General Circulation

So far in this chapter we have limited our discussion to the zonally symmetric component of the general circulation. For a planet with a longitudinally uniform surface the time averaged flow should be completely characterized by the zonally symmetric circulation. However, on the earth the presence of continent–ocean heating contrasts and large-scale topography provides strong forcing for longitudinally asymmetric time mean motions. Such motions, which are often referred to as *stationary waves*, consist both of geographically fixed features which appear on long-term mean maps, and seasonally reversing systems (monsoons).

10.7.1 MIDLATITUDE JET STREAMS

The most significant of the long-term mean zonally asymmetric features in the Northern Hemisphere are the Asian and North American jet streams and their associated storm tracks. The existence of these two jets can be inferred from the January mean 500-mb geopotential height field shown in Fig. 6.3. Note the strong meridional gradients in height associated with the troughs centered just east of the Asian and North American continents (the same features can be seen in annual mean charts, although with somewhat reduced intensity). The zonal flow associated with these semipermanent troughs is illustrated in Fig. 10.13. In addition to the two intense jet cores in the Western Pacific and Western Atlantic there is a third weaker jet centered over North Africa and the Middle East. Figure 10.13 shows dramatically

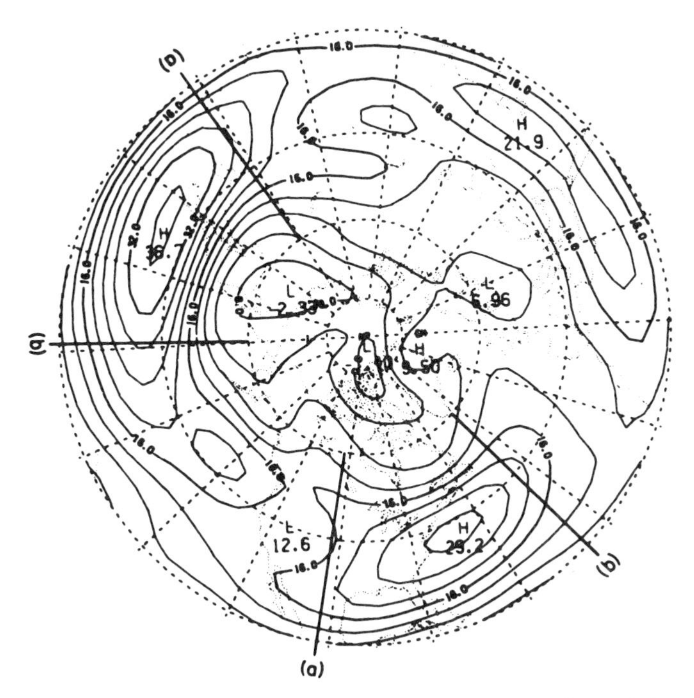

Fig. 10.13 Nine-winter average 500-mb zonal wind speed (contour interval 4 m s^{-1}). Lines designated by letters (a) and (b) refer to locations of cross sections in Fig. 10.14. (After Blackmon *et al.*, 1977. Reproduced with permission of the American Meteorological Society.)

the large deviations from zonal symmetry in the jet stream structure. In midlatitudes the zonal wind speed varies by nearly a factor of three between the core of the Asian jet and the low wind speed area in Western North America. A satisfactory theory of the general circulation must be able to account for the existence of the zonally asymmetric jet structure as well as the zonally averaged winds.

The existence of the Asian and North American jets can be understood in a general sense as simply reflecting a thermal wind balance consistent with the very strong meridional temperature gradients which occur in winter near the eastern edges of the Asian and North American continents due to the contrast between the warm water to the southeast and cold land to the Northwest. However, a satisfactory description of the jet streams must account for the westerly acceleration which air parcels experience as they enter the jet, and the deceleration as they leave the jet core. To understand the momentum budget in the jet streams we consider the zonal component of the momentum equation (2.24) which with the aid of the definition of the geostrophic wind can be written as

$$\frac{du}{dt} = f(v - v_g) \equiv f v_a \tag{10.57}$$

where v_a is the meridional component of the ageostrophic wind. From (10.57) we see that the westerly acceleration ($du/dt > 0$) which air parcels experience as they enter the jet can only be provided by a poleward ageostrophic wind component ($v_a > 0$), and conversely, the easterly acceleration which air parcels experience as they leave the jet requires an equatorward ageostrophic motion. This divergent secondary circulation is illustrated in Fig. 10.14. Note that this secondary circulation is thermally direct upstream of the jet core. A magnitude of $v_a \sim 2 - 3 \text{ m s}^{-1}$ is required to account for the observed zonal wind acceleration. This circulation is an order of magnitude stronger than the zonal mean indirect cell (Ferrel cell) which prevails in midlatitudes. Downstream of the jet core, on the other hand, the secondary circulation is thermally indirect, but much stronger than the zonally averaged Ferrel cell. This secondary circulation is just the type required to balance the strong poleward eddy heat flux (through adiabatic cooling in the region of heat flux convergence) and to balance the eddy momentum flux convergence in the upper troposphere (through the Coriolis force due to equatorward motion). Finally, it is interesting to note that the vertical motion pattern on the poleward (cyclonic shear) side of the jet is similar to that associated with deep transient baroclinic eddies in the sense that subsidence occurs to the west of the stationary trough associated with the jet, and ascent occurs east of the trough (see, for example, Fig. 6.11).

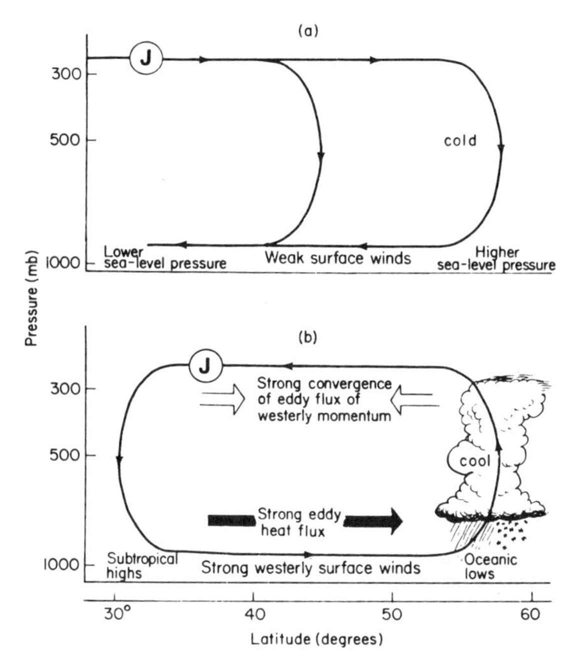

Fig. 10.14 Meridional cross sections showing relationship between the time mean secondary meridional circulation (continuous thin lines with arrows) and the jet stream (denoted by J) at locations (a) upstream and (b) downstream from the jet stream cores. (After Blackmon *et al.*, 1977. Reproduced with permission of the American Meteorological Society.)

10.7.2 Tropical Monsoons

The term "monsoon" is commonly used in a rather general sense to designate any seasonally reversing circulation system. Most tropical regions are influenced to some extent by monsoons. However, the most extensive monsoon circulation by far is the complex circulation associated with the tropical region of Asia. This monsoon completely dominates the climate of the Indian subcontinent, producing warm wet summers and cool dry winters. An idealized model of the structure of the summer monsoons is indicated in Fig. 10.15. The basic drive for this monsoon circulation is provided by the contrast in the thermal properties of the land and sea surfaces. Since land has a small heat capacity compared to ocean, the absorption of solar radiation raises the surface temperature over the land much more rapidly than over the ocean. This surface warming leads to enhanced cumulus convection, and hence to latent heat release which produces warm temperatures throughout the troposphere. Thus, as indicated in Fig. 10.15, the 100-20 kPa (1000–200-mb) thickness becomes larger over the land than over the ocean. As a

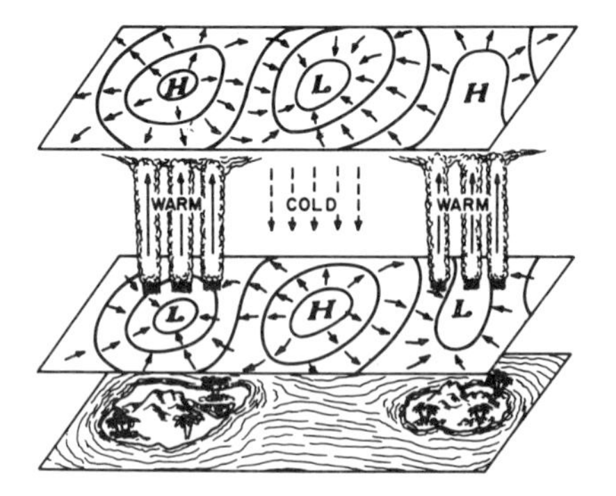

Fig. 10.15 Schematic representation of a summer monsoon circulation. Solid lines represent geopotential contours at 1000 mb (lower plane) and 200 mb (upper plane). Short solid arrows indicate cross-isobar flow. Vertical arrows indicate direction of vertical motion in the midtroposphere. (After Wallace and Hobbs, 1977.)

result there is a pressure gradient force at the upper levels directed from the land to the ocean. The divergent wind which develops in response to this pressure gradient (shown by the small arrows in Fig. 10.15) causes a net mass transport out of the air column above the continent and thereby generates a surface low over the continent (sometimes called a *thermal low*). A compensating convergent wind then develops at low levels. This low-level flow produces a convergence of moisture which serves to maintain the cumulus convection which is the primary energy source for the monsoon circulation.

The low-level convergence and upper-level divergence over the continent constitute a secondary circulation which generates cyclonic vorticity at the lower levels and anticyclonic vorticity at the upper levels. Thus, the vorticity adjusts towards geostrophic balance. From Fig. 10.15 it is clear that there is a positive correlation between the vertical motion and temperature fields. Therefore, the monsoon circulation converts eddy potential energy to eddy kinetic energy, just as midlatitude baroclinic eddies do.

Unlike the case of baroclinic eddies, however, the *primary* energy cycle of the monsoons does not involve the zonal mean potential or kinetic energy. Rather, eddy potential energy is generated directly by diabatic heating (latent and radiative heating); the eddy potential energy is converted to eddy kinetic energy by a thermally direct secondary circulation; and the eddy kinetic energy is frictionally dissipated. (A portion of the eddy kinetic energy may be converted to zonal kinetic energy.) In a dry atmosphere monsoon circulations would still exist; however, they would be much weaker than the observed monsoons. The presence of cumulus convection and its concomitant latent heat release greatly amplifies the eddy potential energy generation and makes the summer monsoons among the most important features of the global circulation.

In the winter season the thermal contrast between the land and sea reverses so that the circulation is just opposite to that shown in Fig. 10.15. As a result the continents are cool and dry and the precipitation is found over the relatively warm oceans. Further discussion of the circulation in the tropical regions will be deferred until Chapter 12.

Problems

1. Starting with Eqs. (10.18) and (10.20) derive the mean and eddy kinetic energy equations (10.21) and (10.22).

2. Starting with Eqs. (10.23) amd (10.24) derive the mean and eddy available potential energy equations (10.25) and (10.26).

3. Using the observed data given in Fig. 10.1, compute the time required for each possible energy transformation or loss to restore or deplete the observed energy stores. (A watt equals 1 J s^{-1}.)

4. Derive the expression (10.53) for the "thermal wind" in the dishpan experiments.

5. Compute the surface torque per unit horizontal area exerted on the atmosphere by topography for the following distribution of surface pressure and surface height:

$$p_s = p_0 + \hat{p}\sin kx, \qquad h = \hat{h}\sin(kx - \gamma)$$

where $p_0 = 10^2$ kPa, $\hat{p} = 1$ kPa, $\hat{h} = 2.5 \times 10^3$ m, $\gamma = \pi/6$ rad, and $k = 1/(a\cos\phi)$, where $\phi = \pi/4$ rad is the latitude, and a is the radius of the earth. Express the answer in kg s^{-2}.

6. Using (10.39) and (10.40) show that for a zonal mean flow which is constant in time the potential vorticity flux must vanish if $\bar{R} = d = 0$. Demonstrate that in this case the vorticity flux $\langle v'\zeta'\rangle$ is proportional to the vertical derivative of the horizontal heat flux B.

***7.** It is sometimes useful to divide the mean meridional flow into two parts by defining X^* as follows: $X \equiv X^* + f_0 B/\sigma$. Show that X^* must satisfy the equation

$$\left(\frac{\partial^2}{\partial y^2} + \frac{f_0^2}{\sigma}\frac{\partial^2}{\partial p^2}\right)X^* = -\frac{f_0}{\sigma}\left[\frac{\partial \bar{R}}{\partial y} - \frac{\partial}{\partial p}\langle v'q'\rangle\right]$$

and that X^* thus will vanish if $\langle v'q'\rangle = \bar{R} = 0$.

***8.** Derive the zonal mean form for the governing equations of the two-level model (8.22)–(8.24). Hence, show that in the absence of dissipation

$$\frac{\partial \bar{u}_1}{\partial t} = \langle v_1'\zeta_1'\rangle + f_0\bar{v}_1, \qquad \frac{\partial \bar{u}_3}{\partial t} = \langle v_3'\zeta_3'\rangle - f_0\bar{v}_1$$

Show that the zonal mean omega equation for the two-level model can be written as

$$\left(\frac{\partial^2}{\partial y^2} - 2\lambda^2\right)\bar{\omega}_2 = \frac{f_0}{\sigma\,\Delta p}\left\{\frac{\partial^3}{\partial y^3}\langle v'_2(\psi'_1 - \psi'_3)\rangle - \frac{\partial}{\partial y}(\langle v'_1\zeta'_1\rangle - \langle v'_3\zeta'_3\rangle)\right\}$$

9. Consider a thermally stratified liquid contained in a rotating annulus of inner radius 0.8 m, outer radius 1.0 m, and depth 0.1 m. The temperature at the bottom boundary is held constant at T_0. The fluid is assumed to satisfy the equation of state (10.50) with $\rho_0 = 10^3\ \mathrm{kg\ m^{-3}}$ and $\varepsilon = 2 \times 10^{-4}\ \mathrm{K^{-1}}$. If the temperature increases linearly with height along the outer radial boundary at a rate of 1°C/cm and is constant with height along the inner radial boundary, determine the geostrophic velocity at the upper boundary for a rotation rate of $\Omega = 1\ \mathrm{rad\ s^{-1}}$. (Assume that the temperature depends linearly on radius at each level.)

Suggested References

Lorenz, *The Nature and Theory of the General Circulation of the Atmosphere*, is an excellent introduction to the subject which has an extensive discussion of the observational aspects of the general circulation, a historical review of former theories of the general circulation, and a discussion of the current status of general circulation theory. The book also contains a useful bibliography of original papers on the subject.

Newell, *et. al.*, *The General Circulation of the Tropical Atmosphere and Interactions with Extra-tropical Latitudes*, gives a more complete and up-to-date account of the general circulation in the tropical areas than is contained in Lorenz's book.

Hide (1966) reviews a number of types of laboratory experiments with rotating fluids including the rotating annulus experiments. His paper also contains an extensive bibliography.

Methods in Computational Physics, Vol. 17: General Circulation Models of the Atmosphere, contains a number of excellent review articles covering all aspects of general circulation modeling.

Chapter

11 Stratospheric Dynamics

Up to this point our discussion has focused almost exclusively on the dynamics of large-scale motions in the troposphere. The troposphere accounts for almost 85% of the total mass of the atmosphere, and virtually all of the atmospheric water. There can be little doubt that processes originating in the troposphere are primarily responsible for the production of the bulk of weather disturbances. Nevertheless, the meteorology of the stratosphere is important for at least two reasons:

(i) Large-scale motions play an essential role in determining the distribution and variability of ozone. Recent concern about the effects of manmade pollutants on the ozone layer has stimulated much research on the combined radiative, dynamical, and photochemical processes in the ozone layer.

(ii) There is growing evidence that dynamical interactions between the stratosphere and the troposphere involving planetary scale waves may be significant factors in modifying the tropospheric circulation on time scales of days to months. Thus, it is important that the stratosphere be considered in any reasonably complete treatment of dynamic meteorology.

11.1 The Observed Mean Structure and Circulation of the Stratosphere

Synoptic scale tropospheric weather disturbances generally decay rapidly with height above the tropopause. Above the lowest several kilometers of the stratosphere the flow is completely dominated by the zonal mean circulation and planetary waves. The overall structure of the zonal mean temperature and zonal wind profiles at the solstices for the atmosphere below ~100 km is shown in Fig. 11.1.

From Fig. 11.1a it is clear that the mean vertical temperature profile provides a convenient standard to divide this region of the atmosphere into three layers: the troposphere, the stratosphere, and the mesosphere. In the troposphere and the mesosphere the temperature generally decreases with height while in the stratosphere the temperature increases with height except in the winter at high latitudes. The thermal structure of the troposphere is maintained by a balance between infrared radiative cooling, vertical transport of sensible and latent heat by small-scale convection, and large-scale heat transport by synoptic scale eddies. The net result is a mean temperature profile with a lapse rate of about 6°C km^{-1}. In the stratosphere, on the other hand, infrared radiative cooling is in the mean balanced primarily by radiative heating due to the absorption of solar ultraviolet radiation in the ozone layer. As a result of the solar heating in the ozone layer, the temperature in the stratosphere increases with height in the mean to a maximum at the *stratopause* near 50 km. Above the stratopause in the mesosphere the ozone heating rate decreases with height and so does the equilibrium temperature.

The radiative heating in the ozone layer also accounts approximately for the latitudinal temperature distribution in the stratosphere. As shown in Fig. 11.1a the temperature is a maximum in the summer polar region and a minimum in the winter polar region, as would be expected from the latitudinal dependence of the solar radiative flux at the solstices. This overall radiative heat balance is, however, modified above 60 km, where temperatures increase from the summer pole to the winter pole. Other notable features of the mean temperature profile are the extremely cold tropical tropopause and the midlatitude temperature maximum in the lower stratosphere in the winter hemisphere.

The mean zonal wind profile shown in Fig. 11.1b indicates an approximate thermal wind balance with the temperature profile of Fig. 11.1a. The main features, consistent with the temperature gradient from winter pole to summer pole, are an easterly jet in the summer hemisphere and a westerly jet in the winter hemisphere, with maximum wind speeds occurring near the 60-km level. Of particular interest is the high latitude westerly jet in the lower stratosphere in the winter season. This so-called *polar night jet* is intimately involved in the sudden stratospheric warmings which will be discussed in Section 11.4.

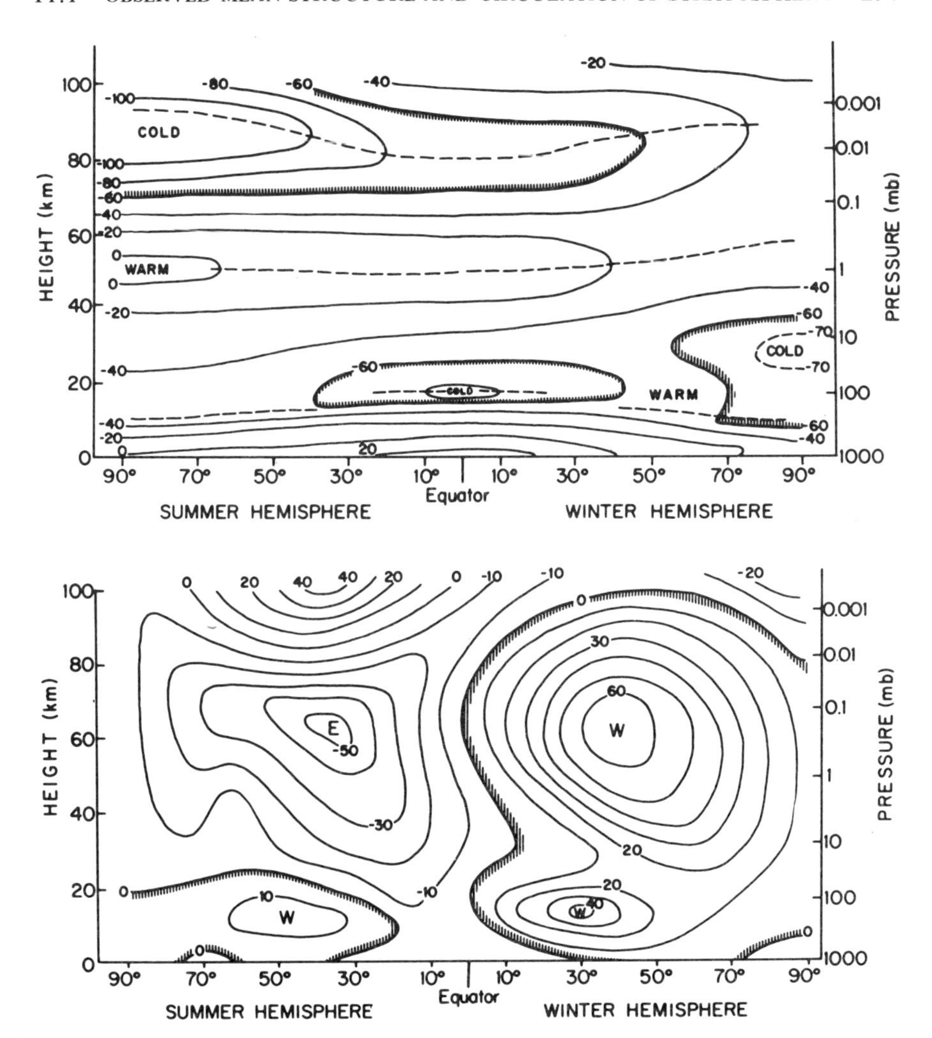

Fig. 11.1 Mean meridional cross sections of (a) temperature in degrees Celsius, and (b) zonal wind in meters per second at the time of the solstices. Dashed lines indicate levels of the tropopause, stratopause, and mesopause. (Courtesy of R. J. Reed.)

The eddy motions in the stratosphere consist primarily of ultralong quasi-stationary planetary waves which are confined to the winter hemisphere. These waves are not generated by in situ instabilities. Rather, they appear to be produced by vertical propagation of planetary waves forced in the troposphere by orography and land–sea heating contrasts. As will be shown in Section 11.3, stationary waves can only propagate vertically when the stratospheric winds are westerly. Thus, as shown in Fig. 11.2, in the Northern

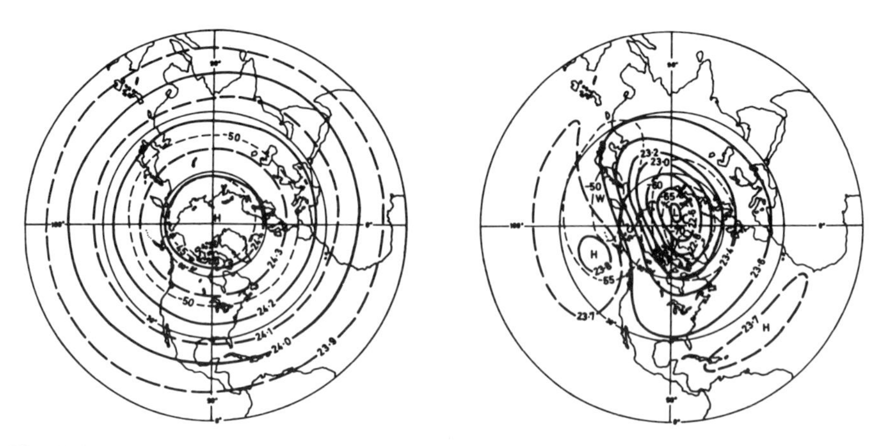

Fig. 11.2 Mean 30-mb geopotential height (solid lines, km) and temperature (dashed lines, °C) for Northern Hemisphere summer (left) and winter (right) averaged for 1958–1961. (After Hare, 1968.)

Hemisphere the seasonal mean zonal vortex is almost completely undisturbed during the summer, but is highly distorted during the winter due to a stationary wave disturbance centered over the Gulf of Alaska.

In addition to these stationary waves, the winter circulation features irregularly varying transient planetary waves which may occasionally interact to produce the so-called sudden stratospheric warmings. Finally, in the equatorial stratosphere there exists a special class of waves whose interaction with the zonal mean flow produces the "quasi-biennial oscillation" discussed in Section 11.6.

11.2 The Energetics of the Lower Stratosphere

In Section 10.2 we discussed the energy cycle of the troposphere. In the troposphere on the average the eddy motion is maintained primarily by baroclinic conversion of energy from zonal mean available potential energy to eddy available potential energy which is in turn partly converted to eddy kinetic energy. The zonal mean available potential energy is maintained in the presence of these eddy conversions through differential radiative heating, The annual mean energy cycle of the lower stratosphere, schematically illustrated in Fig. 11.3, is dramatically different from the tropospheric cycle. In the 30–10-kPa (30–100-mb) layer, cooling to space by infrared emission dominates over solar absorption by ozone. Because the rate of infrared emission increases with the temperature of the emitter, cooling to space tends to reduce the horizontal temperature gradients, and thus acts as a sink for both the zonal mean and eddy available potential energies. The energy of the

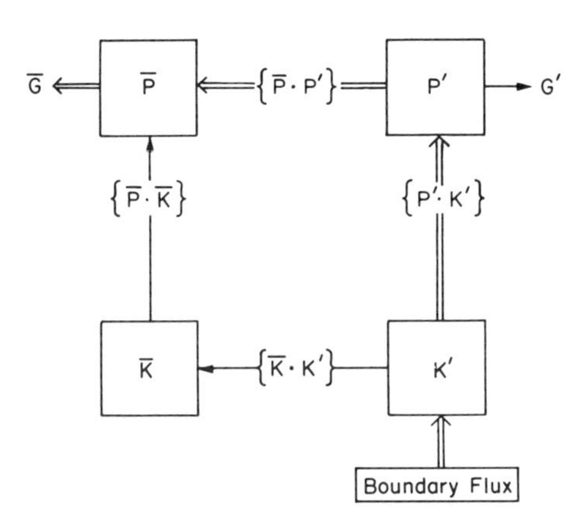

Fig. 11.3 Schematic illustration of the annual mean energy cycle in the lower stratosphere (30–100 mb).

region is maintained against this dissipation by the vertical flux of energy from the troposphere. Theory suggests that only planetary waves of very long wavelength can propagate significant amounts of energy vertically, and this is indeed consistent with the observed predominance of long waves in the stratosphere.

Figure 11.3 shows that in the lower stratosphere the energy cycle is reversed from that of the troposphere. In the annual mean the midlatitude stratosphere is warmer than the equatorial stratosphere. However, the eddies in the stratosphere transport heat poleward just as do the eddies in the troposphere. Thus, the eddy heat transport in the stratosphere is *up the temperature gradient*. The lower stratosphere then (in the mean) is thermodynamically equivalent to a refrigerator. The eddies remove heat from the cold equatorial region and deposit it in the warm midlatitude region, thereby maintaining the reversed mean temperature gradient against radiative damping. This "refrigerator" is driven by the tropospheric heat engine through the vertical propagation of kinetic energy. As shown in Fig. 11.3, the primary energy transformations are from eddy kinetic energy to eddy potential energy due to the sinking of warm air relative to cold air in the eddies and the simultaneous transformation of eddy available potential energy to zonal mean available potential energy by the poleward advection of warm air and equatorward advection of cold air in the eddies. Associated with this poleward eddy heat flux is an indirect mean meridional circulation consisting of rising motion in the cold equatorial and polar regions, and subsidence at midlatitudes. As a result of this mean meridional circulation, there is a small conversion from zonal mean kinetic energy to zonal mean available potential energy.

In summary, from an energetic standpoint the lower stratosphere in the annual mean acts as a rather passive layer which is primarily driven by leakage of energy from the troposphere.

11.3 Vertically Propagating Planetary Waves

In Section 11.1 we pointed out that the predominant eddy motions in the stratosphere are vertically propagating quasi-stationary planetary waves, and that these waves are confined to the winter hemisphere. In order to understand the absence of synoptic scale motions, and the confinement of planetary waves to the winter hemisphere, it is necessary to examine the conditions under which planetary waves can propagate vertically.

To analyze planetary wave propagation in the stratosphere it is convenient to write the equations of motion in a coordinate system in which the vertical coordinate is proportional to the logarithm of pressure. In this system, like the isobaric system, density does not explicitly appear. However, in the *log-pressure* system the vertical coordinate is approximately equal to the geometric height and the static stability parameter is nearly constant with height. The log-pressure system thus combines some of the best features of both the isobaric and height coordinates.

In the log-pressure system, the vertical coordinate is defined as $z^* \equiv -H \ln(p/p_0)$ where p_0 is a standard reference pressure (usually taken to be 100 kPa) and H is a standard scale height, $H \equiv RT_0/g$, with T_0 a global average temperature. For an isothermal atmosphere at temperature T_0, z^* is exactly equal to height. For an atmosphere with variable temperature, z^* will be only approximately equivalent to the actual height. The vertical velocity in this coordinate system is defined as

$$w^* \equiv \frac{dz^*}{dt} \tag{11.1}$$

The horizontal momentum equation in the log-pressure system is the same as that in the isobaric system:

$$\frac{d\mathbf{V}}{dt} + f\mathbf{K} \times \mathbf{V} = -\nabla\Phi \tag{11.2}$$

However, the operator d/dt is now defined as

$$\frac{d}{dt} \equiv \frac{\partial}{\partial t} + \mathbf{V} \cdot \nabla + w^* \frac{\partial}{\partial z^*}$$

The hydrostatic equation

$$\frac{\partial \Phi}{\partial p} = -\alpha$$

may be transformed to the log-pressure system by eliminating α with the ideal gas law to get

$$\frac{\partial \Phi}{\partial \ln p} = -RT$$

or upon dividing through by $-H$

$$\frac{\partial \Phi}{\partial z^*} = \frac{RT}{H} \tag{11.3}$$

The continuity equation may also be conveniently obtained by transforming from the isobaric coordinate form

$$\frac{\partial u}{\partial x} + \frac{\partial v}{\partial y} + \frac{\partial \omega}{\partial p} = 0$$

From (11.1) we have

$$w^* = -\frac{H}{p}\frac{dp}{dt} = -\frac{H\omega}{p}$$

from which we see that

$$\frac{\partial \omega}{\partial p} = -\frac{\partial}{\partial p}\left(\frac{pw^*}{H}\right) = \frac{\partial w^*}{\partial z^*} - \frac{w^*}{H}$$

Thus, in log-pressure coordinates the continuity equation becomes simply

$$\frac{\partial u}{\partial x} + \frac{\partial v}{\partial y} + \frac{\partial w^*}{\partial z^*} - \frac{w^*}{H} = 0 \tag{11.4}$$

It is left as a problem for the student to show that the first law of thermodynamics (2.42) can be written in the following form with the aid of (11.3):

$$\left(\frac{\partial}{\partial t} + \mathbf{V}\cdot\nabla\right)\frac{\partial \Phi}{\partial z^*} + w^* N^2 = \frac{\kappa \dot{q}}{H} \tag{11.5}$$

where

$$N^2 \equiv \frac{R}{H}\left(\frac{\partial T}{\partial z^*} + \frac{\kappa T}{H}\right)$$

and $\kappa \equiv R/c_p$. In the stratosphere the buoyancy frequency squared, N^2, is approximately constant with a value $N^2 \simeq 4 \times 10^{-4}\ \mathrm{s}^{-2}$.

It may be shown, using the methods developed in Section 6.2, that for large-scale adiabatic motions on the midlatitude β plane (11.2)–(11.5) can be combined to yield the quasi-geostrophic potential vorticity equation

$$\left(\frac{\partial}{\partial t} + \mathbf{V}_g \cdot \nabla\right) q = 0 \tag{11.6}$$

where

$$q \equiv \nabla^2 \psi + f + e^{z/H} \frac{f_0^2}{N^2} \frac{\partial}{\partial z}\left(e^{-z/H} \frac{\partial \psi}{\partial z}\right)$$

and N^2 has been assumed to be constant. [Compare with Eq. (6.18)]. Here $\psi = \Phi/f_0$ is the geostrophic streamfunction and f_0 is a constant midlatitude reference value of the Coriolis parameter. We now assume that the motion consists of a small amplitude disturbance superposed on a constant zonal mean flow. Thus, letting $\psi = -\bar{u}y + \psi'$ and linearizing (11.6) we find that the perturbation field must satisfy

$$\left(\frac{\partial}{\partial t} + \bar{u} \frac{\partial}{\partial x}\right) q' + \beta \frac{\partial \psi'}{\partial x} = 0 \tag{11.7}$$

where

$$q' \equiv \nabla^2 \psi' + e^{z/H} \frac{f_0^2}{N^2} \frac{\partial}{\partial z}\left(e^{-z/H} \frac{\partial \psi'}{\partial z}\right)$$

We now assume that solutions of (11.7) exist in the form of a single zonal harmonic wave with zonal and meridional wave numbers k and l, respectively, and zonal phase speed c. Thus, ψ' may be expressed in the form

$$\psi'(x, y, z, t) = \Psi(z) e^{i(kx + ly - kct) + z/2H} \tag{11.8}$$

where the factor $e^{z/2H}$ is included to simplify the equation for the vertical dependence. Substituting (11.8) into (11.7) we find that $\Psi(z)$ must satisfy

$$\frac{d^2 \Psi}{dz^2} + m^2 \Psi = 0 \tag{11.9}$$

where

$$m^2 \equiv \frac{N^2}{f_0^2}\left[\frac{\beta}{\bar{u} - c} - (k^2 + l^2)\right] - \frac{1}{4H^2} \tag{11.10}$$

Referring back to Section 7.4 we recall that $m^2 > 0$ is required for vertical propagation, and in that case m is the vertical wavenumber [i.e., solutions of (11.9) have the form $\Psi(z) = Ae^{imz}$ where A is a constant amplitude.]

For stationary waves ($c = 0$) we thus find from (11.10) that vertically propagating modes can exist only for mean zonal flows which satisfy the condition

$$0 < \bar{u} < \beta[(k^2 + l^2) + f_0^2/(4N^2H^2)]^{-1} \equiv U_c \qquad (11.11)$$

where U_c is called the *Rossby critical velocity*. Thus, vertical propagation of stationary waves can occur only in the presence of westerly winds weaker than a critical value which depends on the horizontal scale of the waves. In the summer hemisphere the stratospheric mean zonal winds are easterly so that stationary planetary waves are all trapped. In the winter hemisphere the mean zonal winds are strong and westerly so that according to (11.11) all but the largest scale waves should be trapped.

In the real atmosphere the mean zonal wind is not constant, but depends on latitude and height. However, both observational and theoretical studies suggest that (11.11) still provides a qualitative guide for estimating vertical propagation of planetary waves even though the actual critical velocity may be larger than indicated by the β-plane theory.

11.4 Sudden Stratospheric Warmings

Although the latitudinal temperature gradient in the annual mean is positive from equator to pole in the stratosphere the situation for a winter seasonal mean is quite different. In the normal winter situation the temperature in the lower stratosphere reaches a maximum at about 45° latitude and decreases poleward so that the polar and equatorial stratospheric temperatures are almost equally cold. From thermal wind considerations this cold polar stratosphere requires a zonal vortex with strong westerly shear with height.

Every few years this normal winter pattern of a cold polar stratosphere with a westerly vortex is interrupted in a spectacular manner in midwinter. Within the space of a few days the polar vortex becomes highly distorted and breaks down with an accompanying large-scale warming of the polar stratosphere which can quickly reverse the meridional temperature gradient and create a circumpolar easterly current. In some cases warmings of as much as 40°C in a few days have occurred at the 50-mb level as shown in Fig. 11.4. Numerous studies of the energetics of the sudden warming confirm that enchanced propagation of energy from the troposphere by planetary scale waves, primarily zonal wave numbers 1 and 2, is essential for the development of the warming. Since the sudden warming is observed only in the Northern Hemisphere, it is logical to conclude that topographically forced waves are responsible for the vertical energy propagation. The Southern Hemisphere with its relatively small land masses at middle latitudes has much smaller

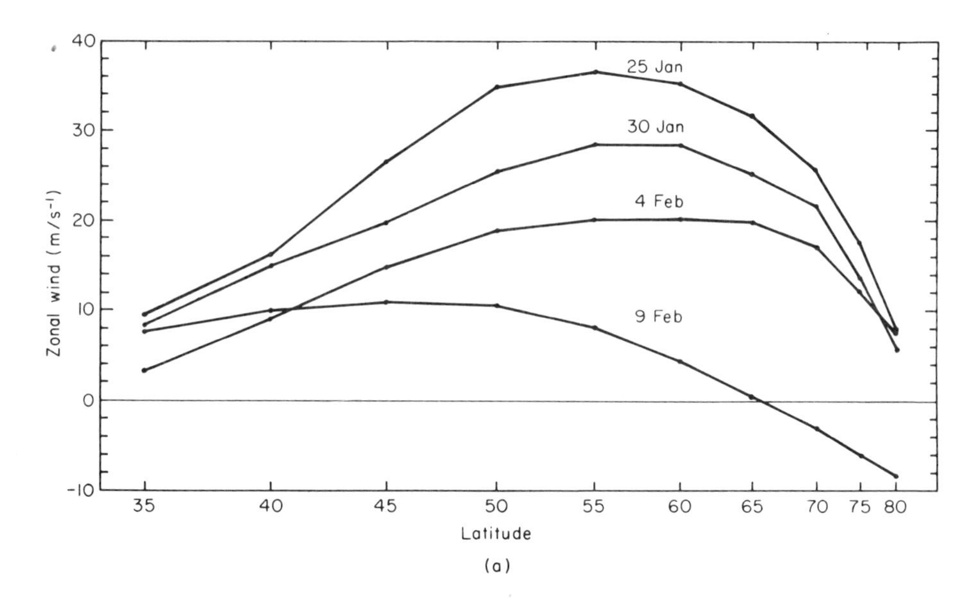

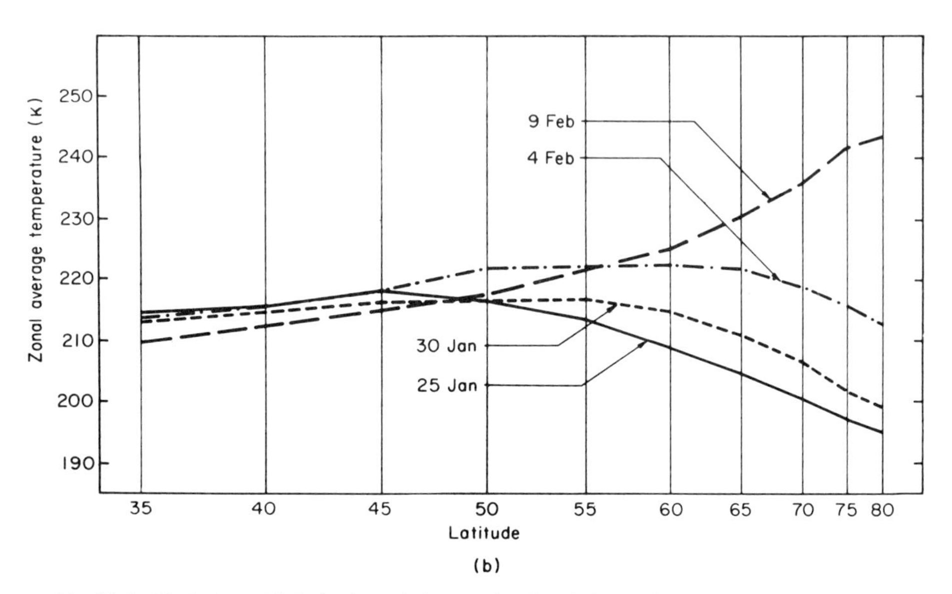

Fig. 11.4 Variation with latitude and time at the 50-mb level of (a) the zonal wind, and (b) the zonal mean temperature during the sudden warming of 1957. (After Reed *et al.*, (1963. Reproduced with permission of the American Meteorological Society.)

amplitude stationary planetary waves. However, even in the Northern Hemisphere, it is apparently only in certain winters that conditions are right to produce the sudden warming.

In Section 10.4 we showed that in order for eddy motions to change the zonal mean circulation there must be a net potential vorticity transport by the eddies. We further showed that for steady nondissipative waves, in the absence of critical lines where $\bar{u} = c$, the potential vorticity transport vanishes. For normal stationary planetary waves in the stratospheric polar night jet this constraint should be at least approximately satisfied, since radiative and frictional dissipation are rather small. Thus, the strong wave–mean flow interaction which occurs during the course of a sudden warming must be due to wave transience and critical line effects. The mean flow deceleration and polar warming which are forced by upward propagating growing waves and by absorption of waves at a critical level can be understood physically by considering an idealized situation in which the waves have vanishing momentum flux. According to (10.42), in the absence of transience, dissipation, or critical lines such waves should have a horizontal heat flux B independent of height. Now the normal stationary planetary waves in the polar night jet have a poleward heat flux ($B > 0$). As shown in the previous section, stationary waves can not propagate vertically in a mean easterly current ($\bar{u} < 0$). Thus stationary waves with $B > 0$ propagating through a region with $\bar{u} > 0$ must have a sudden decrease of B to zero if they encounter a critical level $\bar{u} = 0$. Since as shown in (10.42) the eddy potential vorticity flux depends on the vertical gradient of B, there will be a large poleward potential vorticity flux at the critical level. The resulting easterly acceleration of the mean zonal flow can be physically understood in terms of the mean meridional circulation associated with the eddy heat flux. Because the convergence of the eddy heat flux is to a large extent balanced by adiabatic cooling due to the mean vertical motion (see the discussion above Fig. 10.7), the sharp decrease in the heat flux will induce a large mean meridional flow in the vicinity of the critical level as indicated schematically in Fig. 11.5. The Coriolis torque due to the equatorward mean meridional flow

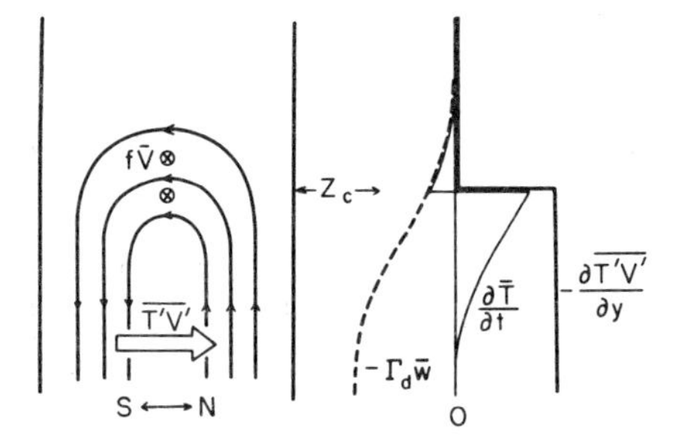

Fig. 11.5 Schematic illustration of the vertical profiles of the northward eddy heat flux, local temperature change, adiabatic heating, and induced meridional circulation for a planetary wave incident on a critical level. (After Matsuno, 1971. Reproduced with permission of the American Meteorological Society.

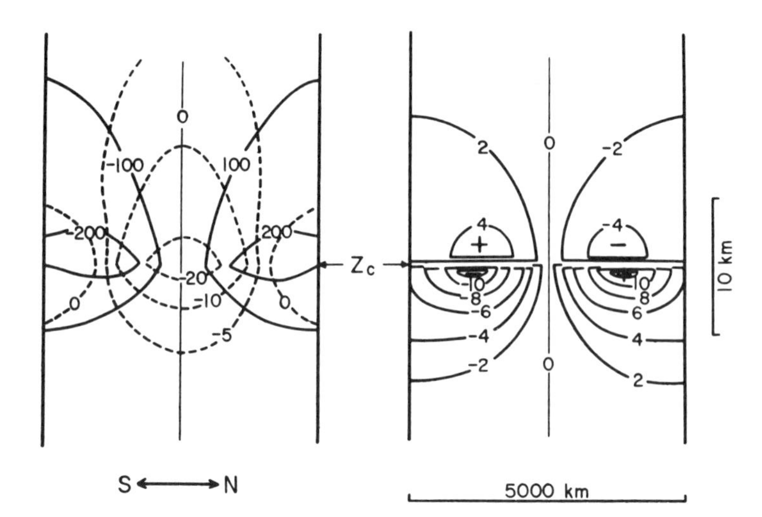

Fig. 11.6 Schematic illustration of zonal mean field changes induced by the meridional circulation of Fig. 11.5. The left side shows changes in geopotential heights (solid lines, m d^{-1}) and change in the mean zonal wind (dashed lines, m s^{-1} d^{-1}). The right side shows temperature changes (°C d^{-1}). The amplitude of the wave incident on the critical level is assumed to be 500 m. (After Matsuno, 1971. Reproduced with permission of the American Meteorological Society.)

will induce a strong easterly acceleration as shown in Fig. 11.6. The thermal wind balance then requires polar warming and equatorial cooling as is also shown in Fig. 11.6. The easterly acceleration will tend to move the $\bar{u} = 0$ line downward so that the region of strong easterly acceleration also moves downward and mean easterlies will gradually fill the polar stratosphere.

In the case of transient growing waves a similar process occurs even in the absence of a critical level. Suppose that the wave amplitude at the tropopause is increased at a time $t = 0$. The wave energy will propagate upward at the vertical group velocity W_g. Thus at a time $t = t_0$ the increased amplitude and its associated increased heat flux will have reached a level $z \simeq W_g t_0$. Above that level the heat flux must decrease with height ($\partial B/\partial z < 0$) so that there must be an easterly zonal flow acceleration and polar warming due to the same processes which occur at a critical level. In the case of transient waves, however, the acceleration and warming are spread out over a deep layer rather than being concentrated at the critical line.

The sudden warmings may thus be understood as consisting of the following sequence of events: (1) planetary zonal wave numbers 1 or 2 become anomalously large in the troposphere; (2) the growing waves propagate into the stratosphere and decelerate the mean zonal winds; (3) the polar night jet weakens and becomes distorted by the growing planetary waves; (4) if the waves are sufficiently strong the mean zonal flow may decelerate sufficiently so that a critical level is formed; (5) further upward transfer of wave energy is then blocked and a very rapid easterly acceleration and polar warming occurs as the critical level moves downward.

In some cases the wave amplitude may be insufficient to lead to formation of a critical line. Such "minor warmings" occur every winter and are generally followed by a quick return to the normal winter circulation. A "major warming," in which the mean zonal flow reverses at least as low as the 30-mb level, seems to occur only about once every couple of years. If the major warming occurs sufficiently late in the winter the westerly polar vortex may not be restored at all before the normal seasonal circulation reversal.

11.5 Waves in the Equatorial Stratosphere

In Section 7.4 we discussed the nature of small-scale internal gravity waves. Planetary scale internal gravity waves differ somewhat from these simple pure gravity modes because they are influenced by the earth's rotation as well as the gravitational stability. Such waves are often referred to as *inertia–gravity* oscillations.

It can be shown that planetary scale gravity waves generally will be "trapped" (that is, cannot propagate energy vertically) unless the frequency of the wave is greater than the Coriolis frequency. Thus at middle latitudes waves with periods in the range of several days are generally not able to propagate significantly into the stratosphere. However, as the equator is approached, the decreasing Coriolis frequency allows these longer period waves to become untrapped and propagate vertically.

The theory predicts that these long-period vertically propagating waves generally must have very short vertical wavelengths. Therefore, even if they existed such waves could not be resolved by conventional radiosonde data. However, there are two important types of vertically propagating waves whose wavelengths are long enough to permit easy detection. These are the eastward traveling atmospheric *Kelvin wave*, and the westward traveling mixed *Rossby–gravity wave*. The horizontal distributions of pressure and velocity characteristic of these waves is shown in Fig. 11.7. The Kelvin wave has a distribution of pressure and zonal velocity which is symmetric about the equator, and has essentially *no meridional velocity* component. The mixed Rossby–gravity wave, on the other hand, has a distribution of pressure and zonal velocity which is antisymmetric about the equator, and has a distribution of meridional velocity which is symmetric.

The dynamics of these waves can easily be deduced theoretically by applying the linear perturbation technique introduced in Chapter 7. It is

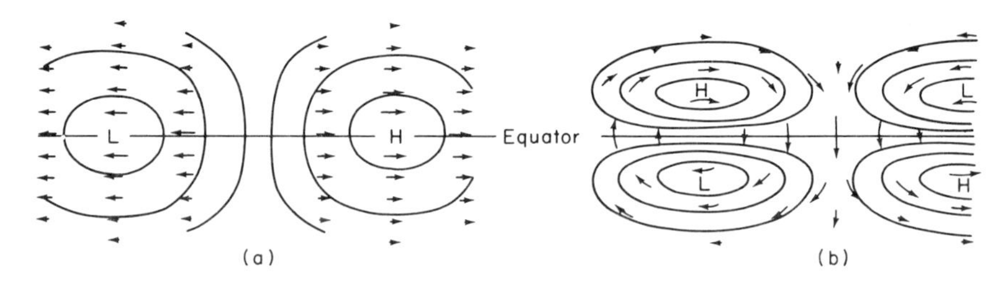

Fig. 11.7 Velocity and pressure distributions in the horizontal plane for (a) Kelvin waves, and (b) mixed Rossby–gravity waves. (After Matsuno, 1966.)

convenient to use the governing equations in log-pressure coordinates (11.2), (11.4), and (11.5) referred to an *equatorial β plane* in which the Coriolis parameter is approximated as

$$f = \frac{2\Omega y}{a} \equiv \beta y$$

where y is the distance from the equator. Thus, β is the rate at which f changes with latitude at the equator. We assume as a basic state an atmosphere at rest with no diabatic heating.[1] If the perturbations are assumed to be zonally propagating waves, we can write

$$\begin{aligned} u &= u'(y, z^*)e^{i(kx - \nu t)} \\ v &= v'(y, z^*)e^{i(kx - \nu t)} \\ w^* &= w^{*\prime}(y, z^*)e^{i(kx - \nu t)} \\ \Phi &= \Phi'(y, z^*)e^{i(kx - \nu t)} \end{aligned} \tag{11.12}$$

From (11.2), (11.4), and (11.5) we obtain the following set of linearized perturbation equations:

$$-i\nu u' - \beta y v' = -ik\Phi' \tag{11.13}$$

$$-i\nu v' + \beta y u' = -\frac{\partial \Phi'}{\partial y} \tag{11.14}$$

$$iku' + \frac{\partial v'}{\partial y} + \left(\frac{\partial}{\partial z^*} - \frac{1}{H}\right)w^{*\prime} = 0 \tag{11.15}$$

$$-i\nu \frac{\partial \Phi'}{\partial z^*} + N^2 w^{*\prime} = 0 \tag{11.16}$$

[1] Adding a constant zonal mean flow merely doppler shifts the frequency of the waves.

For the Kelvin wave case, this set can be considerably simplified. Setting $v' \equiv 0$ and eliminating $w^{*\prime}$ between (11.15) and (11.16) we obtain

$$-i\nu u' = -ik\Phi' \tag{11.17}$$

$$\beta y u' = -\frac{\partial \Phi'}{\partial y} \tag{11.18}$$

$$\left(\frac{\partial}{\partial z^*} - \frac{1}{H}\right)\frac{\partial \Phi'}{\partial z^*} + \frac{k}{\nu} N^2 u' = 0 \tag{11.19}$$

Using (11.17) to eliminate Φ' in (11.18) and (11.19), we obtain two independent equations which the field of u' must satisfy:

$$\beta y u' = \frac{-\nu}{k}\frac{\partial u'}{\partial y} \tag{11.20}$$

and

$$\left(\frac{\partial}{\partial z^*} - \frac{1}{H}\right)\frac{\partial u'}{\partial z^*} + \frac{k^2}{\nu^2} N^2 u' = 0 \tag{11.21}$$

Equation (11.20) determines the meridional distribution of u' and (11.21) determines the vertical distribution.

It may be easily verified that (11.20) has the solution

$$u' = u_0(z^*)\exp\{-\beta y^2 k/2\nu\} \tag{11.22}$$

If we assume that $k > 0$, then $\nu > 0$ corresponds to an eastward-moving wave. In that case (11.22) indicates that the field of u' will have a Gaussian distribution about the equator with an e-folding width given by

$$Y_L = \left|\frac{2\nu}{\beta k}\right|^{1/2} \tag{11.23}$$

For a westward-moving wave ($\nu < 0$) the solution (11.22) increases in amplitude exponentially away from the equator. This solution cannot satisfy reasonable boundary conditions at the poles and must, therefore, be rejected. Therefore, there exists only an *eastward*-propagating atmospheric Kelvin wave.

The equation for the vertical dependence of u' (11.21) is (except for the factor $-1/H$ which appears due to the exponential decay of density with height) simply the vertical structure equation for an internal gravity wave discussed previously in Section 7.4. Solutions can be written in the form

$$u_0(z^*) = \exp(z^*/2H)[C_1\exp(i\lambda z^*) + C_2\exp(-i\lambda z^*)] \tag{11.24}$$

with

$$\lambda^2 \equiv \frac{N^2k^2}{\nu^2} - \frac{1}{4H^2}$$

Here the constants C_1 and C_2 are to be determined by appropriate boundary conditions. For $\lambda^2 > 0$ the solution (11.24) is in the form of a vertically propagating wave. For waves in the equatorial stratosphere which are forced by disturbances in the troposphere, the energy propagation (that is, group velocity) must have an upward component. Therefore, according to the arguments of Section 7.4 the phase velocity must have a downward component. Hence, the constant $C_1 = 0$ in (11.24), and the Kelvin wave has a structure in the x,z plane shown in Fig. 11.8a which is identical to that of the eastward-propagating pure internal gravity wave shown in Fig. 7.9. In summary, the atmospheric Kelvin wave is an eastward-moving wave in which the zonal velocity and meridional pressure fields are in exact geostrophic balance so that on the equatorial β plane the meridional velocity is identically zero and the wave can propagate as an ordinary two-dimensional internal gravity wave in the x,z plane.

A similar analysis is possible for the mixed Rossby–gravity mode. For this case, we must use the full perturbation equations (11.13)–(11.16). After considerable algebraic manipulation, it can be shown that the solution corresponding to the pattern shown in Fig. 11.7 is

$$\begin{Bmatrix} u' \\ v' \\ \Phi' \end{Bmatrix} = \Psi(z^*) \begin{Bmatrix} +i\beta y(1 + kv/\beta)/v \\ 1 \\ +ivy \end{Bmatrix} \exp\left[\frac{-(1 + kv/\beta)\beta^2 y^2}{2v^2}\right] \tag{11.25}$$

where the vertical structure $\Psi(z^*)$ of the three variables is given by

$$\Psi(z^*) = \exp(z^*/2H)[C_1 \exp(i\lambda_0 z^*) + C_2 \exp(-i\lambda_0 z^*)] \tag{11.26}$$

with

$$\lambda_0^2 \equiv N^2 \frac{k^2}{v^2}\left(1 + \frac{\beta}{vk}\right)^2 - \frac{1}{4H^2}$$

and the constants C_1 and C_2 again to be determined by the boundary conditions.

It is obvious from (11.25) that for the mixed Rossby–gravity mode v' has a Gaussian distribution about the equator. The e-folding width of the oscillation is in this case

$$Y_L = \left|\frac{2v}{\beta(\beta/v + k)}\right|^{1/2} \tag{11.27}$$

This solution is valid for westward-propagating waves ($v < 0$) provided that

$$1 + \frac{kv}{\beta} > 0$$

or, noting that $s = ka$ where s is the number of wavelengths around a latitude circle, this condition may be written as

$$|\nu| < \frac{2\Omega}{s} \tag{11.28}$$

For frequencies which do not satisfy (11.28), the wave amplitude will not decay away from the equator and it is not possible to satisfy boundary conditions at the pole.

For upward energy propagation the mixed Rossby–gravity waves must have downward phase propagation just like the Kelvin waves. Thus $C_2 = 0$ is required in (11.26). The resulting wave structure in the x,z plane at a latitude north of the equator is shown in Fig. 11.8b. Of particular interest is the fact

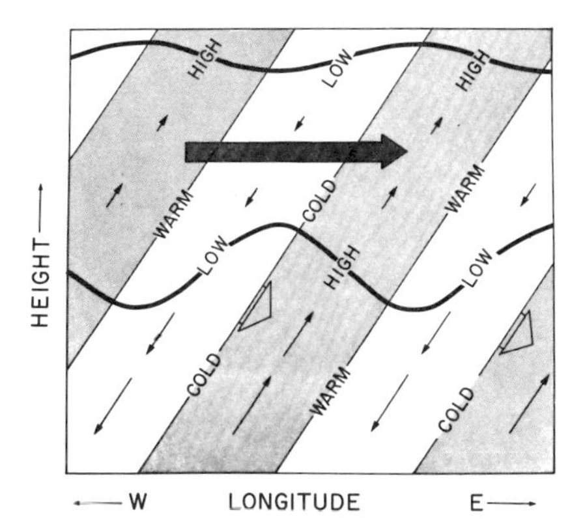

Fig. 11.8a Longitude–height section along the equator showing pressure, temperature, and wind perturbations for a thermally damped Kelvin wave. Heavy wavy lines indicate material lines, short blunt arrows show phase propagation. Areas of high pressure are shaded. Length of the small thin arrows is proportional to the wave amplitude which decreases with height due to damping. The large shaded arrow indicates the net mean flow acceleration due to the wave stress divergence.

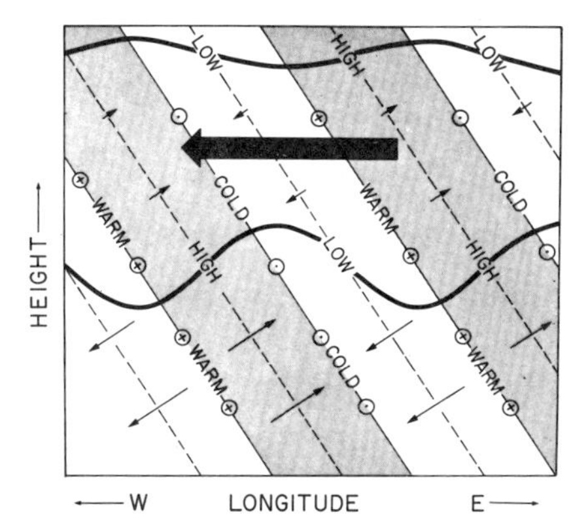

Fig. 11.8b Longitude–height section along a latitude circle north of the equator showing pressure, temperature, and wind perturbations for a thermally damped mixed Rossby–gravity wave. Areas of high pressure are shaded. Small arrows indicate zonal and vertical wind perturbations with length proportional to the wave amplitude. Meridional wind perturbations are shown by arrows pointed into the page (northward) and out of the page (southward). The large shaded arrow indicates the net mean flow acceleration due to the wave stress divergence.

that poleward-moving air is correlated with positive temperature perturbations so that the eddy heat flux $\langle v'T' \rangle$ is positive—the mixed Rossby–gravity waves remove heat from the equatorial region.

Both Kelvin wave and mixed Rossby–gravity wave modes have been identified in observational data from the equatorial stratosphere. The observed Kelvin waves have periods in the range of 12–20 d and appear to be primarily of zonal wave number $s = 1$ (that is, one wavelength spans 360° longitude). The corresponding observed phase speeds of these waves relative to the ground are in the range of $v/k \sim 30$ m s^{-1}. In applying our theoretical formulas for the meridional and vertical scales, however, we must use the doppler shifted phase speed. Assuming a mean zonal wind of -10 m s^{-1} so that the doppler shifted phase speed is $\hat{c} \sim 40$ m s^{-1} we find from (11.23) that $Y_L \simeq 2000$ km. This agrees well with observational indications that the Kelvin waves have significant amplitude only within about 20° of the equator. Knowledge of the observed phase speed also allows us to compute the theoretical vertical wavelength of the Kelvin wave. Assuming that $N^2 = 4 \times 10^{-4}$ s^{-2} we find from (11.24) that

$$\frac{2\pi}{\lambda} \simeq \frac{2\pi\hat{c}}{N} \simeq 12 \quad \text{km}$$

which agrees well with the vertical wavelengths deduced from observations.

An example of zonal wind oscillations caused by the passage of Kelvin waves at a station near the equator is shown in the form of a time–height section in Fig. 11.9. During the observational period shown in the figure the westerly phase of the quasi-biennial oscillation is descending so that at each level there is a general increase in the mean zonal wind with increasing time. However, superposed on this secular trend is a large fluctuating component with a period between speed maxima of about 12 days and a vertical wavelength (computed from the tilt of the oscillations with height) of about 10–12 km. Observations of the temperature field for the same period reveal that the temperature oscillation leads the zonal wind oscillation by one-quarter cycle (that is, maximum temperature occurs one-quarter period prior to maximum westerlies), which is just the phase relationship required for Kelvin waves (see Fig. 11.8a). Furthermore, additional observations from other stations indicate that these oscillations do propagate eastward at about 30 m s^{-1}. Therefore there can be little doubt that the observed oscillations are Kelvin waves.

The existence of the mixed Rossby–gravity mode has also been confirmed in observational data from the equatorial Pacific. This mode is most easily identified in the meridional wind component, since from (11.25) v' is a maximum at the equator for the mixed Rossby–gravity mode. The observed

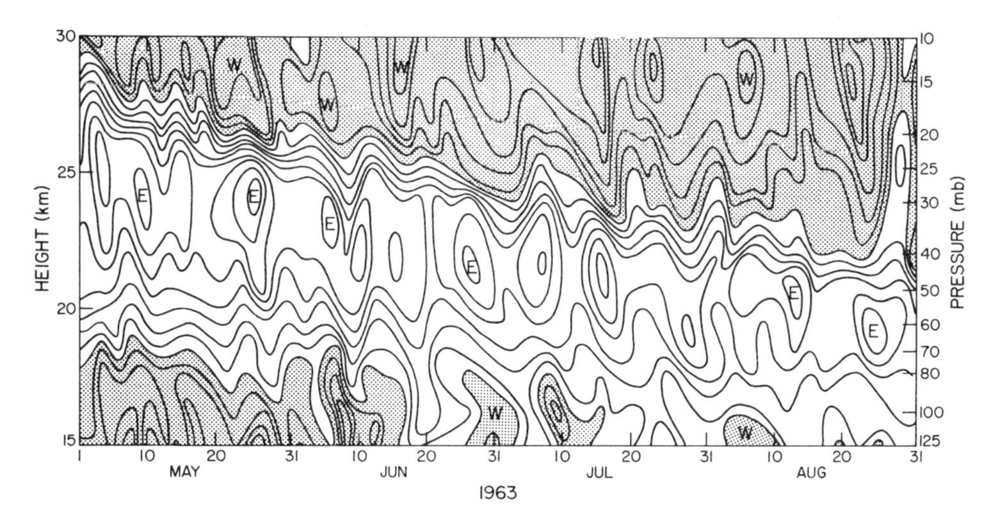

Fig. 11.9 Time–height section of zonal wind at Kwajalein (9°N latitude). Isotachs at intervals of 5 ms^{-1}. Westerlies are shaded. (After Wallace and Kousky, 1968. Reproduced with permission of the American Meteorological Society.)

waves of this mode have periods in the range of 4–5 d and propagate westward at about 20 $\mathrm{m\,s}^{-1}$. The horizontal wavelength appears to be about 10,000 km, corresponding to zonal wave number $s = 4$. The observed vertical wavelength is about 6 km, which agrees with the theoretically derived wavelength (11.26) within the uncertainties of the observations. These waves also appear to have significant amplitude only within about 20° of the equator, which is consistent with the theoretical *e*-folding width (11.27).

At present it appears that both the Kelvin waves and the mixed Rossby-gravity waves are excited by oscillations in the large-scale convective heating pattern in the equatorial troposphere. Although these waves do not contain much energy compared to typical tropospheric disturbances, they are the predominant disturbances of the equatorial stratosphere, and through their vertical energy and momentum transport play a crucial role in the general circulation of the stratosphere.

11.6 The Quasi-Biennial Oscillation

The search for periodicities in atmospheric motions has a long history. However, aside from the externally forced diurnal and annual components and their harmonics, no evidence of truly periodic oscillations has been found. Probably the phenomenon which comes closest to exhibiting periodic behavior is the quasi-biennial oscillation in the mean zonal winds of the equatorial stratosphere. This oscillation has the following observed features: Zonally symmetric easterly and westerly wind regimes alternate regularly with a period varying from about 24 to 30 months. Successive regimes first appear above 30 km, but propagate downward at a rate of 1 km month^{-1}. The downward propagation occurs without loss of amplitude between 30 and 23 km, but there is rapid attenuation below 23 km. The oscillation is symmetric about the equator with a maximum amplitude of about 20 m s^{-1}, and a half-width of about 12° latitude. This oscillation is best depicted by means of a time–height section of the zonal wind speed at an equatorial station as shown in Fig. 11.10. It is apparent from the figure that the vertical shear of the wind is quite strong at the level where one regime is replacing the other. Because this oscillation is zonally symmetric and symmetric about the equator, and has very small mean meridional and vertical motions associated with it, the zonal wind is in geostrophic balance nearly all the way to the equator. Thus, there must be a relatively strong meridional temperature gradient in the vertical shear zone to satisfy the thermal wind balance.

The main factors which a theoretical model of the quasi-biennial oscillation must explain are the approximate biennial periodicity, the downward propagation without loss of amplitude, and the occurrence of zonally symmetric westerlies at the equator. Since a zonal ring of air in westerly motion at the equator has an angular momentum greater than that of the earth, no symmetric advection process could account for the westerly phase of the oscillation. Therefore, there must be an eddy momentum source to produce the westerly accelerations in the downward-propagating westerly shear zone. Both observational and theoretical studies have confirmed that vertically propagating equatorial waves—the Kelvin and mixed Rossby–gravity modes—provide the zonal momentum sources necessary to drive the quasi-biennial oscillation. From Fig. 11.8a it is clear that Kelvin waves with upward energy propagation transfer westerly momentum upward (i.e., u' and w' are positively correlated so that $\langle u'w' \rangle > 0$). Thus, the Kelvin waves can provide the needed source of *westerly* momentum. Referring to Fig. 11.8b it is apparent that $\langle u'w' \rangle > 0$ also for the mixed Rossby–gravity mode. However, the mixed Rossby–gravity mode has a strong positive horizontal heat flux, $\langle v'T' \rangle > 0$. By the mechanism discussed in Section 10.4, this heat flux induces a mean meridional circulation which through the Coriolis acceleration $f\bar{v} < 0$ produces a net *easterly* acceleration.

It was pointed out in Section 11.4 that quasi-geostrophic wave modes do not produce any net mean flow acceleration unless the waves are transient, damped, or encounter a critical level. Similar considerations also apply to the equatorial Kelvin and mixed Rossby–gravity modes. Theoretical calculations indicate that equatorial stratospheric waves are damped primarily

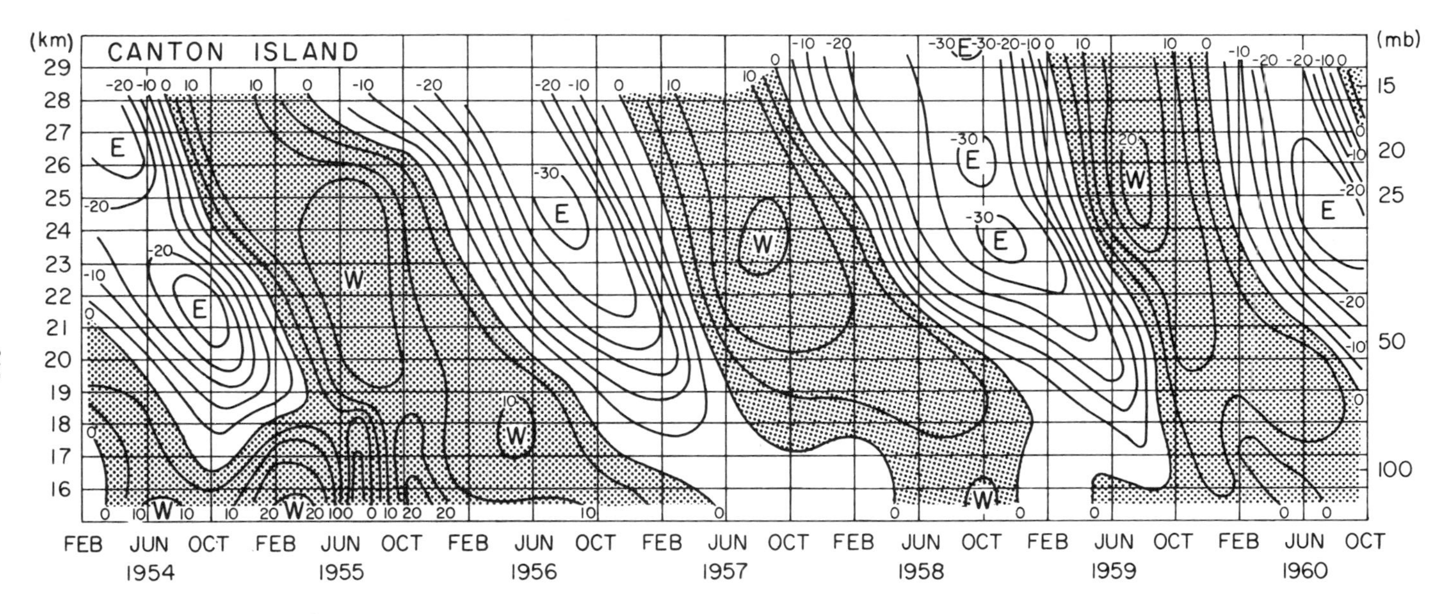

Fig. 11.10 Time–height section for Canton Island (2°46′S, 171°43′W), February 1954–October 1960. Isopleths are monthly mean zonal wind components in meters per second. Negative values denote easterly winds. (After Reed and Rogers, 1962. Reproduced with permission of the American Meteorological Society.)

by infrared radiation to space which tends to smooth out the temperature perturbations associated with the waves. Moreover, this damping is strongly dependent on the doppler shifted frequency of the waves. As the doppler shifted frequency decreases, the vertical component of group velocity also decreases, and there is a longer time available for the energy to be damped for a given vertical propagation distance. Thus, the westerly Kelvin waves tend to be damped rapidly in westerly shear zones below their critical levels thereby transferring westerly momentum to the mean flow and causing the westerly shear zone to descend. Similarly, the easterly mixed Rossby–gravity waves are damped in easterly shear zones, thereby causing an easterly acceleration and descent of the easterly shear zone. We conclude that the quasi-biennial oscillation is indeed excited primarily by vertically propagating equatorial wave modes through the mechanism of radiative damping which causes the waves to decay in amplitude with height and thus to transfer momentum to the mean zonal flow.

This process of wave–mean flow interaction can be elucidated by considering the heavy wavy lines in Figs. 11.8a and 11.8b. These lines indicate the vertical displacement of horizontal lines of fluid parcels (*material* lines) by the vertical velocity field associated with the waves. (Note that the maximum upward displacement occurs $\frac{1}{4}$ cycle after the maximum upward velocity.) In the Kelvin wave case (Fig. 11.8a) positive pressure perturbations coincide with negative material surface slopes. Thus, the fluid below a wavy material line exerts an eastward directed pressure force on the fluid above. Since the wave amplitude decreases with height this force is larger for the lower of the two material lines in the figure. There will thus be a net *westerly* acceleration of the block of fluid contained between the two wavy material lines shown in Fig. 11.8a. For the mixed Rossby–gravity wave, on the other hand, positive pressure perturbations coincide with positive slopes of the material lines. There is thus a westward directed force exerted by the fluid below the lines on the fluid above. In this case the result is a net *easterly* acceleration of the fluid contained between the two wavy material lines shown in the figure. Thus, by considering the stresses acting across a material surface corrugated by the waves it is possible to deduce the mean flow acceleration caused by the waves without reference to the momentum and/or heat fluxes due to the waves.

How such a mechanism can cause a mean zonal flow oscillation when equal amounts of easterly and westerly momentum are carried upward across the tropopause by the waves can be seen qualitatively by referring to Fig. 11.10. For example, in February 1957 upward-propagating mixed Rossby–gravity waves would have had very low doppler shifted frequencies in the 18–22-km region and hence could not propagate to higher levels. The Kelvin waves, on the other hand, could propagate relatively undisturbed into the region above 25 km where they would have been absorbed in the westerlies and produced a westerly acceleration and descent of the shear zone. A year later in February 1958, the opposite situation prevailed; Kelvin waves would have been absorbed in the westerlies below 22 km while the mixed Rossby–gravity waves would have propagated well into the easterlies above 24 km where they would have been absorbed and produced an easterly acceleration and descent of the easterly shear zone. In this manner the mean zonal wind is forced to oscillate back and forth between westerlies and easterlies with a period which depends primarily on the vertical momentum transport and other properties of the waves *not* on the annual cycle of solar heating.

11.7 The Ozone Layer

Above 25 km in the stratosphere oxygen atoms are photodissociated by solar ultraviolet radiation of wavelengths less than 242 nm. The resulting oxygen atoms rapidly combine with molecular oxygen to form ozone (O_3). Ozone itself strongly absorbs solar ultraviolet radiation at wavelengths below 300 nm. As a result there is a strong radiative heat input into the stratosphere which causes the observed temperature inversion between the tropopause and the stratopause. In addition, the absorption of solar ultraviolet radiation by O_3 protects the biosphere from biologically harmful ultraviolet radiation. The existence of the ozone layer is thus crucial not only to the general circulation of the atmosphere, but to the very existence of life itself.

It should be emphasized that this essential atmospheric component is indeed a *trace* substance. Even at the peak of the ozone layer the ozone mixing ratio is only about 10 ppm (parts per million) by volume. If the entire atmospheric ozone content were brought to standard temperature and pressure (0°C, 101.325 kPa) the thickness of the ozone column would only be about 0.3 cm! The concentration of ozone is highly variable both in space and time. This variability depends not only on the distribution of ozone sources and sinks, but also importantly on transport of ozone by all scales of atmospheric motions. This dependence of ozone concentration on atmospheric motions is illustrated in Fig. 11.11. As indicated in the figure ozone is formed primarily in tropical latitudes above 25 km. By dividing local ozone concentrations by the ozone formation rate at each latitude and height it is possible to compute an ozone "replacement time." The heavy dashed and solid contours in Fig. 11.11 indicate the 4-month and 10-year replacement time boundaries, respectively. Poleward and below the 10-year replacement time contour the ozone is "detached" from the source region (i.e., the concentration is completely controlled by air motions). The fact that the maximum

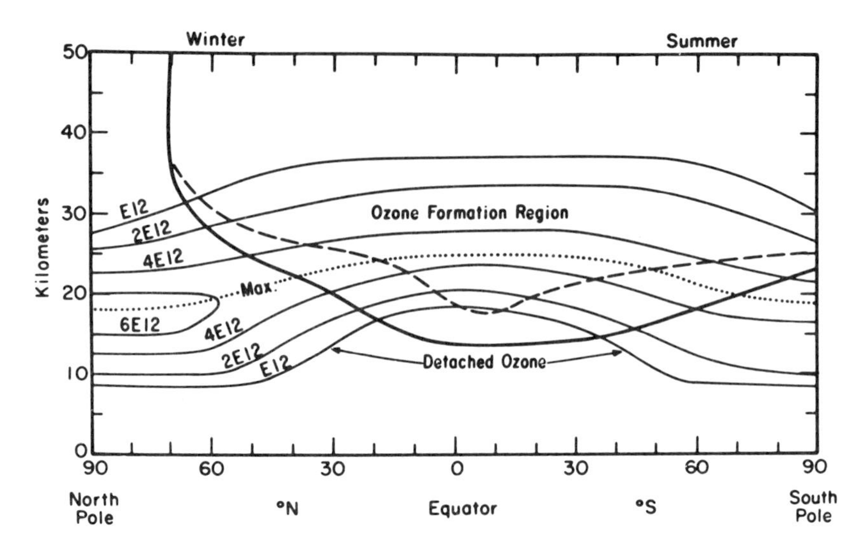

Fig. 11.11 Longitudinal mean ozone concentration in units of molecules cm^{-3} (4E12 means 4×10^{12}) for mid-January. See text for further explanation. (After Climatic Impact Committee, 1975.)

ozone concentration occurs in the detached region below 20 km near the winter pole indicates clearly the predominent role which quasi-horizontal transport by air motions must play in accounting for the climatological ozone distribution.

A complete understanding of the ozone budget requires not only a knowledge of the source distribution and transport by air motions, but also of the sinks. In recent years great progress has been made in our understanding of the photochemistry of the ozone layer. It is now recognized that ozone is catalytically destroyed by a complex series of reactions involving (among other species) NO, ClO, and HO. The concentration of these ozone destroying trace constituents is in itself controlled partly by the mass exchange between the stratosphere and troposphere.

The overall tropospheric–stratospheric mass exchange is illustrated schematically in Fig. 11.12. The cross tropopause exchange in middle latitudes is thought to be dominated by frontal scale processes associated with the upper troposphere jet stream. Studies of radioactive tracers and dynamical tracers (potential temperature and potential vorticity) indicate that considerable stratospheric air is mixed into the troposphere by intrusions of stratospheric air which occur in conjunction with upper-level frontogenesis. These intrusions, which occur in thin layers of ~100-km horizontal and ~1-km vertical scale are quickly destroyed by irreversible vertical mixing in

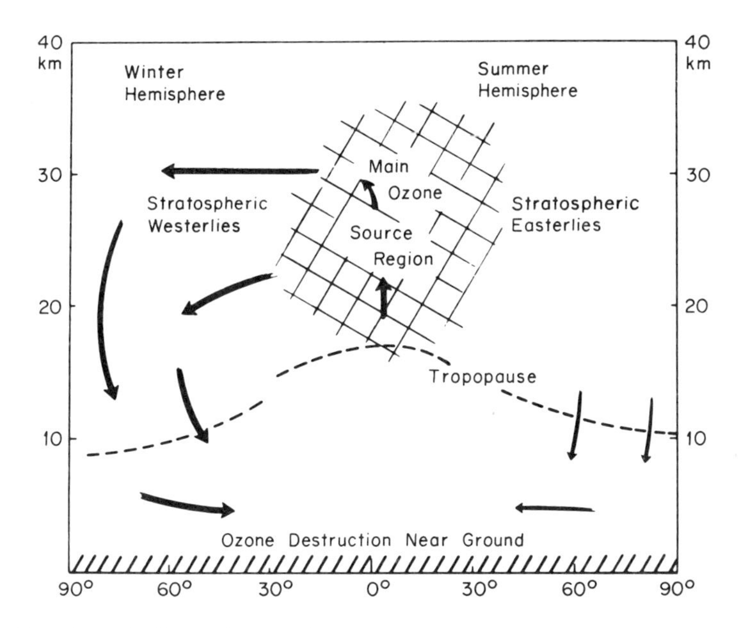

Fig. 11.12 Schematic illustration showing sources, sinks, and mass transport of ozone. (After Dütsch, 1971.)

the troposphere. A large fraction of the stratospheric air mass is injected into the troposphere by this process each year. However, most of this air probably originates in the lowest few kilometers of the stratosphere. Although some tropospheric air no doubt is mixed into the stratosphere by slow meridional circulations associated with the jet stream, the primary transport of mass from the troposphere into the stratosphere takes place in the equatorial region in conjunction with the upper branch of the tropical Hadley cell. Since air entering the stratosphere across the equatorial tropopause is cooled to temperatures less than 200 K, the water vapor mixing ratio for such air must be less than about 4.5×10^{-6} (the saturation mixing ratio for $T = 193$ K and $z = 17$ km). Observations indicate that stratospheric water vapor mixing ratios are generally in the range of 2×10^{-6} to 4×10^{-6}, which is consistent with a tropical injection. If significant tropospheric air were to enter the stratosphere in midlatitudes, where the tropopause temperature exceeds 220 K, the stratosphere would be much wetter than observed.

According to Fig. 11.12 the overall mean mass flow within the stratosphere is directed poleward and is downward outside the tropics. This pattern of mass flow accounts for the observed distribution of ozone shown in Fig. 11.11. The high concentration in the polar winter is accounted for by the strong poleward and downward mass flow at that season. The mass flow shown in

Fig. 11.11 should not be confused with the eulerian mean meridional circulation which at high latitudes in winter is directed upward and equatorward (see Fig. 11.5). This apparent paradox may be resolved by recalling that planetary waves are very active in the winter stratosphere and that these waves produce eddy fluxes which contribute to the transport of trace species. Thus, the *net* transport depends on the combined effects of the eddies and the mean meridional flow.

The situation is closely analogous to the zonal mean heat balance discussed in Section 10.4, where it was shown that net heating or cooling results from a small difference between the eulerian mean eddy heat flux convergence, the adiabatic temperature change due to the eulerian mean meridional circulation, and the diabatic heating. Similar considerations apply to ozone and other strongly vertically stratified tracers.

A conservation equation for the ozone mixing ratio can be written in a form similar to the thermodynamic energy equation (10.40) provided that we divide the ozone mixing ratio into its horizontal average at each level $(q_i)_o$ and the deviation from this average q_i, and utilize the fact that below the ozone maximum the mean gradient is sufficiently strong that $d(q_i)_o/dz \gg dq_i/dz$. Now, following the motion of an individual parcel the rate of change of mixing ratio must be due to the net effect of all sources and sinks S. Thus,

$$\frac{d}{dt}[q_i + (q_i)_o] = S \tag{11.29}$$

Assuming, consistent with the quasi-geostrophic theory, that the horizontal winds are quasi-nondivergent, (11.29) can be approximated below the ozone maximum as

$$\left(\frac{\partial}{\partial t} + u\frac{\partial}{\partial x} + v\frac{\partial}{\partial y}\right)q_i + w\frac{d(q_i)_o}{dz} \simeq S \tag{11.30}$$

which is similar in form to the first law of thermodynamics (11.5).

Noting that $\partial u/\partial x + \partial v/\partial y \simeq 0$ we can rewrite (11.30) in flux form and average zonally to obtain

$$\frac{\partial \bar{q}_i}{\partial t} = -\frac{\partial}{\partial y}\langle q_i'v'\rangle - \bar{w}\frac{d(q_i)_o}{dz} + \bar{S} \tag{11.31}$$

where $\bar{q}_i$ and q_i' designate the zonal mean and deviation from the zonal mean, respectively. Thus, in the long-term mean (11.31) indicates that the zonal mean ozone concentration is controlled by a balance between the net source $\bar{S}$, the transport due to the horizontal eddy motions $\langle q_i'v'\rangle$, and vertical advection by the eulerian mean meridional motion.

As has been previously pointed out there is a tendency for the transports by the eddy motions and the mean meridional cell to cancel each other especially in high latitudes during the winter. Thus in order to study the *net* transport there are some advantages in adopting a lagrangian point of view. If we follow the motion of an individual air parcel we know that in the tropical lower stratosphere there is a net radiative heating due to the absorption of solar radiation by the ozone. Thus individual fluid parcels must be moving upward toward higher potential temperature surfaces. In high latitudes, on the other hand, there is a net radiative cooling (especially in the winter hemisphere) and individual parcels must be moving downward toward lower potential temperature. Thus, by considering the stratospheric heat balance we obtain the simple mass flow picture shown in Figure 11.12. It then becomes obvious that in high latitudes during the winter the eddy flux convergence must dominate over the advection by the eulerian mean meridional motion in order that the net lagrangian mass motion be downward.

Problems

1. Show that the first law of thermodynamics (2.42) can be written in the form (11.5) in log-pressure coordinates.

2. Suppose that temperature increases linearly with height in the layer between 20 and 50 km at a rate of 2 K km^{-1}. If the temperature is 200 K at 20 km, find the value of the scale height H for which the log-pressure height z^* will coincide with the actual height at 50 km. (Assume that z^* coincides with actual height at 20 km and let g be a constant.)

3. Find the Rossby critical velocities for zonal wave numbers 1, 2, and 3 (i.e., for 1, 2, and 3 wavelengths around a latitude circle) assuming a β plane centered at 45°N, scale height $H = 7$ km, buoyancy frequency $N = 2 \times 10^{-2}\ s^{-1}$, and infinite meridional scale ($l = 0$).

4. Find the geopotential and vertical velocity fluctuations for a Kelvin wave of zonal wave number 1, phase speed 40 m s^{-1}, and zonal velocity perturbation amplitude 5 m s^{-1}. Let $N^2 = 4 \times 10^{-4}\ s^{-2}$.

5. For the situation of Problem 4 compute the vertical momentum flux $M \equiv \rho_s \langle uw \rangle$, where $\rho_s = e^{-z/H}$ and the angle brackets denote a zonal average. Show that M is constant with height.

6. Determine the form for the vertical velocity perturbation for the mixed Rossby–gravity wave corresponding to the u', v', and Φ' perturbations given in (11.25).

7. For a mixed Rossby–gravity wave of zonal wave number 4 and phase speed $-20\ \mathrm{m\ s^{-1}}$ determine the latitude at which the vertical momentum flux $\rho_s\langle uw\rangle$ is a maximum.

8. Show that for a mean zonal wind symmetric about the equator the condition for geostrophic balance in the vicinity of the equator is $\beta\bar{u} = -\partial^2\Phi/\partial y^2$. Use this relationship to estimate the amplitude of the temperature oscillation associated with the quasi-biennial oscillation assuming a vertical shear of $20\ \mathrm{m\ s^{-1}}$ over a depth of 5 km and a half-width of 12° latitude.

Suggested References

Holton (1975) gives an advanced treatment of the dynamics of the stratosphere.

Climatic Impact Committee, *Environmental Impact of Stratospheric Flight*, gives an excellent review of the photochemical, radiative, and dynamical aspects of the ozone layer.

Chapter

12 Tropical Motion Systems

Throughout the previous chapters of this book, we have emphasized the circulation systems of the extratropical regions (that is, the regions poleward of about 30° latitude). This emphasis has not been due to any lack of interesting motion systems in the tropics, but is a result, rather, of the relatively primitive state of knowledge concerning the dynamics of tropical circulations. Most meteorologists live in the middle latitudes of the Northern Hemisphere, and the best data coverage is also in that region. For these reasons, there has historically been a much greater effort devoted to the dynamics of extratropical systems. These efforts have culminated in the development of the quasi-geostrophic theory, which, as we have tried to emphasize in this book, provides a basic theoretical framework for understanding the dynamics of middle latitude synoptic systems and their role in the general circulation of the atmosphere. It is primarily due to the existence of a reasonably satisfactory theory that middle latitude synoptic systems have been emphasized in this book.

However, it is in the tropics that the majority of the solar energy which drives the atmospheric heat engine is absorbed by the earth and transferred to the atmosphere. Therefore, an understanding of the general circulation of the tropics must be regarded as a fundamental goal of dynamic meteorology. Furthermore, since the tropics cover half the surface of the earth, development of forecasting models for the tropics is an important goal in itself. In addition, as mentioned in the previous chapter, the tropical and middle latitude atmosphere are coupled so that for long-term numerical forecasts of the motions at middle latitudes accurate forecasts in the tropics are also required.

Unfortunately, there is as yet no single unifying theory for tropical motions comparable to the quasi-geostrophic theory for middle latitude motions. Rapid progress is being made in the understanding of tropical motion systems. However, the problems involved are in some ways much more difficult than in middle latitudes. We have seen that outside the tropics the primary energy source for synoptic disturbances is the zonal available potential energy associated with the strong latitudinal temperature gradient. In fact, observations indicate that diabatic heating due to latent heat release and radiative heating is as a rule only a secondary energy source in mid-latitude systems. In the tropics, on the other hand, the storage of available potential energy is very small due to the very small temperature gradients. Latent heat release appears to be the primary energy source, at least for those disturbances which originate within the equatorial zone. Most latent heat release in the tropics occurs in convective cloud systems. These convective clouds are generally embedded in large-scale circulations. But the convective elements themselves are inherently cumulus scale circulations. Thus, there is a strong interaction between the cumulus scale convection and large-scale circulations which is of primary importance for understanding tropical motion systems.

Of course, the tropical atmosphere and middle latitude atmosphere cannot be treated in complete isolation from each other. In the subtropical regions ($\sim 30°$ latitude) circulation systems characteristic of both tropical and extratropical regions may be observed depending on the season and geographical location. To keep the discussion as simple as possible we will, therefore, in this chapter focus primarily on the zone well equatorward of 30° where the influence of middle latitude systems should be a minimum.

12.1 Scale Analysis of Tropical Motions

Despite the uncertainties involved with the interaction between the convective and synoptic scales, some information on the character of synoptic scale motions in the tropics can be obtained through the methods of scale analysis. The scaling arguments can be carried out most conveniently if the governing equations are written in the log-pressure coordinate system. Referring to Section 11.3 we recall that the momentum equation, hydrostatic equation, continuity equation, and thermodynamic energy equation may be written as follows in the log-pressure system:

$$\frac{d\mathbf{V}}{dt} + f\mathbf{k} \times \mathbf{V} = -\nabla\Phi \tag{12.1}$$

$$\frac{\partial \Phi}{\partial z^*} = \frac{RT}{H} \tag{12.2}$$

$$\frac{\partial u}{\partial x} + \frac{\partial v}{\partial y} + \frac{\partial w^*}{\partial z^*} - \frac{w^*}{H} = 0 \tag{12.3}$$

$$\left(\frac{\partial}{\partial t} + \mathbf{V} \cdot \nabla\right) T + w^* S = \frac{1}{c_p} \dot{q} \tag{12.4}$$

Here $\dot{q}$ is the diabatic heating rate per unit mass and

$$S = \frac{\partial T}{\partial z^*} + \frac{RT}{c_p H} = \frac{T}{\theta} \frac{\partial \theta}{\partial z^*} = \frac{N^2 H}{R}$$

is the static stability parameter. For the tropical troposphere, $S \simeq 3°\mathrm{C\,km^{-1}}$ is nearly constant.

We now compare the magnitudes of the various terms in (12.1)–(12.4) for synoptic scale motions in the tropics. We define characteristic scales for the various field variables as follows:

$U \sim 10\ \mathrm{m\,s^{-1}}$	horizontal velocity scale
W	vertical velocity scale
$L \sim 10^6\ \mathrm{m}$	horizontal length scale
D	depth scale
$\Delta\phi$	geopotential fluctuation scale
$L/U \sim 10^5\ \mathrm{s}$	time scale for advection

We have here assumed magnitudes for the horizontal length and velocity scales which are typical for observed values in synoptic systems both in the tropics and at middle latitudes. We now wish to show how the corresponding characteristic scales for vertical velocity and geopotential fluctuations are limited by the dynamic constraints imposed by continuity, momentum balance, and heat energy balance.

We first note that an upper limit on W is imposed by the continuity equation (12.3). Thus, following the discussion of Section 4.5,

$$\frac{\partial u}{\partial x} + \frac{\partial v}{\partial y} \lesssim \frac{U}{L}$$

But for motions with a vertical scale less than or equal to the scale height $(D \lesssim H)$,

$$\frac{\partial w^*}{\partial z^*} - \frac{w^*}{H} \sim \frac{W}{D}$$

so that the vertical velocity scale must satisfy

$$W \lesssim \frac{D}{L} U \qquad (12.5)$$

We can estimate the magnitude of the geopotential fluctuation by scaling the terms in the momentum equation. For this purpose, it is convenient to compare the magnitude of the horizontal inertial acceleration,

$$\mathbf{V} \cdot \nabla \mathbf{V} \sim \frac{U^2}{L}$$

with each of the other terms in (12.1) as follows:

$$\left| \frac{\partial \mathbf{V}/\partial t}{(\mathbf{V} \cdot \nabla \mathbf{V})} \right| \sim 1 \qquad (12.6)$$

$$\left| w^* \frac{\partial \mathbf{V}/\partial z^*}{(\mathbf{V} \cdot \nabla \mathbf{V})} \right| \sim \frac{WL}{UD} \lesssim 1 \qquad (12.7)$$

$$\left| \frac{f\mathbf{k} \times \mathbf{V}}{(\mathbf{V} \cdot \nabla \mathbf{V})} \right| \sim \frac{fL}{U} = Ro^{-1} \qquad (12.8)$$

$$\left| \frac{\nabla \Phi}{(\mathbf{V} \cdot \nabla \mathbf{V})} \right| \sim \frac{\Delta \Phi}{U^2} \qquad (12.9)$$

We have previously shown that in middle latitudes where $f \sim 10^{-4}\ \mathrm{s}^{-1}$, the Rossby number Ro is small so that to a first approximation the Coriolis force and pressure gradient force terms balance. In that case $\Delta\Phi \sim fUL$. In the equatorial region, however, $f \lesssim 10^{-5}\ \mathrm{s}^{-1}$ and the Rossby number is of order unity or greater. Therefore, it is not appropriate to assume that the Coriolis force term balances the pressure gradient. In fact from (12.6)–(12.9) we see that in general the pressure gradient must be balanced by the inertial acceleration so that $\Delta\Phi \sim U^2 \sim 100\ \mathrm{m}^2\ \mathrm{s}^{-2}$. Thus, the geopotential perturbations associated with equatorial synoptic scale disturbances will be an order of magnitude smaller than those for midlatitude systems of similar scale.

This constraint on the geopotential fluctuations in the tropics has profound consequences for the structure of synoptic scale tropical motion systems. These consequences can be easily understood by applying scaling arguments to the thermodynamic energy equation. It is first necessary to obtain an estimate of temperature fluctuations. From the hydrostatic approximation (12.2) we have

$$T = \frac{H}{R}\frac{\partial \Phi}{\partial z^*} \sim \frac{H}{D}\frac{\Delta\Phi}{R} \tag{12.10}$$

Thus, for systems whose vertical scale is comparable to the scale height,

$$T \sim \frac{U^2}{R} \sim 0.3^\circ\mathrm{C}$$

Therefore, deep tropical systems are characterized by practically negligible temperature fluctuations. Referring to the thermodynamic energy equation, we find that for such systems

$$\left(\frac{\partial}{\partial t} + \mathbf{V}\cdot\nabla\right)T \sim 0.3^\circ\mathrm{C\ d^{-1}}$$

In the absence of precipitation the diabatic heating is primarily due to long-wave radiation which tends to cool the troposphere at a rate of $\dot{q}/c_p \sim -1^\circ\mathrm{C\ d^{-1}}$. Since the actual temperature fluctuations are small, this radiative cooling must be approximately balanced by adiabatic warming due to subsidence. Thus, to a first approximation (12.4) becomes

$$w^* S \simeq \frac{\dot{q}}{c_p} \tag{12.11}$$

For the tropical troposphere, $S \sim 3^\circ\mathrm{C\ km^{-1}}$ and the vertical motion scale must satisfy

$$W \sim \frac{\dot{q}}{Sc_p} \sim 0.3 \quad \mathrm{cm\ s^{-1}} \tag{12.12}$$

Therefore, in the absence of precipitation the vertical motion is constrained to be even smaller than in extratropical synoptic systems of a similar scale. Therefore, to a rather high degree of accuracy the horizontal motion field is governed by the vorticity equation

$$\left(\frac{\partial}{\partial t} + \mathbf{V}_\psi\cdot\nabla\right)(\zeta + f) + (\zeta + f)\nabla\cdot\mathbf{V} = 0 \tag{12.13}$$

which may be obtained by taking the vertical component of the curl of (12.1), neglecting terms involving w^*, and approximating $\mathbf{V}$ by $\mathbf{V}_\psi$ in the advection term.

For the midlatitude case, we have seen that since $\zeta \ll f$ the divergence term can be approximated as simply $f\nabla\cdot\mathbf{V}$. In the tropics $\zeta \sim f$ so that the relative vorticity cannot be neglected compared to f in the divergence term. However, the terms

$$\left(\frac{\partial}{\partial t} + \mathbf{V}_\psi \cdot \nabla\right)\zeta \sim \frac{U^2}{L^2} \sim 10^{-10} \quad \mathrm{s}^{-1}$$

and

$$\mathbf{V}_\psi \cdot \nabla f \sim U\beta \sim 10^{-10} \quad \mathrm{s}^{-1}$$

have the same magnitudes in the tropics as in middle latitudes. But since in the tropics $(\zeta + f) \sim 10^{-5}\ \mathrm{s}^{-1}$ and for deep systems $\nabla \cdot \mathbf{V} \sim W/H \sim 0.3 \times 10^{-6}\ \mathrm{s}^{-1}$ we see that outside precipitation zones

$$(\zeta + f)\nabla \cdot \mathbf{V} \sim 10^{-11} \quad \mathrm{s}^{-1}$$

so that to a first approximation (12.13) becomes simply

$$\left(\frac{\partial}{\partial t} + \mathbf{V}_\psi \cdot \nabla\right)(\zeta + f) = 0 \tag{12.14}$$

Thus, in the absence of condensation heating, tropical motions in which the vertical scale is comparable to the scale height of the atmosphere must be *barotropic*. Such disturbances cannot convert potential energy to kinetic energy. They must, therefore, be driven by lateral coupling either to mid-latitude systems or to precipitating tropical disturbances, or by barotropic conversion of mean flow kinetic energy.

For precipitating synoptic systems in the tropics, the above scaling considerations require considerable modification. Precipitation rates in such systems are typically in the range of 2 cm d^{-1}. This implies condensation of $m_w = 20$ kg water for an atmospheric column of 1-m^2 cross section. For a latent heat of condensation $L_c \simeq 2.5 \times 10^6\ \mathrm{J\ kg^{-1}}$; this precipitation rate implies an addition of heat energy to the atmospheric column of

$$m_w L_c \sim 5 \times 10^7\ \mathrm{J\ m^{-2}\ d^{-1}}$$

If this heat is distributed uniformly over the entire atmospheric column of mass $p_0/g \simeq 10^4\ \mathrm{kg\ m^{-2}}$, then the average heating rate per unit mass of air is

$$\frac{\dot{q}}{c_p} \simeq \frac{L_c m_w}{c_p(p_0/g)} \sim 5^\circ\mathrm{C\ d^{-1}}$$

In reality the condensation heating due to deep convective clouds is not distributed evenly over the entire vertical column, but is a maximum between

300 and 400 mb, where the heating rate may be $\sim 10°\text{C d}^{-1}$. In this case the approximate thermodynamic energy equation (12.11) implies that the vertical motion on the synoptic scale in precipitating systems must have a magnitude of order $W \sim 3 \text{ cm s}^{-1}$ in order that the adiabatic cooling can balance the condensation heating in the 300–400-mb layer.

Therefore, the average vertical motion in precipitating disturbances in the tropics is an order of magnitude larger than the vertical motion outside the disturbances. As a result the flow in these disturbances has a relatively large divergent component so that the vorticity generation by the divergence term is comparable in magnitude to the other terms in (12.13). However, the vorticity equation in the form (12.13) turns out to be inadequate to describe the vorticity budget of precipitating tropical disturbances. Although our crude scaling arguments indicate that the divergence term is the same order in magnitude as the vorticity advection a more careful analysis shows that in the trough region where ζ and $\nabla \cdot \mathbf{V}$ are both large and the vorticity advection is small the term $(\zeta + f)\, \nabla \cdot \mathbf{V}$ can not be balanced by $d(\zeta + f)/dt$. There is growing evidence that in such disturbances the vertical transport of vorticity (momentum) by cumulus convection is an essential part of the vorticity (momentum) budget. This, of course, introduces a complicated problem of interaction between the cumulus scale convection and the synoptic scale. On the other hand, temperature fluctuations are observed to remain small in precipitating tropical systems (at least above the planetary boundary layer). Thus, the approximate thermodynamic energy balance remains in the simple form (12.11) so that the synoptic scale vertical motion, and hence the divergence field, can be diagnostically computed if the synoptic scale distribution of diabatic heating is known. This matter will be discussed further in Section 12.2.6.

All the above scaling arguments have been based on the assumption that the depth scale of tropical disturbances is comparable to the scale height of the atmosphere. There is, however, one important class of large-scale tropical motions for which this assumption does not apply. These are the vertically propagating planetary scale gravity waves which were discussed in Section 11.5. Such vertically propagating waves have vertical scales which are generally small compared to the scale height. Hence, since

$$\nabla \cdot \mathbf{V} \sim \frac{W}{D}$$

these waves tend to have large horizontal divergence fields even though the vertical velocities remain quite small. Such waves are important for transporting momentum and energy into the stratosphere, but probably have little influence on tropospheric weather systems.

12.2 Cumulus Convection

The scaling arguments of the previous section suggest that to a large extent the structure of synoptic disturbances in the tropics must be determined by the field of heating provided by latent heat release in areas of active precipitation. Observations indicate that most precipitation in tropical systems is of convective origin and is concentrated in a comparatively small number of very deep vigorous cumulus convection cells. These so-called "hot towers" occupy only a small fraction of the area of the tropics even in active disturbances. Therefore, to understand the dynamics of synoptic scale systems in the tropics, it is necessary to elucidate the nature of the interaction between the cumulus scale motions and the synoptic scale disturbances in which the cumulus motions are embedded. This in turn requires an understanding of the dynamics of cumulus scale motions themselves.

The subject of convective motions is extremely complex to treat theoretically. Part of this difficulty results from the fact that convective motions are generally nonhydrostatic, nonsteady turbulent motions which require the full three-dimensional equations of motion for their description. In this section we, therefore, concentrate primarily on the *thermodynamic* aspects of moist convection.

12.2.1 Equivalent Potential Temperature

We have previously applied the parcel method to discuss the vertical stability of a dry atmosphere. We found that the stability of a dry parcel with respect to a vertical displacement depends on the lapse rate of potential temperature in the environment such that a parcel displacement is neutrally stable provided that $\partial\theta/\partial z = 0$. The same condition also applies to parcels in a moist atmosphere when the relative humidity is less than 100%. However, if a parcel of moist air is forced to rise, it will eventually become saturated and a further rise will cause condensation and latent heat release. If the environmental lapse rate is greater than the moist adiabatic lapse rate, the parcel may then become buoyant relative to its surroundings and be accelerated upward. The criterion for this type of parcel instability, which is called *conditional instability*, can be expressed conveniently in terms of a quantity called *equivalent potential temperature*. Equivalent potential temperature, designated by θ_e, is the potential temperature that a parcel of air would have if all its moisture were condensed out and the resultant latent heat used to warm the parcel. The temperature of an air parcel can be brought to its equivalent potential value by raising the parcel from its original level until all the water vapor in the parcel has condensed and fallen out, then compressing the parcel adiabatically to a pressure of 1000 mb. Since the condensed water is assumed to fall out, the temperature increase during the compression will be at the dry adiabatic rate and the parcel will arrive back at its original level with a temperature which is higher than its original temperature. The process is, therefore, irreversible. Ascent of this type, in which all condensation products are assumed to fall out, is called *pseudoadiabatic ascent*. (It is not a truly adiabatic process because the liquid water which falls out carries a small amount of heat with it.)

A complete derivation of the mathematical expression relating θ_e to the other variables of state is rather involved and will be relegated to Appendix D. However, for most purposes, it is sufficient to use an approximate expression for θ_e which can be immediately derived from the entropy form of the first law of thermodynamics (2.46). If we let q_s denote the mass of water vapor per unit mass of dry air in a saturated parcel (q_s is called the saturation *mixing ratio*), then the rate of diabatic heating per unit mass is

$$\dot{q} = -L_c \frac{dq_s}{dt}$$

where L_c is the latent heat of condensation. Thus, from the first law of thermodynamics

$$c_p \frac{d \ln \theta}{dt} = -\frac{L_c}{T} \frac{dq_s}{dt} \tag{12.15}$$

For a saturated parcel undergoing pseudoadiabatic ascent the rate of change in q_s following the motion is much larger than the rate of change in T or L_c. Therefore,

$$d \ln \theta \simeq -d\left(\frac{L_c q_s}{c_p T}\right) \tag{12.16}$$

Integrating (12.16) from the initial state (θ, q_s) to a state where $q_s \simeq 0$ we obtain

$$\ln\left(\frac{\theta}{\theta_e}\right) \simeq -\frac{L_c q_s}{c_p T}$$

where θ_e, the potential temperature in the final state, is approximately the equivalent potential temperature defined above. Thus, θ_e for a saturated parcel is given by

$$\theta_e \simeq \theta \exp\left(\frac{L_c q_s}{c_p T}\right) \tag{12.17}$$

The expression (12.17) may also be used to compute θ_e for an unsaturated parcel provided that the temperature used in the formula is the temperature which the parcel would have if expanded adiabatically to saturation. Thus, equivalent potential temperature is conserved for a parcel during both dry adiabatic and pseudoadiabatic changes of state.

In many applications tropical meteorologists prefer to use the *moist* static energy $h \equiv s + L_c q$, where $s \equiv c_p T + gz$ is the *dry static energy*. It may be shown (Problem 7) that $c_p T\, d \ln \theta \simeq ds$ so that $c_p T\, d \ln \theta_e \simeq dh$. Hence, moist static energy is approximately conserved when θ_e is conserved.

12.2.2 The Pseudoadiabatic Lapse Rate

The first law of thermodynamics for a moist adiabatic process (12.15) can be used to derive a formula for the rate of change of temperature with respect to height for a saturated parcel undergoing pseudoadiabatic ascent. Using the definition of θ (2.44) we can rewrite (12.15) as

$$\frac{d \ln T}{dz} - \frac{R}{c_p}\frac{d \ln p}{dz} = -\frac{L_c}{c_p T}\frac{dq_s}{dz}$$

which with the aid of the hydrostatic equation and equation of state can be expressed as

$$\frac{dT}{dz} + \frac{g}{c_p} = -\frac{L_c}{c_p}\frac{dq_s}{dT}\frac{dT}{dz}$$

Thus, following the ascending saturated parcel we have

$$\Gamma_s \equiv -\frac{dT}{dz} = \Gamma_d/[1 + (L_c/c_p)dq_s/dT] \tag{12.18}$$

where $\Gamma_d = g/c_p$ is the dry adiabatic lapse rate, and Γ_s is the *pseudoadiabatic lapse rate*. Since dq_s/dT is positive Γ_s is always less than Γ_d. In the midtroposphere Γ_s is about 6°C km^{-1}.

12.2.3 Conditional Instability

We showed in Section 2.7.3 that for dry adiabatic motions the atmosphere is statically stable provided that the lapse rate is less than the dry adiabatic lapse rate (i.e., the potential temperature increases with height). If the lapse rate Γ lies between the dry adiabatic and pseudoadiabatic values ($\Gamma_s < \Gamma < \Gamma_d$) the atmosphere is stably stratified with respect to dry abiabatic displacements but unstable with respect to pseudoadiabatic displacements. Such a situation is referred to as *conditional instability* (i.e., the instability is conditional to *saturation* of the air parcel).

The conditional stability criterion can also be expressed in terms of the gradient of θ_e^*, where θ_e^* is defined as the equivalent potential temperature of a hypothetically saturated atmosphere which has the thermal structure of the actual atmosphere. Thus,

$$d \ln \theta_e^* = d \ln \theta + d(L_c q_s / c_p T) \tag{12.19}$$

where T is the actual temperatue, *not* the temperature after adiabatic expansion to saturation as in (12.17). To derive this condition we consider the motion of a saturated parcel in an environment in which the potential temperature is θ_0 at the level z_0. At the level $z_0 - \delta z'$ the undisturbed environmental air thus has potential temperature

$$\theta_0 - \frac{\partial \theta}{\partial z} \delta z'$$

Suppose a saturated parcel which has the environmental potential temperature at $z_0 - \delta z'$ is raised to the level z_0. When it arrives at z_0 the parcel will have potential temperature

$$\theta_1 = \left(\theta_0 - \frac{\partial \theta}{\partial z} \delta z' \right) + \delta \theta$$

where $\delta\theta$ is the change in potential temperature due to ascent through vertical distance $\delta z'$. Assuming a pseudoadiabatic ascent we see from (12.16) that

$$\frac{\delta \theta}{\theta} \simeq -\delta \left(\frac{L_c q_s}{c_p T} \right) \simeq -\frac{\partial}{\partial z} \left(\frac{L_c q_s}{c_p T} \right) \delta z'$$

so that the buoyancy of the parcel when it arrives at z_0 is proportional to

$$\frac{\theta_1 - \theta_0}{\theta_0} \simeq -\left[\frac{1}{\theta} \frac{\partial \theta}{\partial z} + \frac{\partial}{\partial z} \left(\frac{L_c q_s}{c_p T} \right) \right] \delta z' \tag{12.20}$$

Now the first term on the right in (12.20) refers to conditions in the environment, while the second refers to the saturated parcel. However, provided that the parcel temperature is not too different from that of the environment, the two terms on the right may be combined with the aid of (12.19) to provide an expression for the buoyancy in terms of θ_e^*:

$$\frac{\theta_1 - \theta_0}{\theta_0} \simeq -\frac{\partial \ln \theta_e^*}{\partial z} \delta z'$$

Now, the saturated parcel will be positively buoyant at z_0 provided $\theta_1 > \theta_0$. Thus the conditional stability criterion for a *saturated* parcel is

$$\frac{\partial \theta_e^*}{\partial z} \begin{cases} < 0 & \text{conditionally unstable} \\ = 0 & \text{saturated neutral} \\ > 0 & \text{absolutely stable} \end{cases} \tag{12.21}$$

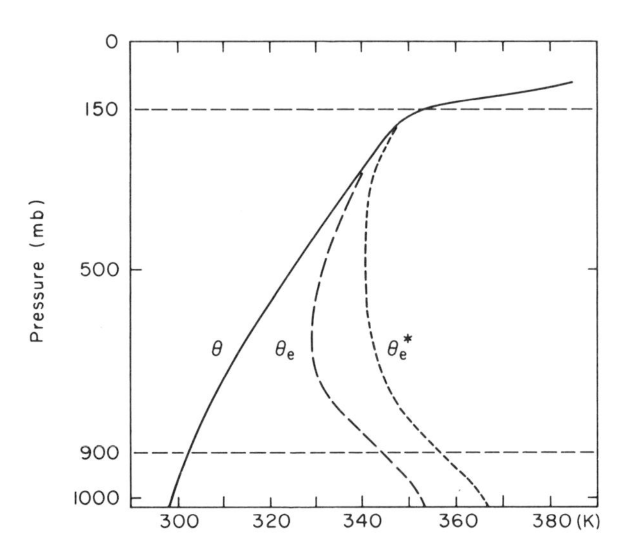

Fig. 12.1 Typical sounding in the tropical atmosphere showing the vertical profiles of potential temperature θ, equivalent potential temperature θ_e, and the equivalent potential temperature θ_e^* of a hypothetically saturated atmosphere with the same temperature at each level. (After Ooyama, 1969. Reproduced with permission of the American Meteorological Society.)

In Fig. 12.1 the vertical profiles of θ, θ_e, and θ_e^* for a typical tropical sounding are shown. It is obvious from the figure that the mean tropical atmosphere is conditionally unstable in the lower and middle troposphere. However, this observed profile does not imply that convective overturning will spontaneously occur in the tropics. The release of conditional instability requires not only $\partial\theta_e^*/\partial z < 0$, but also a saturated atmosphere, and the mean relative humidity in the tropics is well below 100%. Thus, low-level convergence with its resultant forced ascent or vigorous vertical turbulent mixing in the boundary layer is required to produce saturation. The amount of ascent necessary to produce a positively buoyant parcel can be estimated simply from Fig. 12.1. A parcel rising pseudoadiabatically from a level $z_0 - \delta z'$ will conserve the value of θ_e characteristic of the environment at $z_0 - \delta z'$. Now, the buoyancy of a parcel depends only on the difference in density between the parcel and the environment. Thus, in order to compute the buoyancy of the parcel at z_0, it is not correct simply to compare θ_e of the environment at z_0 to $\theta_e(z_0 - \delta z')$ because if the environment is unsaturated the difference in θ_e for the parcel and the environment may be due primarily to the difference in mixing ratios, not to any temperature (density) difference. To estimate the buoyancy of the parcel $\theta_e(z_0 - \delta z')$ should instead be compared to $\theta_e^*(z_0)$, which is the equivalent potential temperature which the environment at z_0 would have if it were isothermally brought to saturation. The parcel will thus become buoyant when $\theta_e(z_0 - \delta z') > \theta_e^*(z_0)$, for then the parcel temperature will exceed the temperature of the environment. From Fig. 12.1 we see that θ_e for a parcel raised from 1000 mb will intersect the θ_e^* curve just above 900 mb, whereas a parcel raised from any level much above 900 mb will not intersect θ_e^* no matter how far it is forced to ascend. It is for this reason that *low-level convergence* is required to initiate convective overturning in the tropical atmosphere. Only air near the surface has a sufficiently high value of θ_e to become buoyant when it is forcibly raised. Of course, convergence at higher levels may play an important role in maintaining the convection by adding substantial moisture to the system.

12.2.4 The Slice Method

In the parcel method the environment is assumed to remain undisturbed by the convective parcels. In reality the rising motion in the convection must be compensated by subsidence in the environment if an overall mass balance is to be maintained. For this reason, the parcel method tends to overestimate the degree of instability of the atmosphere. More significantly, it is the compensating subsidence and adiabatic warming in the environment which is the direct cause of the large-scale heating associated with cumulus convection in synoptic scale disturbances. The simplest manner in which to take account of this adjustment in the environment is the so-called *slice method*. In this method it is assumed that on any horizontal plane the upward mass flux in the convection cells is just balanced by downward mass flux in the environment immediately surrounding the convection cells. This mass balance may be expressed as follows: We let ρ' and w' be the density and vertical velocity in the convection cells and ρ and w be the corresponding fields in the environment. If the fractional horizontal area occupied by the convection cells is designated by a, then for mass balance

$$\rho' w' a = \rho w(1 - a) \tag{12.22}$$

Now suppose that in a time increment δt the convective parcels rise an amount $\delta z'$ and the environment sinks by δz. We can write

$$w' \simeq \frac{\delta z'}{\delta t}, \qquad w \simeq \frac{\delta z}{\delta t}$$

Substituting into (12.22) and neglecting the small difference between ρ' and ρ we obtain

$$a\,\delta z' \simeq (1 - a)\,\delta z \tag{12.23}$$

We now consider the difference at a level z_0 between the temperatures of a saturated parcel which has risen from the level $z_0 - \delta z'$ and the environmental air which has sunk from the level $z_0 + \delta z$. Letting θ_0 be the initial

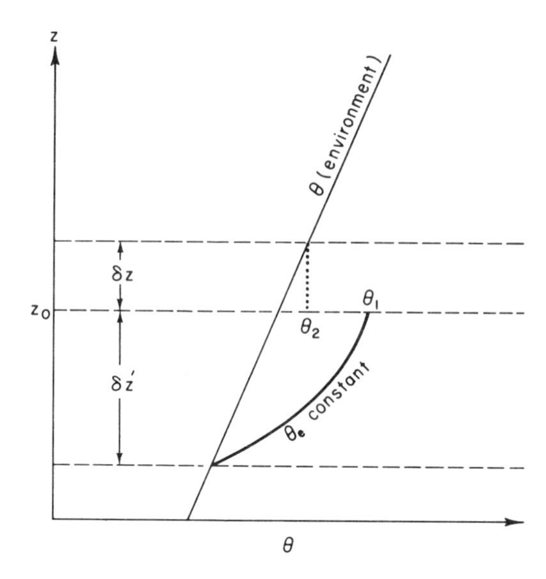

Fig. 12.2 Graphical computation of parcel buoyancy using the slice method.

undisturbed environmental potential temperature at z_0, we have for the potential temperature of the environment at level $z_0 + \delta z$

$$\theta = \theta_0 + \frac{\partial \theta}{\partial z}\delta z$$

The descending environmental air heats dry adiabatically. It thus preserves its original potential temperature and arrives at z_0 with potential temperature

$$\theta_2 = \theta_0 + \frac{\partial \theta}{\partial z}\delta z \tag{12.24}$$

The rising parcel, on the other hand, arrives at z_0 with potential temperature θ_1 given by (12.20). The buoyancy at z_0 is thus proportional to $\theta_1 - \theta_2$, which as shown in Fig. 12.2 is a smaller buoyancy than the $\theta_1 - \theta_0$ given by the parcel method. From (12.20) and (12.24) we find that

$$\frac{\theta_1 - \theta_2}{\theta} = -\frac{\partial \ln \theta_e^*}{\partial z}\delta z' - \frac{\partial \theta}{\partial z}\delta z \tag{12.25}$$

Therefore, with the aid of (12.23) we obtain the stability criterion

$$\frac{\partial \ln \theta_e^*}{\partial z} + \frac{\partial \ln \theta}{\partial z}\left(\frac{a}{1-a}\right)\begin{cases} <0 & \text{conditionally unstable} \\ =0 & \text{neutral} \\ >0 & \text{stable} \end{cases} \tag{12.26}$$

Observations indicate that in general active convection cells occupy only a small percent of the total area even in synoptic disturbances. Thus, $a/(1 - a) \ll 1$ and the slice stability does not in practice differ much from the ordinary moist parcel stability. However, even for $a \simeq 0.01$ an average vertical velocity of 1 m s^{-1} in the cumulus towers implies a compensating subsidence of 1 cm s^{-1} which is easily sufficient to account for the small temperature changes observed in tropical disturbances.

12.2.5 Entrainment

In the previous subsection it was assumed that the rising convective parcels do not mix with the environment. In reality, however, rising saturated parcels tend to be diluted by *entraining*, or mixing in, some of the relatively dry environmental air. If the air in the environment is unsaturated, some of the liquid water in the rising parcel must be evaporated to maintain saturation in the convection cell as air from the environment is entrained. The evaporative cooling caused by entrainment will reduce the buoyancy of the convective parcel. Thus, the equivalent potential temperature in an entraining convection cell will decrease with height rather than remaining constant. The magnitude of the entrainment effect can be determined by application of the thermodynamic energy equation to a mass M of saturated cloud air and a mass dM of entrained air. In entropy form the appropriate approximate energy equation for the cloud air alone is (12.16). Evaporation of liquid water to bring the entrained air to saturation will cause a diabatic heating of $-(q_s - \bar{q})L_c\,dM$, where $\bar{q}$ is the mixing ratio of the environment. In addition an amount of heat, $c_p(T - \bar{T})\,dM$ is required to raise the temperature of the environmental air $\bar{T}$ to the cloud temperature. Thus the approximate thermodynamic energy equation for this case is

$$-M\,d\left(\frac{L_c q_s}{T}\right) - \frac{(q_s - \bar{q})L_c\,dM}{T} = Mc_p\,d\ln\theta + c_p\frac{(T - \bar{T})}{T}\,dM \quad (12.27)$$

where we have neglected the heat accession by the liquid water. From the definition of θ_e (12.17) we can write (12.27) as

$$-M\,d\ln\theta_e = dM\left[(q_s - \bar{q})\frac{L_c}{c_p T} + \frac{(T - \bar{T})}{T}\right]$$

or

$$\frac{d\ln\theta_e}{dz} = -\frac{d\ln M}{dz}\left[\frac{L_c}{c_p T}(q_s - \bar{q}) + \frac{T - \bar{T}}{T}\right] \quad (12.28)$$

For there to be positive buoyancy $T > \bar{T}$, and since $q_s \geq \bar{q}$ the terms in square brackets will be positive and if the entrainment is positive ($d\ln M/dz > 0$)

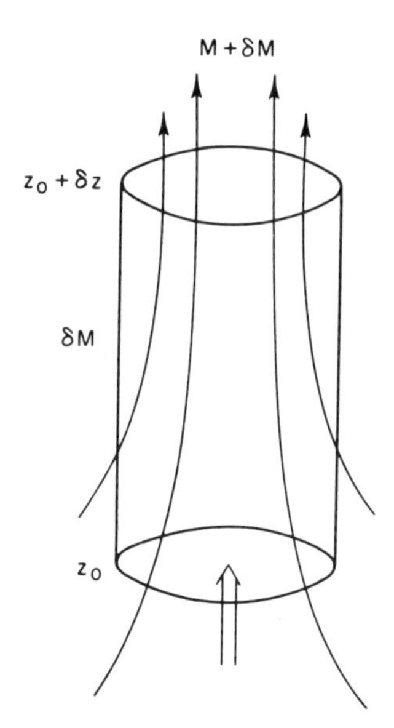

Fig. 12.3 An entraining jet model of cumulus convection.

the equivalent potential temperature of the cloud air must decrease with height. It is observed that in typical cumulus clouds $T - \overline{T} \sim 1°\text{C}$. Thus, the first term in the brackets in (12.28) dominates. If the last term is neglected, (12.28) yields the maximum entrainment possible to just maintain neutral buoyancy for given profiles of θ_e, q_s, and $\bar{q}$.

Under certain restrictive assumptions, it is possible to use (12.28) together with the vertical momentum equation and continuity equation to predict the vertical temperature profile and maximum height of penetration for a cumulus cell given the temperature and specific humidity profiles in the environment. If the convection cell is assumed to be a steady-state entraining jet, then the momentum balance in a slice of depth δz and cross-sectional area S as shown in Fig. 12.3 may be computed as follows: We let

$$M = S\rho w \tag{12.29}$$

be the vertical mass flux at the level $z = z_0$. Then the vertical mass flux at $z = z_0 + \delta z$ may be written

$$M + \delta M = M + \frac{dM}{dz}\delta z$$

The vertical *momentum* flux is thus wM at z_0 and

$$\left(w + \frac{dw}{dz}\delta z\right)\left(M + \frac{dM}{dz}\delta z\right)$$

at $z_0 + \delta z$. Thus, the net divergence of momentum from the volume element $S\,\delta z$ is

$$\left(w + \frac{dw}{dz}\,\delta z\right)\left(M + \frac{dM}{dz}\,\delta z\right) - wM \simeq \frac{d}{dz}(wM)\,\delta z$$

For steady-state motion, the momentum divergence must be balanced by production of vertical momentum by the buoyancy force. It was shown in Chapter 2 [see Eq. (2.51)] that the buoyancy force per unit mass is approximately

$$g\,\frac{(T - \bar{T})}{T}$$

Thus, the momentum balance for the mass $\rho S\,\delta z$ becomes

$$\frac{d}{dz}(wM)\,\delta z = g\,\frac{(T - \bar{T})}{T}\,S\rho\,\delta z$$

or after eliminating S with (12.29)

$$\frac{d}{dz}(wM) = \frac{M}{w}\,g\,\frac{(T - \bar{T})}{T} \tag{12.30}$$

Equation (12.30) can also be written in the form

$$\frac{1}{2}\frac{d}{dz}(w^2) = g\left(\frac{T - \bar{T}}{T}\right) - \frac{d\ln M}{dz}\,w^2 \tag{12.31}$$

which indicates that the rate of kinetic energy increase with height is given by the kinetic energy production by buoyancy minus the kinetic energy loss due to entrainment.

Provided that the cross section S of the convective cell is known as a function of height the system (12.28)–(12.30) may be regarded as a simultaneous set of equations for the determination of the vertical profiles of M, w, and T given $\bar{q}$ and $\bar{T}$. This model can to some extent duplicate observed conditions in cumulus clouds; however, it by no means constitutes a complete theory of cumulus convection because the cross section S must be empirically specified. In addition the entraining jet model assumes a steady state which is certainly not valid for real cumulus convection.

12.2.6 Heating by Condensation

The manner in which the atmosphere is heated by condensation of water vapor depends crucially on the nature of the condensation process. In particular, it is necessary to differentiate between latent heat release through large-scale vertical motion (that is, synoptic scale forced uplift) and the latent heat release due to small-scale deep cumulus convection. The former process, which is generally associated with midlatitude synoptic systems, can be easily incorporated into the thermodynamic energy equation in terms of the synoptic scale field variables. However, since the large-scale heating field resulting from the cooperative action of many cumulonimbus cells involves a complex interaction between the cumulus scale and synoptic scale motions, representation of this type of latent heating in terms of the synoptic scale field variables is much more difficult.

Before considering the problem of condensation heating by cumulus convection, it is worth indicating briefly how the condensation heating by large-scale forced uplift can be included in a prediction model. The approximate thermodynamic energy equation for a pseudoadiabatic process (12.15) states that

$$\frac{d \ln \theta}{dt} \simeq -\frac{L_c}{c_p T}\frac{dq_s}{dt} \tag{12.32}$$

Now the change in q_s following the motion is primarily due to ascent, so that

$$\frac{dq_s}{dt} \simeq \begin{cases} w\dfrac{\partial q_s}{\partial z} & \text{for} \quad w > 0 \\ 0 & \text{for} \quad w < 0 \end{cases} \tag{12.33}$$

and (12.32) can be written in the form

$$\left(\frac{\partial}{\partial t} + \mathbf{V}\cdot\nabla\right)\ln\theta + w\left[\frac{\partial \ln \theta}{\partial z} + \frac{L_c}{c_p T}\frac{\partial q_s}{\partial z}\right] \simeq 0 \tag{12.34}$$

for regions where $w > 0$. But from (12.17)

$$\frac{\partial \ln \theta_e}{\partial z} \simeq \frac{\partial \ln \theta}{\partial z} + \frac{L_c}{c_p T}\frac{\partial q_s}{\partial z}$$

so that (12.34) can be written in a form valid for both positive and negative vertical motion as

$$\left(\frac{\partial}{\partial t} + \mathbf{V}\cdot\nabla\right)\theta + w\Gamma_e \simeq 0 \tag{12.35}$$

where Γ_e is an *equivalent static stability* defined by

$$\Gamma_e = \begin{cases} \dfrac{\theta}{\theta_e}\dfrac{\partial \theta_e}{\partial z} & \text{for} \quad q > q_s \quad \text{and} \quad w > 0 \\ \dfrac{\partial \theta}{\partial z} & \text{for} \quad q < q_s \quad \text{or} \quad w < 0 \end{cases}$$

Thus, in the case of condensation due to large scale forced ascent ($\Gamma_e > 0$) the thermodynamic energy equation has essentially the same form as for adiabatic motions except that the static stability is replaced by the equivalent static stability. As a consequence, the local temperature changes induced by the forced ascent will be smaller than for the case of forced dry ascent with the same lapse rate.

If, on the other hand, $\Gamma_e < 0$, the atmosphere is conditionally unstable and the condensation will occur primarily through cumulus convection. In that case (12.33) is still valid but the vertical velocity must be that of the individual cumulus updrafts, not the synoptic scale w. Thus, a simple formulation of the thermodynamic energy equation in terms of only the synoptic scale variables is not possible. However, we can still simplify the thermodynamic energy equation to some extent. We recall from Section 12.1 [see Eq. (12.11)] that due to the smallness of temperature fluctuations in the tropics the adiabatic cooling and diabatic heating terms must approximately balance. Thus, (12.32) becomes approximately[1]

$$w \frac{\partial \ln \theta}{\partial z} \simeq -\frac{L_c}{c_p T} \frac{dq_s}{dt} \tag{12.36}$$

The synoptic scale vertical velocity w which appears in (12.36) is the average of very large vertical motions in the active convection cells and small vertical motions in the environment. Thus, if we let w' be the vertical velocity in the convective cells and $\bar{w}$ the vertical velocity in the environment, we have

$$w = aw' + (1 - a)\bar{w} \tag{12.37}$$

where a is the fractional area occupied by the convection. With the aid of (12.33) we can then write (12.36) in the form

$$w \frac{\partial \ln \theta}{\partial z} = -\frac{L_c}{c_p T} aw' \frac{\partial q_s}{\partial z} \tag{12.38}$$

The problem is then to express the condensation heating term on the right in (12.38) in terms of the synoptic scale field variables.

This problem of *parameterizing* the effect of the cumulus convection is one of the most challenging areas in tropical meteorology. A simple approach which has been used successfully in some theoretical studies[2] is based on the fact that, since the storage of water in the clouds is rather small, the total

[1] This simplification is primarily useful in diagnostic studies. In forecast models it is necessary to retain $\partial \ln \theta / \partial t$ in order to predict the (small) changes in temperature resulting from the slight imbalance in (12.36).

[2] See Stevens and Lindzen (1978).

vertically integrated heating rate due to condensation must be approximately proportional to the net precipitation rate:

$$-\int_{z_c}^{z_T}\left(\rho a w' \frac{\partial q_s}{\partial z}\right) dz = P \tag{12.39}$$

where z_c and z_T represent the cloud base and cloud top heights, respectively, and P is the precipitation rate (kg m^{-2} s^{-1}).

Since relatively little moisture goes into changing the atmospheric vapor mixing ratio, the net precipitation rate must approximately equal the moisture convergence into an atmospheric column plus surface evaporation

$$P = -\int_0^{z_m} \nabla \cdot (\rho q \mathbf{V})\, dz + E \tag{12.40}$$

where E is the evaporation rate (kg m^{-2} s^{-1}) and z_m is the top of the moist layer ($z_m \simeq 2$ km in the equatorial Western Pacific). Substituting into (12.40) from the approximate continuity equation for q,

$$\nabla \cdot (\rho q \mathbf{V}) + \frac{\partial}{\partial z}(\rho q w) \simeq 0$$

we obtain

$$P = \rho(z_m) w(z_m) q(z_m) + E \tag{12.41}$$

Using (12.41) we can relate the vertically averaged heating rate to the synoptic scale variables $w(z_m)$ and $q(z_m)$.

We still, however, need to determine the distribution of the heating in the vertical. The most common approach is to use an empirically determined vertical distribution based on observations. In that case (12.34) can be written as

$$\left(\frac{\partial}{\partial t} + \mathbf{V} \cdot \nabla\right) \ln \theta + w \frac{\partial \ln \theta}{\partial z} = \frac{L_c}{\rho c_p T} \eta(z)[\rho(z_m) w(z_m) q(z_m) + E] \tag{12.42}$$

where $\eta(z) = 0$ for $z < z_c$ and $z > z_T$ and $\eta(z)$ for $z_c \le z \le z_T$ is a weighting function which must satisfy

$$\int \eta(z)\, dz = 1$$

Recalling that the diabatic heating must be approximately balanced by adiabatic cooling as indicated in (12.38) we see from (12.42) that $\eta(z)$ will have a vertical structure similar to that of the large scale vertical mass flux, ρw. Observations indicate that for many tropical synoptic scale disturbances $\eta(z)$ reaches its maximum at about the 40 kPa (400 mb) level, consistent with the divergence pattern shown in Fig. 12.7.

The above formulation is designed to model average tropical conditions. In reality the vertical distribution of diabatic heating is determined by the local distribution of cloud heights. Thus the cloud height distribution is clearly a key parameter in cumulus parameterization. A sophisticated scheme for relating the cloud height distribution to synoptic scale variables has been developed by Arakawa and Schubert (1974). Discussion of this scheme is beyond the scope of this text.

12.2.7 Cumulus Friction

It was pointed out in Section 12.1 that scaling arguments suggest that in precipitating tropical disturbances it is necessary to consider not only the heating due to cumulus convection, but also the vertical momentum (vorticity) transport by the cumulus. Observational studies of equatorial waves indicate that (at least in troughs and ridges, where vorticity advection is small) the vorticity budget can be satisfied only if this so-called *cumulus friction* acts to nearly balance the vorticity generation due to the divergence term. Thus, the vertical profile of cumulus friction should qualitatively resemble that of the divergence field shown in Fig. 12.7.

A simple theoretical formulation of cumulus friction has been derived by Schneider and Lindzen (1976). They showed that this effect could be parameterized by adding a force in the momentum equation (12.1) of the form

$$\mathbf{F}_c = -\frac{1}{\rho}\frac{\partial}{\partial z}[M_c(\mathbf{V} - \mathbf{V}_c)] \tag{12.43}$$

where M_c is the vertical cloud mass flux, $\mathbf{V}$ is the large scale horizontal velocity, and $\mathbf{V}_c$ is a measure of the horizontal velocity within the convective clouds. Since most of the mass flux within deep cumulus originates at cloud base, and the time scale for ascent within the clouds is $\lesssim 2$ h, $\mathbf{V}_c$ can be reasonably approximated by using the large-scale horizontal velocity at cloud base. It is then only necessary to parameterize the vertical cloud mass flux M_c in order to have a theoretical representation of cumulus friction in terms of the large scale variables. For some purposes it is probably sufficient to simply let M_c be proportional to the precipitation rate P.

12.3 The Observed Structure of Large-Scale Motions in the Equatorial Zone

In the first section of this chapter we have presented scaling arguments which indicate that the large-scale motions in the equatorial zone are driven primarily by latent heat release. Observations indicate that this latent heating occurs primarily through cumulonimbus convection rather than large-scale forced ascent. For this type of driving force to be effective, an interaction between the cumulus scale and large-scale motions is necessary in which the large-scale convergence provides moisture for the convection and the cumulus cells act cooperatively to provide a large-scale heat source. Because of the special nature of this driving force, as well as the smallness of the Coriolis parameter, large-scale equatorial motion systems have certain distinctive characteristic structural features which are quite different from those of midlatitude systems.

12.3.1 The Intertropical Convergence Zone

In the traditional view, the tropical general circulation was thought to consist of a direct Hadley circulation in which the air in the lower troposphere in both hemispheres moved equatorward toward the *intertropical convergence zone* (ITCZ) where by continuity considerations it was forced to rise and move poleward, thus transporting heat away from the equator in the upper troposphere. However, this simple model of large-scale overturning is not consistent with the observed vertical profile of θ_e. As indicated in Fig. 12.1, the equivalent static stability in the tropics is positive above about 600 mb. Thus, a large-scale upward mass flow would be up the gradient of θ_e in the upper troposphere and would actually cool the upper troposphere in the region of the ITCZ [see Eq. (12.35)]. Such a circulation could not generate potential energy and would not, therefore, satisfy the heat balance in the equatorial zone.

In fact it appears that the only way in which heat can effectively be brought from the surface to the upper-troposphere in the ITCZ is through pseudo-adiabatic ascent in the cores of large cumulonimbus clouds (often referred to as "hot towers"). For such motions, the air parcels approximately conserve θ_e and can therefore arrive in the upper troposphere with a moderate temperature excess. Thus, the heat balance of the equatorial zone can be accounted for, at least qualitatively, provided that the vertical motion in the ITCZ is confined primarily to the updrafts of individual convective cells. Riehl and Malkus (1958) have estimated that only 1500–5000 individual "hot towers" would need to exist simultaneously around the globe to account for the required vertical heat transport in the ITCZ.

This view of the ITCZ as a narrow zonal band of vigorous cumulus convection has been confirmed beyond doubt by recent observations, particularly satellite cloud photos. Observations indicate that within the ITCZ precipitation greatly exceeds the moisture supplied by evaporation from the ocean surface below. Thus, much of the vapor necessary to maintain the convection in the ITCZ must be supplied by the converging tradewind flow in the lower troposphere. In this manner the large-scale flow provides the latent heat supply for the convection and the convective heating in turn produces the large-scale pressure field which maintains the low-level inflow.

The above description of the ITCZ is actually oversimplified. In reality, the ITCZ over the oceans rarely appears as a long unbroken band of heavy convective cloudiness, and it almost never is found right at the equator. Rather, it is usually made up of a number of distinct *cloud clusters*, with scales of the order of a few hundred kilometers, which are separated by regions of relatively clear skies. The strength of the ITCZ is also quite variable in both space and time. It is particularly persistent and well defined over the Pacific and Atlantic between about 5 and 10°N latitude, and occasionally appears in the Pacific between 5 and 10°S. That the ITCZ is centered away from the equator is clearly demonstrated by Fig. 12.4 which shows a 15-d average of satellite cloud brightness photographs for the Pacific. The dry zone along the equator is a particularly striking feature.

12.3.2 Equatorial Wave Disturbances

The cloud clusters which are observed along the ITCZ are actually in general just precipitation zones associated with weak wave disturbances which propagate westward along the ITCZ. That such westward-propagating disturbances exist and are responsible for a large part of the cloudiness in the ITCZ can be easily seen by viewing time–longitude sections constructed from daily satellite pictures cut into thin zonal strips. An example is shown in Fig. 12.5. The well-defined bands of cloudiness which slope from right to left down the page define the locations of the cloud clusters as a function of time. Clearly much of the cloudiness in the 5–10°N latitude zone of the Pacific is associated with westward moving disturbances. The slope of the cloud lines in Fig. 12.5 implies a westward propagation speed of about 8–10 m s^{-1}. The longitudinal separation of the cloud bands is about 3000–4000 km, corresponding to a period range of about 4–5 d for this type of disturbance.

Evidence from diagnostic studies indicates that these westward-propagating disturbances are driven by the release of latent heat in the convective precipitation areas accompanying the waves. The vertical structure of this type of disturbance is shown in schematic form in Fig. 12.6. Since the approximate thermodynamic energy equation (12.36) requires that the vertical motion be proportional to the diabatic heating, the maximum large-scale vertical velocities occur in the convective zone. By continuity there must thus be convergence at low levels in the convection zone and divergence in the upper troposphere. Hence, provided that the absolute vorticity is positive (negative) for Northern (Southern) Hemisphere disturbances there will be cyclonic vorticity generation in the lower troposphere and anticyclonic vorticity generation in the upper troposphere due to the $(\zeta + f)\,\nabla \cdot \mathbf{V}$ term

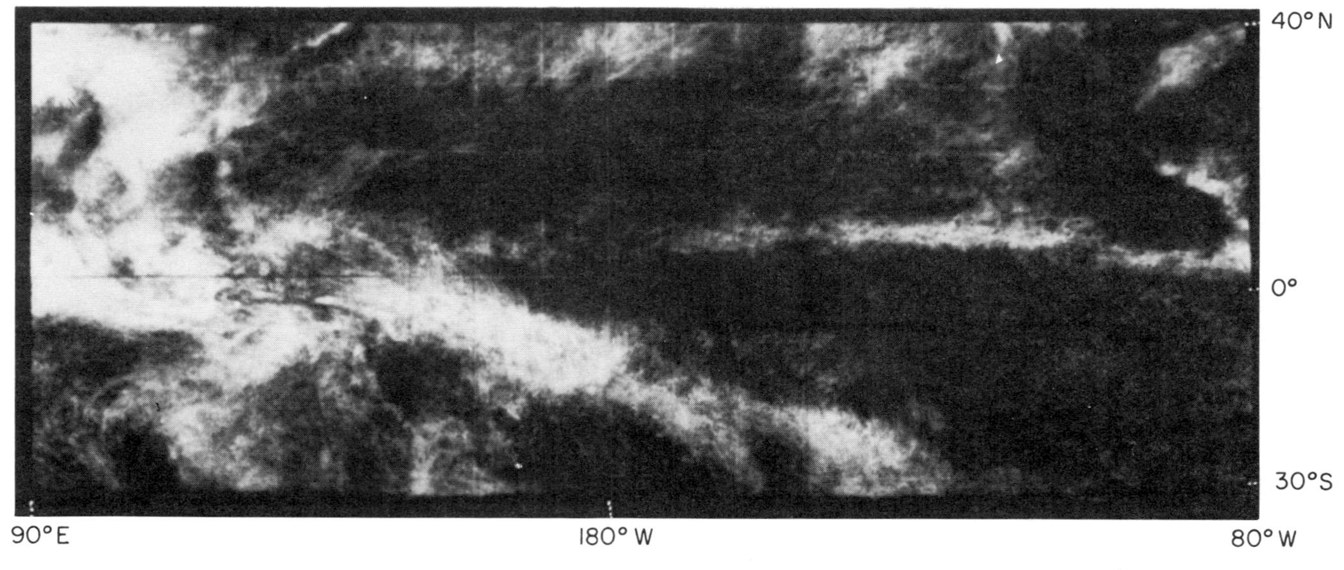

Fig. 12.4 15-d satellite cloud brightness averages for the equatorial Pacific, 16–31 January 1967 (Mercator projection). (After Kornfield *et al.*, 1967. Reproduced with permission of the American Meteorological Society.)

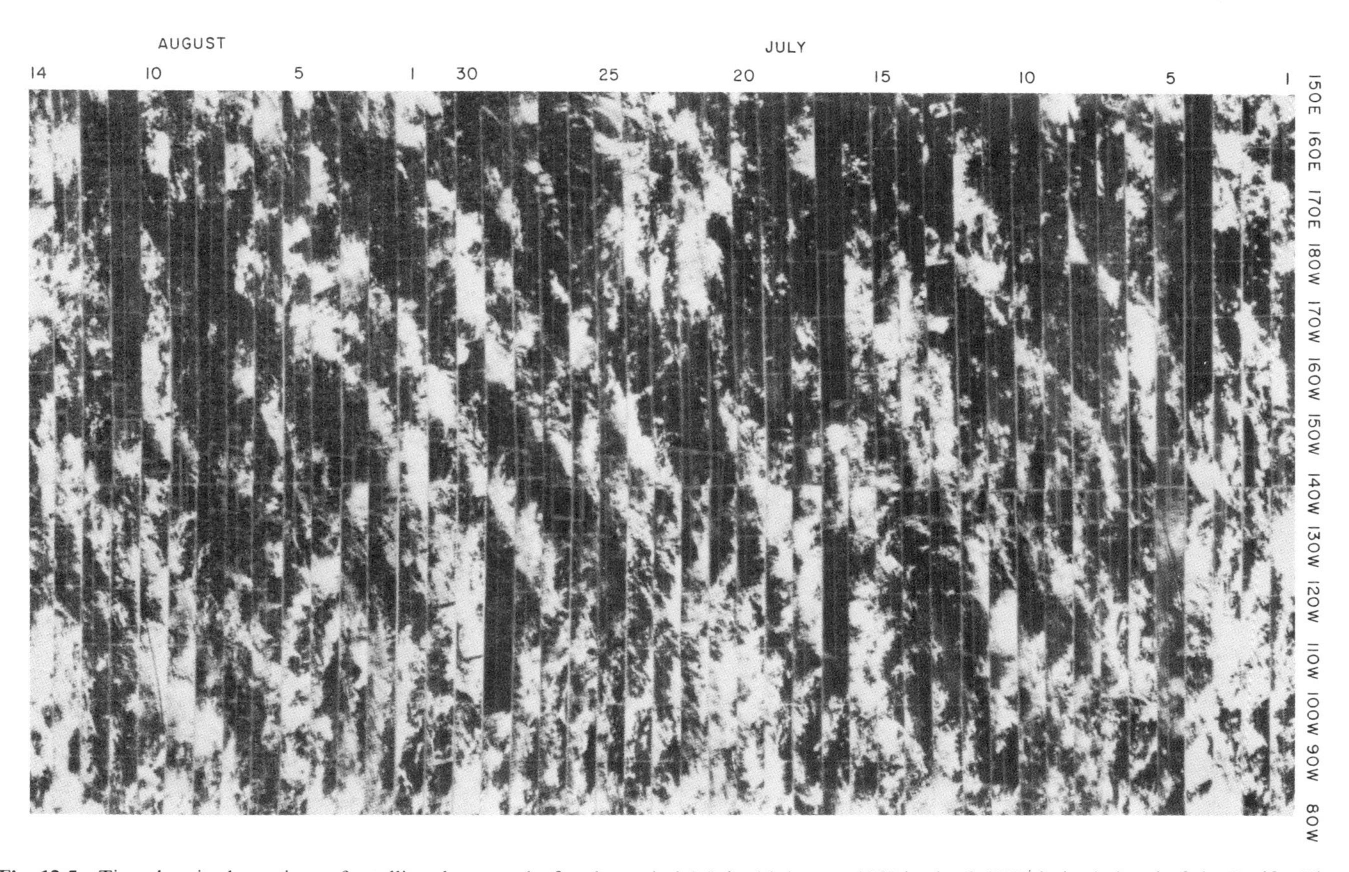

Fig. 12.5 Time–longitude sections of satellite photographs for the period 1 July–14 August 1967 in the 5–10°N latitude band of the Pacific. The westward progression of the cloud clusters is indicated by the bands of cloudiness sloping down the page from right to left. (After Chang, 1970. Reproduced with permission of the American Meteorological Society.)

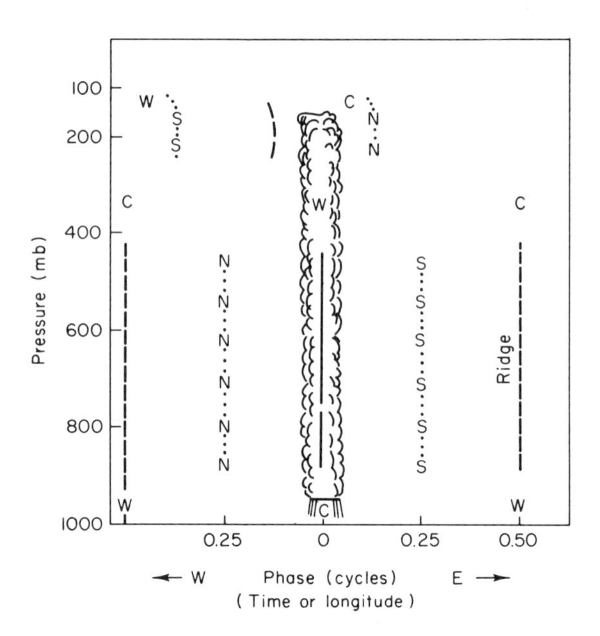

Fig. 12.6 Schematic model for equatorial wave disturbances showing trough axis (solid line)j ridge axis (dashed lines), axes of northerly and southerly wind components (dotted lines). Regions of warm and cold air designated by W and C, respectively. (After Wallace, 1971.)

in the vorticity equation. The mass–velocity adjustment process thus tends to generate a pressure minimum (trough) at low levels.[3] Thus, the thickness (or layer mean temperature) in the convective zone must be greater than in the surrounding environment. As previously mentioned, however, the vorticity generated by the divergence term in such systems is much greater than is actually realized. The difference must be accounted for by the vorticity damping due to cumulus friction.

The strongest convective activity in these waves occurs where the mid-tropospheric temperatures are warmer than average (although generally by less than 1°C). The correlation between temperature and the diabatic heating rate is therefore positive and from the approximate thermodynamic energy equation (12.11) we find that

$$\langle w^* T \rangle S \simeq \langle \dot{q} T \rangle / c_p$$

[3] The terms "trough" and "ridge" as used by tropical meteorologists designate pressure minima and maxima, respectively, just as in midlatitudes. However, in the easterlies of the Northern Hemisphere tropics the zonal mean pressure increases with latitude so that the pattern of isobars depicting a tropical trough will resemble the pattern associated with a ridge in middle latitudes (i.e., there is a poleward deflection of the isobars).

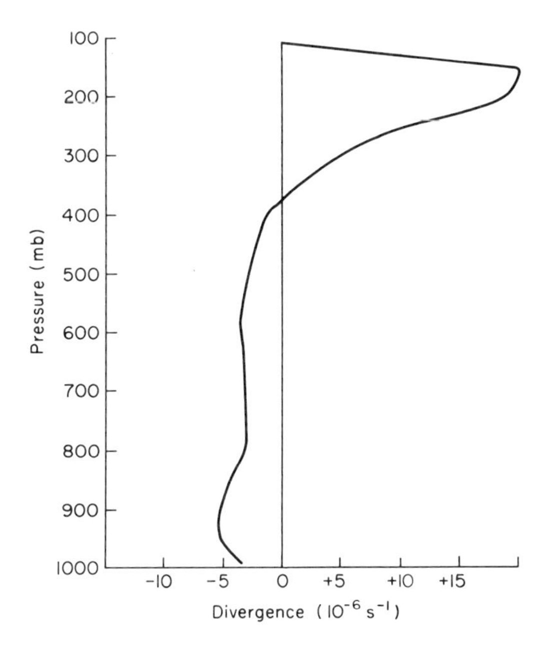

Fig. 12.7 Vertical profile of 4°-square area average divergence based on composites of many equatorial disturbances. (Adapted from Williams, 1971.)

where the angle brackets indicate an average over one wavelength. Thus, the potential energy generated by the diabatic heating is immediately *converted* to kinetic energy through the $\langle w^* T \rangle$ correlation. There is, in this approximation, no storage in the form of available potential energy. The energetics of these disturbances, therefore, differs remarkably from that of midlatitude baroclinic systems in which the available potential energy greatly exceeds the kinetic energy.

A typical vertical profile of divergence in the precipitation zone of this type of disturbance is shown in Fig. 12.7. Convergence is not limited to low-level frictional inflow in the planetary boundary layer, but extends up to nearly 400 mb, which is the height where the hot towers achieve their maximum buoyancy. The deep convergence implies that there must be substantial entrainment of midtropospheric air into the convective cells. Because the midtropospheric air is relatively dry, this entrainment will require considerable evaporation of liquid water to bring the mixture of cloud and environment air to saturation. It will thus reduce the buoyancy of the cloud air, and may in fact produce negatively buoyant convective downdrafts if there is sufficient evaporative cooling. However, in the large cumulonimbus clouds present in equatorial waves the central core updrafts are protected from entrainment by the surrounding cloud air so that they can penetrate nearly to the tropopause without dilution by environmental air. It is these undiluted cores which constitute the "hot towers" referred to in the previous section. Since the hot towers are responsible for most of the vertical heat transport in the ITCZ, and the wave disturbances contain most of the active convective precipitation areas along the ITCZ, it is obvious that the equatorial waves play an essential role in the general circulation of the atmosphere.

12.3.3 African Wave Disturbances

The considerations of the previous subsection are valid for ITCZ disturbances over most regions of the tropics. However, in the region of North Africa local effects due to surface conditions create a unique situation which requires separate discussion. During the Northern Hemisphere summer intense surface heating over the Sahara generates a strong positive temperature gradient in the lower troposphere between the equator and about 25°N. The resulting easterly thermal wind is responsible for the existence of a strong easterly jet core near 650 mb centered near 16°N as shown in Fig. 12.8.

Synoptic scale disturbances are observed to form and propagate westward in the cyclonic shear zone to the south of this jet core. Occasionally such disturbances are progenitors of tropical storms and hurricanes in the Western Atlantic. The average wavelength of the observed African wave disturbances is about 2500 km and the westward propagation speed is about 8 m s^{-1},

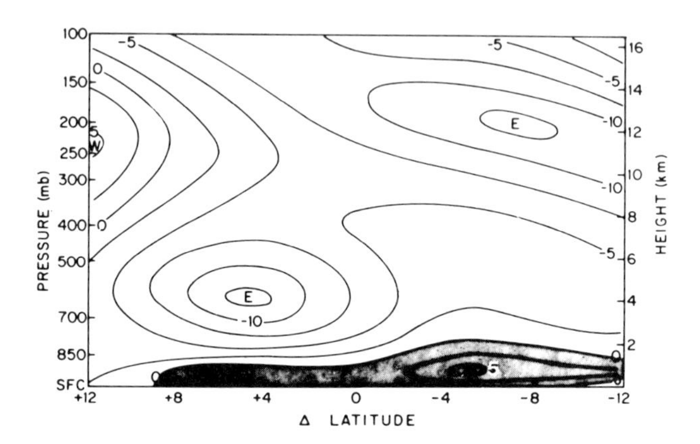

Fig. 12.8 The mean zonal wind distribution in the North African region (30°W to 10°E longitude) for the period 23 August 1974 to 19 September 1974. Latitude is shown relative to latitude of maximum disturbance amplitude at 700 mb (about 12°N). The contour interval is 2.5 m s^{-1}. (After Reed *et al.*, 1977. Reproduced with permission of the American Meteorological Society.)

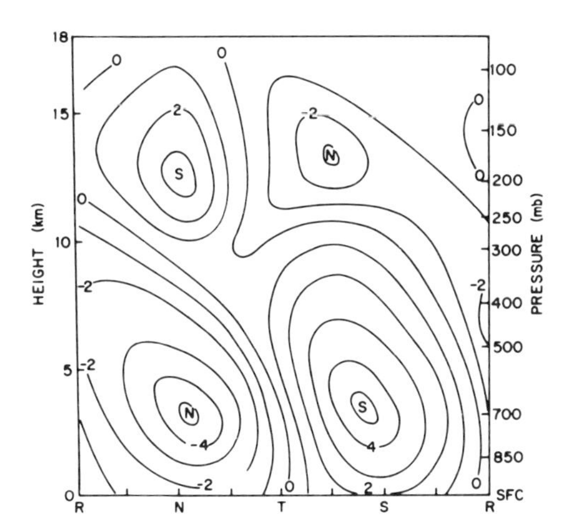

Fig. 12.9 Vertical cross section along the reference latitude of Fig. 12.8 showing perturbation meridional velocities ms^{-1}. R, N, T, S, refer to ridge, northwind, trough, and southwind sectors of the wave respectively. (After Reed *et al.*, 1977. Reproduced with permission of the American Meteorological Society.)

implying a period of about 3.5 d. The disturbances have horizontal velocity perturbations which reach maximum amplitude at the 650-mb level, as indicated in Fig. 12.9. Although there is considerable organized convection associated with these waves they do not appear to be primarily driven by latent heat release, but depend, rather, on barotropic and baroclinic conversions of energy from the easterly jet. The energetics of such disturbances will be discussed further in the following section.

12.4 The Origin of Equatorial Disturbances

Some tropical disturbances probably originate as midlatitude baroclinic waves which move equatorward and gradually assume tropical characteristics. However, there can be little doubt that most synoptic scale disturbances in the equatorial zone originate in situ. Baroclinic instability cannot account for the bulk of these disturbances because, except in a few isolated regions such as North Africa and India, the meridional temperature gradient is very small. Thus, there is an insufficient source of available potential energy to account for the development and maintenance of equatorial disturbances.

Two processes which have been suggested as possible mechanisms for initiating equatorial disturbances are *barotropic instability* of the mean zonal flow due to lateral shear, and *conditional instability of the second kind*

(CISK) due to organized convection driven by moisture convergence in the boundary layer. Neither of these mechanisms can be readily tested with a simple analytic normal modes stability analysis such as the one which we carried out for baroclinic instability. Therefore, we are limited here to a fairly qualitative discussion.

12.4.1 Barotropic Instability

According to the scale analysis of Section 12.1, in the absence of condensation vertical motions must be small in the tropics. Thus, to a first approximation the flow is governed by the barotropic vorticity equation[4]

$$\left(\frac{\partial}{\partial t} + \mathbf{V}\cdot\nabla\right)(\zeta + f) = 0 \tag{12.44}$$

We now assume that the flow consists of a small barotropic perturbation superposed on a zonal current which depends only on latitude. Thus, we let

$$u = \bar{u}(y) + u', \qquad v = v'$$

Since the flow is quasi-nondivergent, we can express the perturbed motion in terms of a perturbation streamfunction ψ' by letting

$$u' = -\frac{\partial\psi'}{\partial y}, \qquad v' = \frac{\partial\psi'}{\partial x}$$

we can then write the linearized perturbation form of (12.44) as

$$\left(\frac{\partial}{\partial t} + \bar{u}\frac{\partial}{\partial x}\right)\nabla^2\psi' + \left(\beta - \frac{d^2\bar{u}}{\partial y^2}\right)\frac{\partial\psi'}{\partial x} = 0 \tag{12.45}$$

The quantity

$$\beta - \frac{d^2\bar{u}}{dy^2} = \frac{d}{dy}(f + \bar{\zeta})$$

is merely the latitudinal gradient of the basic state absolute vorticity. We now assume following the technique introduced in Chapter 7 that solutions of (12.45) can be represented in terms of zonally propagating harmonic waves in the form

$$\psi'(x, y, t) = \psi(y)e^{ik(x-ct)} \tag{12.46}$$

[4] An exception is the African wave disturbance discussed in Section 12.3.3 in which baroclinicity can not be neglected. However, even in this case (12.44) provides a reasonable qualitative model near the 700-mb level.

where $\psi = \psi_r + i\psi_i$ is a complex function of y alone. Substituting from (12.46) into (12.45) we obtain

$$(\bar{u} - c)\left(\frac{d^2\psi}{dy^2} - k^2\psi\right) + \left(\beta - \frac{d^2\bar{u}}{dy^2}\right)\psi = 0 \tag{12.47}$$

which is an ordinary second-order differential equation in $\psi(y)$. As boundary conditions for (12.47), it is usually assumed that the perturbed flow is confined to a zonal channel with walls at $y = \pm L$ so that

$$\psi(y) = 0 \quad \text{at} \quad y = \pm L \tag{12.48}$$

The presence of the physically unrealistic walls at $y = \pm L$ should not significantly affect the results for perturbations whose amplitudes are small near the walls. For a given distribution of $\bar{u}(y)$ it turns out that (12.47) will have solutions which satisfy (12.48) only for certain values of the phase speed c. In cases where c is complex with a positive imaginary part we see from (12.46) that the perturbation amplitude will grow exponentially in time.

In practice it is not a simple matter to determine the solutions of (12.47) for a particular profile $u(y)$ because the coefficients are not constant. However, it is possible to obtain *necessary* conditions for the existence of instability by application of simple integral considerations. Dividing through in (12.47) by $(\bar{u} - c)$ we have

$$\frac{d^2\psi}{dy^2} - \left(k^2 - \frac{\beta - d^2\bar{u}/dy^2}{\bar{u} - c}\right)\psi = 0 \tag{12.49}$$

If the phase speed is complex, $c = c_r + ic_i$, then $(\bar{u} - c)^{-1}$ is also complex and has real and imaginary parts

$$\delta_r = \frac{u - c_r}{(u - c_r)^2 + c_i^2}, \qquad \delta_i = \frac{c_i}{(u - c_r)^2 + c_i^2}$$

Equation (12.49) can then be separated into real and imaginary parts as follows:

$$\frac{d^2\psi_r}{dy^2} - \left[k^2 - \left(\beta - \frac{d^2\bar{u}}{dy^2}\right)\delta_r\right]\psi_r - \left(\beta - \frac{d^2\bar{u}}{dy^2}\right)\delta_i\psi_i = 0 \tag{12.50}$$

$$\frac{d^2\psi_i}{dy^2} - \left[k^2 - \left(\beta - \frac{d^2\bar{u}}{dy^2}\right)\delta_r\right]\psi_i + \left(\beta - \frac{d^2\bar{u}}{dy^2}\right)\delta_i\psi_r = 0 \tag{12.51}$$

Multiplying (12.50) by ψ_i, (12.51) by ψ_r, and subtracting the latter from the former, we obtain

$$\psi_i\frac{d^2\psi_r}{dy^2} - \psi_r\frac{d^2\psi_i}{dy^2} - \left(\beta - \frac{d^2\bar{u}}{dy^2}\right)\delta_i(\psi_i^2 + \psi_r^2) = 0$$

which can also be written as

$$\frac{d}{dy}\left(\psi_i \frac{d\psi_r}{dy} - \psi_r \frac{d\psi_i}{dy}\right) = \left(\beta - \frac{d^2\bar{u}}{dy^2}\right)\delta_i(\psi_r^2 + \psi_i^2) \tag{12.52}$$

Integrating (12.52) with respect to y and applying the boundary conditions

$$\psi_r = \psi_i = 0 \quad \text{at} \quad y = \pm L$$

we find that the terms on the left integrate to zero so that we are left with the integral condition

$$\int_{-L}^{+L}\left(\beta - \frac{d^2\bar{u}}{dy^2}\right)\delta_i|\psi|^2\, dy = 0 \tag{12.53}$$

where

$$|\psi|^2 = \psi_r^2 + \psi_i^2$$

Now, for an unstable perturbation to exist requires that $\delta_i > 0$ (i.e., $c_i > 0$). Since $|\psi|^2 \geq 0$ everywhere in the domain, the integral condition in (12.53) can be satisfied for an unstable wave only if $\beta - d^2\bar{u}/dy^2$ *changes sign* somewhere in the region $-L < y < L$. Thus, a *necessary* condition for barotropic instability is that the gradient of absolute vorticity of the mean current must vanish somewhere in the region, that is,

$$\beta - \frac{d^2\bar{u}}{dy^2} = 0 \qquad \text{somewhere} \tag{12.54}$$

In Fig. 12.10 the absolute vorticity profile for the African easterly jet shown in Fig. 12.8 is plotted. The shaded region indicates the area in which

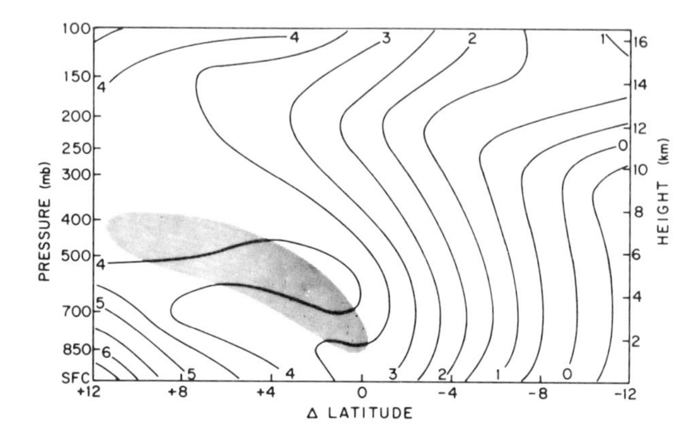

Fig. 12.10 Absolute vorticity (units $10^{-5}\ s^{-1}$) corresponding to the mean wind field of Fig. 12.8. Shading shows region where $\beta - \bar{u}_{yy}$ is negative. (After Reed *et al.*, 1977. Reproduced with permission of the American Meteorological Society.)

the vorticity gradient is negative. Thus it is clear that the African jet satisfies the necessary condition (12.54) for barotropic instability.[5]

Numerical solutions of the perturbation equation (12.49) for the 700-mb profile $\bar{u}(y)$ observed in the African jet indicate that the jet is in fact barotropically unstable and that the most unstable disturbance mode has characteristics qualitatively in accord with the observed waves. Thus, barotropic instability appears to be the mechanism responsible for the generation of African wave disturbances. It should be mentioned, however, that observations indicate that baroclinic energy conversions are also of importance in the energetics of the African waves.

Although barotropic instability provides a satisfactory mechanism for the generation of African waves, and may also play a role in the equatorial Pacific, it must be noted that barotropically unstable disturbances can be maintained only if the shear of the mean zonal flow remains unstable so that the waves can extract energy from the mean flow. Since tropical disturbances are observed to exist in the absence of strong lateral shear it is unlikely that barotropic instability is the primary energy source for the maintenance of the waves over most of the oceanic equatorial zone. Finally, it should be mentioned that barotropic instability is not a uniquely tropical phenomenon. The Coriolis parameter only appears in the barotropic vorticity equation in differentiated form as β. Thus, the equator has no special significance for barotropic flow. Barotropic instability is also possible in the vicinity of the midlatitude jet stream. However, at middle latitudes baroclinic instability is generally the more important mechanism.

12.4.2 Conditional Instability of the Second Kind

As we have previously pointed out, the release of latent heat in cumulus convection is almost certainly the primary energy source for the maintenance of finite-amplitude equatorial wave disturbances. It might be thought, therefore, that such waves are merely a direct result of the release of conditional instability in a saturated atmosphere with a super moist-adiabatic lapse rate $(\partial\theta_e^*/\partial z < 0)$. However, numerous theoretical studies have shown that conditional instability produces maximum growth rates for motions on the scale of individual cumulus clouds. Therefore, ordinary conditional instability cannot explain the synoptic scale organization of the motion. Observations indicate, moreover, that the mean tropical atmosphere is not saturated even in the planetary boundary layer. Thus, a parcel must undergo a considerable

[5] It should be noted here that the profile shown in Fig. 12.10 is not a *zonal* mean, but is, rather, a *time* mean for a limited longitudinal domain. Provided that the longitudinal scale of variation of this time mean zonal flow is large compared to the scale of the disturbances, the time mean flow may be regarded as a locally valid basic state for linear stability calculations.

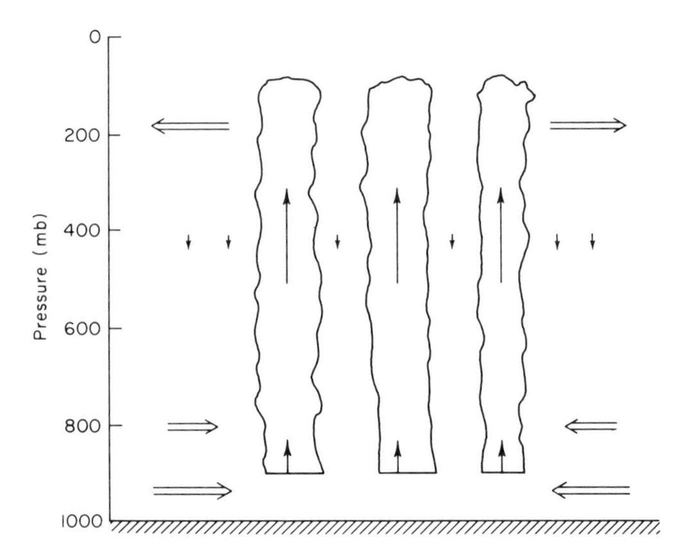

Fig. 12.11 The relationship between forced convection and low-level synoptic scale convergence.

amount of forced ascent before it becomes positively buoyant. Such forced ascent will occur in an organized manner only in regions of low-level convergence. The cumulus convection and the large-scale motion must then be viewed as cooperatively interacting. The cumulus supplies the heat necessary to drive the large-scale disturbance, and the large-scale disturbance produces the moisture convergence necessary to drive the cumulus convection. This interaction process is indicated schematically in Fig. 12.11.

When the cooperative interaction between the cumulus convection and a large-scale perturbation leads to unstable growth of the large scale system the process is referred to as *conditional instability of the second kind*,[6] or CISK. The primary difficulty in testing the CISK mechanism theoretically is that the heating field due to the cumulus clouds must be expressed in terms of the large-scale field variables. Otherwise, it would be necessary to model the detailed structure of all the cumulus elements in the disturbance. This is just the cumulus parameterization problem discussed in Sections 12.2.6 and 12.2.7. Models for the large-scale heating by cumulus convection similar to that given in (12.42) have been used to study the CISK mechanism theoretically. There are two primary classes of such models: In the first type the low-level convergence field which supplies the moisture for the cumulus clouds is produced by boundary layer frictional inflow into a surface trough.

[6] This term was coined by Charney and Eliassen (1964). The acronym CISK was first used by Rosenthal and Koss (1968).

This mechanism has been quite successful in explaining the development of intense tropical storms and hurricanes. In the second type, often called *wave-CISK*, the low-level convergence is simply the convergent velocity field associated with the inviscid wave itself (which may be modified by boundary layer frictional effects). Many theoretical studies have been directed toward understanding the dynamics of wave-CISK. Such efforts have, however, failed to provide an explanation for the observed scales of tropical wave disturbances. Unlike the baroclinic instability theory for midlatitude disturbances, wave-CISK does not indicate maximum growth rates for synoptic scale disturbances. Rather, wave-CISK models indicate that small-scale gravity wave disturbances should be the most rapidly amplifying. Nevertheless, the wave-CISK process may be crucial for the dynamics of tropical synoptic waves. In a wave-CISK model which included the cumulus friction effects discussed in Section 12.2.7, Stevens and Lindzen (1978) found that for parameters appropriate to observed synoptic wave disturbances, the wave-CISK modes with cumulus friction tend to be *neutrally* stable. Thus, if such modes were excited in certain locations (e.g., over North Africa) by barotropic instability, they would be able to maintain themselves over the tropical oceans almost indefinitely through the wave-CISK energy source.

12.5 Tropical Cyclones

Tropical cyclones are intense vortical storms which develop over the tropical oceans in regions of very warm surface water. These storms, which are called hurricanes in the Atlantic and typhoons in the Pacific, owe their existence to a very effective interactive coupling between the cumulus scale motion and the large-scale wind field.

In tropical cyclones the horizontal scale of the region where convection is strong is typically about 100 km in radius. Maximum tangential wind speeds in these storms range typically from 50–100 m s^{-1}. For such high velocities and relatively small scales, the centrifugal force term cannot be neglected compared to the Coriolis force. Thus, to a first approximation the radial force balance in a steady-state hurricane satisfies the gradient wind relationship, not geostrophic balance. Motion on the hurricane scale does, however, remain in hydrostatic balance although motion in the individual cumulus towers is not in hydrostatic equilibrium. Contrary to the situation for weak tropical wave disturbances which we have previously discussed, there do exist large-amplitude temperature fluctuations in hurricanes. The magnitude of these temperature fluctuations can be estimated by scaling arguments similar to those used in Section 12.1.

We first require a relationship between the mass field and tangential velocity field for an axially symmetric hurricane vortex. The gradient wind balance (3.10) can be expressed in cylindrical isobaric coordinates as

$$\frac{v^2}{r} + fv = \frac{\partial \Phi}{\partial r} \tag{12.55}$$

where r is the radial distance from the axis of the storm (positive outward) and v is the tangential velocity (positive for cyclonic flow). We can eliminate Φ in (12.55) with the aid of the hydrostatic equation in the log-pressure coordinate system,

$$\frac{\partial \Phi}{\partial z^*} = \frac{RT}{H}$$

to obtain a relationship between the radial temperature gradient and the vertical shear of the tangential wind:

$$\frac{\partial v}{\partial z^*}\left(f + \frac{2v}{r}\right) = \frac{R}{H}\frac{\partial T}{\partial r} \tag{12.56}$$

Now the cyclonic flow in a hurricane is observed to decrease quite rapidly with height from its maximum values in the lower troposphere. Thus, above the boundary layer $\partial v/\partial z^* < 0$, which by (12.56) implies that $\partial T/\partial r < 0$ and the temperature maximum must occur at the center of the storm. This is consistent with the observation that hurricanes are *warm core* systems, and of course is necessary if the cumulus heating is to generate kinetic energy.

If we let the vertical scale of the system be equal to the scale height H, the tangential velocity scale be $U \sim 50 \text{ m s}^{-1}$, the horizontal scale be $L \sim 100$ km, and assume that $f \sim 5 \times 10^{-5} \text{ s}^{-1}$, we find from (12.56) that the radial temperature fluctuation must have a magnitude

$$\Delta T \sim \frac{LH}{R}\left(\frac{U}{H}\right)\left(f + \frac{2U}{L}\right) \sim 10°C$$

Therefore, the radial advection of temperature cannot be neglected in the thermodynamic energy equation. The approximate expression of the thermodynamic energy equation (12.36) which was valid for weak tropical disturbances cannot be used in modeling the dynamics of a fully developed hurricane.

The origin of tropical cyclones is still a matter of controversy. It is not clear under what conditions a weak tropical disturbance can be transformed into a hurricane. Although there are many tropical disturbances each year, only rarely does one develop into a hurricane. Thus, the development of a hurricane must require rather special conditions. Theoretical investigations of this problem to date have nearly always assumed the existence of a small cylindrically symmetric disturbance, and examined the conditions under which unstable amplification of the disturbance can occur. The results of those investigations show that under favorable conditions of moisture supply, the CISK mechanism can lead to rapid development of a hurricane scale disturbance. In this amplification process friction must be thought of as playing an energy *producing* role. The kinetic energy loss due to frictional convergence in the boundary layer is more than balanced by the energy generated from the latent heating due to the moisture which the frictional convergence supplies to the cumulus. The first study of this type was due to Charney and Eliassen (1964) who used a model with two levels in the vertical. In their model the heating field is related to the large-scale variables by assuming that all moisture which converges into a vertical column is condensed out as rain and that the latent heat released is distributed in the vertical in proportion to the heat release for moist adiabatic ascent. The boundary layer is assumed to be an Ekman layer so that the boundary layer convergence is proportional to the vorticity at the top of the layer. Even such a simplified model results in a rather complicated mathematical analysis. However, Charney and Eliassen did obtain solutions in the form of self-amplifying perturbations. For an atmosphere with mean relative humidity of 80%, which corresponds to observed conditions in regions of hurricane genesis, a doubling time of about one day results for a disturbance with a radial scale of about 100 km. Unfortunately the growth rate is even larger for smaller scale disturbances. Inclusion of viscosity or nonlinear effects in the model would no doubt damp the growth rate for the smaller scales. Therefore, the conclusion from Charney and Eliassen's study is that the cooperative interaction of cumulus convection and the large-scale fields can produce an amplifying disturbance on the hurricane scale provided that the mean atmosphere is unsaturated but conditionally unstable.

Several numerical models to simulate the life cycle of hurricanes have also been developed. Nearly all the successful models of this type have used some form of the CISK mechanism as the energy source for the hurricane.[7] One of the most interesting conclusions to come from this work is the confirmation of the observational evidence that the hurricane can maintain itself only in the presence of very warm ocean surface temperatures. For surface temperatures less than 26°C, the converging flow in the boundary layer apparently cannot acquire a high enough equivalent potential temperature to sustain the intense rate of convective heating needed to maintain the hurricane circulation.

[7] The exception is the model of Rosenthal (1978) in which the convective elements are explicitly resolved by the grid system.

Finally, it should be said that most theoretical studies of the hurricane have been limited to axially symmetric models. However, observed precipitation patterns in the hurricane are hardly symmetric. Rather the convection tends to concentrate along *spiral bands* radiating from the center of the storm. Furthermore, the environment in which hurricanes develop is far from symmetric. Thus, current modeling efforts are being directed towards development of three-dimensional models which can simulate the important role of asymmetry in the development and evolution of hurricanes.

Problems

1. An air parcel at 920 mb with temperature 20°C is saturated (mixing ratio 16 g kg^{-1}). Compute θ_e for the parcel.

2. Suppose an entraining cumulus updraft has a vertical mass flux M which increases exponentially with height according to $M = e^{z/H}$ where $H = 8$ km. If the updraft speed is 3 m s^{-1} at 2-km height, what is its value at a height of 8 km assuming that the updraft has zero net buoyancy?

3. Suppose that the relative vorticity at the top of the Ekman layer is $\zeta = 2 \times 10^{-5}\ \text{s}^{-1}$. Let the eddy viscosity coefficient be $K = 10\ \text{m}^2\ \text{s}^{-1}$, the Coriolis parameter $5 \times 10^{-5}\ \text{s}^{-1}$, and the water vapor mixing ratio at the top of the Ekman layer 12 g kg^{-1}. Use the method of Section 12.2.6 to estimate the precipitation rate due to moisture convergence in the Ekman layer.

***4.** Show by multiplying (12.50) by ψ_r, (12.51) by ψ_i and adding, that for unstable waves ($c_i \neq 0$)

$$\int_{-L}^{+L} [(\beta - d^2\bar{u}/dy^2)|\psi|^2\bar{u}/|\bar{u} - c|^2]\, dy > 0$$

i.e., $\bar{u}$ and the gradient of absolute vorticity must be positively correlated.

5. Consider the following profiles of the mean zonal current:

(a) $$\bar{u}(y) = +u_0 \tanh[l(y - y_0)]$$

(b) $$\bar{u}(y) = -u_0 \sin^2[l(y - y_0)]$$

where u_0, y_0, and l are constants and y is the distance from the equator. Determine the necessary conditions for each profile to be barotropically unstable.

6. The azimuthal velocity component in some hurricanes is observed to have a radial dependence given by $v = V_0(r_0/r)^2$ for distances from the center given by $r \geq r_0$. Letting $V_0 = 50$ m s^{-1} and $r_0 = 50$ km find the total geopotential difference between the far field ($r \to \infty$) and $r = r_0$, assuming gradient wind balance and $f_0 = 5 \times 10^{-5}$ s^{-1}. At what distance from the center does the Coriolis force equal the centrifugal force?

7. Show that for dry adiabatic processes the thermodynamic energy equation may be expressed approximately as $ds/dt \simeq 0$, where $s = c_p T + gz$. Hence, show that for moist adiabatic processes $dh/dt \simeq 0$, where $h = s + L_c q$.

Suggested References

Riehl, *Tropical Meteorology*, although somewhat out of date remains the basic reference on the descriptive aspects of the subject.

Palmén and Newton, *Atmospheric Circulation Systems*, contains a brief but more up-to-date review of observational aspects of tropical circulations.

Wallace (1971) is a detailed review of the structure of synoptic scale tropospheric wave disturbances in the equatorial Pacific.

Anthes (1974) provides an excellent review of observational and theoretical studies of the mature hurricane.

Appendix

Useful Constants and Parameters

Gravitational constant	$G = 6.673 \times 10^{-11}$ N m^2 kg^{-2}
Acceleration due to gravity at sea level	$g_0 = 9.81$ m s^{-2}
Mean radius of the earth	$a = 6.37 \times 10^6$ m
Angular speed of rotation of the earth	$\Omega = 7.292 \times 10^{-5}$ rad s^{-1}
Universal gas constant	$R^* = 8.314 \times 10^3$ J K^{-1} kmol^{-1}
Gas constant for dry air	$R = 287$ J K^{-1} kg^{-1}
Specific heat of dry air at constant pressure	$c_p = 1004$ J K^{-1} kg^{-1}
Specific heat of dry air at constant volume	$c_v = 717$ J K^{-1} kg^{-1}
Ratio of specific heats	$\gamma \equiv c_p/c_v = 1.4$
Molecular weight of water	$m_v = 18.016$ kg kmol^{-1}
Latent heat of condensation at 0°C	$L_c = 2.5 \times 10^6$ J kg^{-1}
Mass of the earth	$M = 5.988 \times 10^{24}$ kg
Standard sea level pressure	$p_0 = 101.325$ kPa
Standard sea level temperature	$T_0 = 288.15$ K
Standard sea level density	$\rho_0 = 1.225$ kg m^{-3}

Appendix

B List of Symbols

Only the principal symbols are listed. Symbols formed by adding primes or subscripted indices are not listed separately. Boldface type indicates vector quantities. Where symbols have more than one meaning, the section where the second meaning is first used is indicated in the list

a	(1) Radius of the earth; (2) inner radius of a laboratory annulus (Section 10.5)
b	Outer radius of a laboratory annulus
c	Phase speed of a wave
c_p	Specific heat of dry air at constant pressure
c_{pv}	Specific heat of water vapor at constant pressure
c_v	Specific heat of dry air at constant volume
c_w	Specific heat of liquid water
d	Grid distance
e	(1) Vapor pressure; (2) internal energy per unit mass (Section 2.6)
e_s	Saturation vapor pressure
f	Coriolis parameter, $2\Omega \sin \phi$
g	Magnitude of gravity
$\mathbf{g}$	Gravity
$\mathbf{g}^*$	Gravitational acceleration

h (1) Depth of fluid layer; (2) moist static energy (Section 12.2.1)
i Square root of minus one
$\mathbf{i}$ Unit vector along the x axis
$\mathbf{j}$ Unit vector along the y axis
$\mathbf{k}$ Unit vector along the z axis
k Zonal wave number
l (1) Eddy mixing length; (2) meridional wave number (Section 7.5)
m (1) A mass element; (2) vertical wave number (Section 7.4)
m_v Molecular weight of water
n Distance in direction normal to a parcel trajectory
$\mathbf{n}$ Unit vector normal to a parcel trajectory
p Pressure
p_s (1) Standard constant pressure; (2) surface pressure in σ-coordinate system (Section 8.7)
q (1) quasi-geostrophic potential vorticity; (2) water vapor mixing ratio (Chapter 12)
$\dot{q}$ Diabatic heating rate per unit mass
q_i Ozone mixing ratio
q_s Saturation mixing ratio
r Radial distance in spherical coordinates
$\mathbf{r}$ A position vector
s (1) Distance along a parcel trajectory: (2) generalized vertical coordinate (Section 1.6.3); (3) zonal wave number (Section 11.5); (4) dry static energy (Section 12.2.1); (5) entropy (Section 2.7)
t Time
$\mathbf{t}$ Unit vector parallel to a parcel trajectory
u_* Friction velocity
u x component of velocity (eastward)
v y component of velocity (northward)
w z component of velocity (upward)
w^* Vertical motion in log-pressure system
x, y, z Eastward, northward, and upward distance, respectively, in spherical coordinates or on a β plane
z^* $\equiv -H \ln(p/p_s)$, vertical coordinate in log-pressure system
$\mathbf{A}$ An arbitrary vector
A Area
A_M, A_T Wave amplitudes
A_Z Eddy mixing coefficient
De Depth of Ekman layer
E_I Internal energy
$\mathbf{F}$ A force
$\mathbf{Fr}$ Frictional force

G	(1) Universal gravitational constant; (2) rate of energy generation by diabatic heating (Section 10.2)
H	Scale height
K	(1) Eddy viscosity coefficient; (2) kinetic energy (Section 10.2)
L	A length scale
M	Mass
N	Buoyancy frequency
P	Available potential energy
R	(1) Gas constant for dry air; (2) distance from the axis of rotation of the earth to a point on the surface of the earth (Section 1.5)
$\mathbf{R}$	Vector in the equatorial plane directed from the axis of rotation to a point on the surface of the earth
R^*	Universal gas constant
S	$\equiv HN^2/R$, stability parameter in log-pressure coordinates
S_p	$\equiv -T\,\partial \ln \theta/\partial p$, stability parameter in pressure coordinates
T	Temperature
T_0	Standard temperature, constant or dependent only on height
U	Horizontal velocity scale
V	Speed in natural coordinates
δV	Volume increment
$\mathbf{U}$	Three-dimensional velocity vector
$\mathbf{V}$	Horizontal velocity vector
W	Vertical motion scale
X	Meridional streamfunction
Z	Geopotential height
α	Specific volume
β	(1) $\equiv df/dy$, variation of the Coriolis parameter with latitude; (2) the angular direction of the wind (Section 3.3)
γ	$\equiv c_p/c_v$, the ratio of the specific heats
ε	(1) Rate of frictional energy dissipation; (2) thermal expansion coefficient of water (Section 10.5)
ζ	Vertical component of relative vorticity
η	(1) Vertical component of absolute vorticity; (2) diabatic heating profile (Section 12.2.6)
θ	Potential temperature
θ_e	Equivalent potential temperature
κ	$\equiv R/c_p$, ratio of gas constant to specific heat at constant pressure
λ	Longitude, positive eastward
μ	(1) Dynamic viscosity coefficient; (2) angular momentum per unit mass (Section 10.3)
ν	(1) Angular frequency of a wave; (2) kinematic viscosity (Section 1.4.3)

ρ Density
σ (1) $\equiv -\alpha\, \partial\theta/\partial p$, static stability parameter in isobaric coordinates; (2) $\equiv p/p_s$, vertical coordinate in σ system (Section 8.7)
τ_z Horizontal frictional stress due to vertical shear
ϕ Latitude
χ Geopotential tendency
ψ Streamfunction
ω Vertical wind component (dp/dt) in isobaric coordinates
Γ $\equiv -dT/dz$, Lapse rate of temperature
Γ_d Dry adiabatic lapse rate
Φ Geopotential
Ω (1) Angular speed of rotation of the earth; (2) angular speed of rotation of laboratory annulus (Section 10.5)
$\mathbf{\Omega}$ Angular velocity of the earth

Appendix

C Vector Analysis

C.1 Vector Identities

The following formulas may be shown to hold where Φ is an arbitrary scalar and **A** and **B** are arbitrary vectors.

$$\nabla \times \nabla\Phi = 0$$

$$\nabla \cdot (\Phi\mathbf{A}) = \Phi\nabla \cdot \mathbf{A} + \mathbf{A} \cdot \nabla\Phi$$

$$\nabla \times (\Phi\mathbf{A}) = \nabla\Phi \times \mathbf{A} + \Phi(\nabla \times \mathbf{A})$$

$$\nabla \cdot (\nabla \times \mathbf{A}) = 0$$

$$(\mathbf{A} \cdot \nabla)\mathbf{A} = \tfrac{1}{2}\nabla(\mathbf{A} \cdot \mathbf{A}) - \mathbf{A} \times (\nabla \times \mathbf{A})$$

$$\nabla \times (\mathbf{A} \times \mathbf{B}) = \mathbf{A}(\nabla \cdot \mathbf{B}) - \mathbf{B}(\nabla \cdot \mathbf{A}) - (\mathbf{A} \cdot \nabla)\mathbf{B} + (\mathbf{B} \cdot \nabla)\mathbf{A}$$

C.2 Integral Theorems

(a) Divergence theorem:

$$\int_A \mathbf{B} \cdot \mathbf{n}\, dA = \int_V \nabla \cdot \mathbf{B}\, dV$$

where V is a volume enclosed by surface A and **n** is a unit normal on A.

(b) Stokes' theorem:

$$\oint \mathbf{B} \cdot d\mathbf{l} = \int_A (\nabla \times \mathbf{B}) \cdot \mathbf{n}\, dA$$

where A is a surface bounded by the line $\mathbf{l}$ and $\mathbf{n}$ is a unit normal of A.

C.3 Vector Operations in Various Coordinate Systems

(a) Cartesian coordinates: (x, y, z)

position vector $\quad \mathbf{r} = \mathbf{i}x + \mathbf{j}y + \mathbf{k}z$

horizontal velocity $\quad \mathbf{V} = \mathbf{i}u + \mathbf{j}v$

$$\nabla\Phi = \mathbf{i}\frac{\partial\Phi}{\partial x} + \mathbf{j}\frac{\partial\Phi}{\partial y} + \mathbf{k}\frac{\partial\Phi}{\partial z}$$

$$\nabla \cdot \mathbf{V} = \frac{\partial u}{\partial x} + \frac{\partial v}{\partial y}$$

$$\mathbf{k} \cdot (\nabla \times \mathbf{V}) = \frac{\partial v}{\partial x} - \frac{\partial u}{\partial y}$$

$$\nabla_h^2\Phi = \frac{\partial^2\Phi}{\partial x^2} + \frac{\partial^2\Phi}{\partial y^2}$$

(b) Cylindrical coordinates: (r, λ, z)

position vector $\quad \mathbf{r} = \mathbf{i}r + \mathbf{k}z$

horizontal velocity $\quad \mathbf{V} = \mathbf{i}u + \mathbf{j}v$

$$\nabla\Phi = \mathbf{i}\frac{\partial\Phi}{\partial r} + \mathbf{j}\frac{1}{r}\frac{\partial\Phi}{\partial\lambda} + \mathbf{k}\frac{\partial\Phi}{\partial z}$$

$$\nabla \cdot \mathbf{V} = \frac{1}{r}\frac{\partial}{\partial r}(ru) + \frac{1}{r}\frac{\partial v}{\partial\lambda}$$

$$\mathbf{k} \cdot (\nabla \times \mathbf{V}) = \frac{1}{r}\frac{\partial}{\partial r}(rv) - \frac{1}{r}\frac{\partial u}{\partial\lambda}$$

$$\nabla_h^2\Phi = \frac{1}{r}\frac{\partial}{\partial r}\left(r\frac{\partial\Phi}{\partial r}\right) + \frac{1}{r^2}\frac{\partial^2\Phi}{\partial\lambda^2}$$

(c) Spherical coordinates: (λ, ϕ, r)

$$\text{position vector} \qquad \mathbf{r} = \mathbf{k}r$$

$$\text{horizontal velocity} \qquad \mathbf{V} = \mathbf{i}u + \mathbf{j}v$$

$$\nabla\Phi = \frac{\mathbf{i}}{r\cos\phi}\frac{\partial\Phi}{\partial\lambda} + \frac{\mathbf{j}}{r}\frac{\partial\Phi}{\partial\phi} + \mathbf{k}\frac{\partial\Phi}{\partial r}$$

$$\nabla\cdot\mathbf{V} = \frac{1}{r\cos\phi}\left[\frac{\partial u}{\partial\lambda} + \frac{\partial}{\partial\phi}(v\cos\phi)\right]$$

$$\mathbf{k}\cdot(\nabla\times\mathbf{V}) = \frac{1}{r\cos\phi}\left[\frac{\partial v}{\partial\lambda} - \frac{\partial}{\partial\phi}(u\cos\phi)\right]$$

$$\nabla_{\mathrm{h}}^2\Phi = \frac{1}{r^2\cos^2\phi}\left[\frac{\partial^2\Phi}{\partial\lambda^2} + \cos\phi\frac{\partial}{\partial\phi}\left(\cos\phi\frac{\partial\Phi}{\partial\phi}\right)\right]$$

Appendix

D The Equivalent Potential Temperature

A mathematical expression for θ_e can be derived by applying the first law of thermodynamics to a mixture of 1 g dry air plus q g of water vapor (q, called the *mixing ratio*, is usually expressed as grams of vapor per kilogram of dry air). If the parcel is not saturated, the dry air satisfies the energy equation

$$c_p \, dT - \frac{d(p-e)}{p-e} RT = 0 \tag{D.1}$$

and the water vapor satisfies

$$c_{pv} \, dT - \frac{de}{e} \frac{R^*}{m_v} T = 0 \tag{D.2}$$

where the motion is assumed to be adiabatic. Here, e is the partial pressure of the water vapor, c_{pv} the specific heat at constant pressure for the vapor, R^* the universal gas constant, and m_v the molecular weight of water. If the parcel is saturated, then condensation of $-dq_s$ g vapor per kilogram dry air will heat the mixture of air and vapor by an amount of $-L_c \, dq_s$, where L_c is the latent heat of condensation. Thus, neglecting the small amount of heat which goes into the liquid water, the saturated parcel must satisfy the energy equation

$$c_p \, dT + q_s c_{pv} \, dT - \frac{d(p-e_s)}{(p-e_s)} RT - q_s \frac{de_s}{e_s} \frac{R^*}{m_v} T = -L_c \, dq_s \tag{D.3}$$

where q_s and e_s are the saturation mixing ratio and vapor pressure, respectively. The quantity de_s/e_s may be expressed in terms of temperature using the Clausius–Clapeyron equation[1]

$$\frac{de_s}{dT} = \frac{m_v L_c e_s}{R^* T^2} \tag{D.4}$$

Substituting from (D.4) into (D.3) and rearranging terms we obtain

$$-L_c\, d\left(\frac{q_s}{T}\right) = c_p \frac{dT}{T} - \frac{R d(p - e_s)}{p - e_s} + q_s c_{pv} \frac{dT}{T} \tag{D.5}$$

If we now define the potential temperature of the dry air θ_d, according to

$$c_p\, d \ln \theta_d = c_p\, d \ln T - R\, d \ln(p - e_s)$$

we can rewrite (D.5) as

$$-L_c\, d\left(\frac{q_s}{T}\right) = c_p\, d \ln \theta_d + q_s c_{pv}\, d \ln T \tag{D.6}$$

However, it may be shown that

$$\frac{dL_c}{dT} = c_{pv} - c_w \tag{D.7}$$

where c_w is the specific heat of liquid water. Using (D.7) to eliminate c_{pv} in (D.6) we obtain

$$-d\left(\frac{L_c q_s}{T}\right) = c_p\, d \ln \theta_d + q_s c_w\, d \ln T \tag{D.8}$$

Neglecting the last term in (D.8) we may integrate from the original state $(p, T, q_s, e_s, \theta_d)$ to a state where $q_s \simeq 0$. The result is

$$c_p \ln\left(\frac{\theta_d}{\theta_e}\right) = -\frac{L_c q_s}{T}$$

where θ_e is the potential temperature in the final state with $q_s \simeq 0$. Therefore, the equivalent potential temperature of a saturated parcel is given by

$$\theta_e = \theta_d \exp(L_c q_s / c_p T) \simeq \theta \exp(L_c q_s / c_p T) \tag{D.9}$$

Equation (D.9) may also be applied to an unsaturated parcel provided that the temperature used is the temperature that the parcel would have if brought to saturation by an adiabatic expansion.

[1] For a derivation, see, for example, Hess (1959, p. 46).

Appendix

E Standard Atmosphere Data[1]

Table E.1 *Geopotential Height as a Function of Pressure*

Pressure (kPa)	Pressure (mb)	Z (km)
100	1000	0.111
90	900	0.988
85	850	1.457
70	700	3.012
60	600	4.206
50	500	5.574
40	400	7.185
30	300	9.164
20	200	11.784
10	100	16.180
5	50	20.576
3	30	23.849
1	10	31.055

[1] After *U.S. Standard Atmosphere*, 1976.

Table E.2 *Standard Atmosphere Temperature, Pressure, and Density as a Function of Geopotential Height*

Z (km)	Temperature (K)	Pressure (kPa)	Density (kg m^{-3})
0	288.15	101.325	1.225
1	281.65	89.874	1.112
2	275.15	79.495	1.007
3	268.65	70.108	0.909
4	262.15	61.640	0.819
5	255.65	54.019	0.736
6	249.15	47.181	0.660
7	242.65	41.060	0.590
8	236.15	35.599	0.525
9	229.65	30.742	0.466
10	223.15	26.436	0.412
12	216.65	19.330	0.311
14	216.65	14.101	0.227
16	216.65	10.287	0.165
18	216.65	7.505	0.121
20	216.65	5.475	0.088
24	220.65	2.930	0.046
28	224.65	1.586	0.025
32	228.65	0.868	0.013

Answers to Selected Problems

Chapter 1

1. $\alpha \simeq \Omega^2 a \sin 2\phi/(2g)$ **2.** 36,000 km **3.** 0.35 N kg^{-1} **4.** 556 m **5.** 1.46 cm
6. 6.22 m s^{-1}
7. lateral force = 10^3 N directed southward, upward force = 2×10^3 N greater for westward travel
8. 7.8 m eastward **10.** 1 rad s^{-1}, 1.22 rad s^{-1}, 0.17 J **12.** 5.536 km, 5.070 km
13. 3°C **14.** 7.987 km **17.** 5.187 km

Chapter 2

1. −2 mb/3 h **2.** −2°C/h **5.** 130-km deflection **6.** 1.02 cm s^{-1}
7. 135 J kg^{-1}, 16.4 m s^{-1}

Chapter 3

1. −1 m km^{-1} **2.** -10^{-3} m s^{-2} **3.** 94 kPa **5.** $V_{\text{grad}}/V_g = 2$
7. North, R_t = 250 km; west, R_t = 500 km; south, $R_t \to \infty$; east, R_t = 500 km
8. North, 10.5 m s^{-1}; west and east, 12.1 m s^{-1}; south, 15 m s^{-1} **10.** 25 m s^{-1}, 34°
11. −1.5°C h^{-1} **12.** 0.5°C/h in the 90–70-kPa layer **13.** 245 km to left
15. North, 9.4 m s^{-1}; west and east, 11.5 m s^{-1}; south, 15 m s^{-1} **17.** 7.66 K
18. −852 km **20.** 2×10^{-5} s^{-1} **21.** 110% **23.** +0.96 cm s^{-1}

Chapter 4

1. $-2 \times 10^7\ \mathrm{m^2\ s^{-1}}$, $-2 \times 10^{-5}\ \mathrm{s^{-1}}$ **2.** $-5.5\ \mathrm{m\ s^{-1}}$ (anticyclonic)
3. $-2.3 \times 10^{-5}\ \mathrm{s^{-1}}$ **4.** $9.4 \times 10^{-5}\ \mathrm{s^{-1}}$, $-8.4 \times 10^{-6}\ \mathrm{s^{-1}}$
5. 0 for annulus, $5 \times 10^{-5}\ \mathrm{s^{-1}}$ for inner circle **6.** $-7.2\ \mathrm{m^2\ s^{-2}}$
8. $h(r) = H + \Omega^2 r^2/2g$ **9.** $2.3\ \mathrm{s^{-1}}$ **11.** $10^{-5}\ \mathrm{s^{-1}}$, $-1.3 \times 10^{-5}\ \mathrm{s^{-1}}$, -1540 km
13. (a) 0.71° equatorward, (b) 138.6 km upward

Chapter 5

4. 0.307 cm, 293 s **5.** $\simeq 1.7$ cm **8.** $2474\ \mathrm{kg\ m^{-1}\ s^{-1}}$, $1750\ \mathrm{kg\ m^{-1}\ s^{-1}}$ **9.** 14 m
11. $w_{max} = 9.62 \times 10^{-3}\ \mathrm{mm\ s^{-1}}$, $v_{max} = -0.75\ \mathrm{mm\ s^{-1}}$ **12.** 1.34×10^7 s
13. $1.55\ \mathrm{mm\ s^{-1}}$

Chapter 6

3. $+2.84 \times 10^{-6}\ \mathrm{s^{-1}}$ **4.** $\omega(p) = (k^2 c^2 p_0/f_0 \pi) \sin(\pi p/p_0) \cos k(x - ct)$
5. $\omega(p) = (c^2 f_0 \pi/\sigma p_0) \sin(\pi p/p_0) \cos k(x - ct)$ **8.** $W_0 = -c f_0 U_0 \pi/(\sigma p_0)$
9. $\omega(p) = (V p_0/f_0 \pi)[k^2(c - U) + \beta] \sin(\pi p/p_0) \cos k(x - ct)$
10. $c = U - \beta[k^2 + f_0^2 \pi^2/\sigma p_0^2]^{-1}$ **11.** $0.01\ \mathrm{Pa\ s^{-1}}$

Chapter 7

2. $B_r = A \cos kx_0$, $B_i = -A \sin kx_0$; $c_r = v/k$, $c_i = \alpha/k$
4. $u' = p'(c - \bar{u})/(\gamma \bar{p})$, $\rho' = p'/(c - \bar{u})^2$ **6.** $u' = (c - \bar{u})h'/H$
7. $u' = -mw'/k$, $p'/\rho_0 = -vmw'/k^2$, $\theta'/\bar{\theta} = -iw'(N^2/vg)$ **8.** $\langle \rho_0 u'w' \rangle = -\rho_0 m |A|^2/(2k)$
11. $u' = [0.66\ \mathrm{m\ s^{-1}}] \cos(kx + mz)$, $w' = -[0.75\ \mathrm{m\ s^{-1}}] \sin(kx + mz)$ **12.** $-24.3\ \mathrm{m\ s^{-1}}$
13. $-0.633\ \mathrm{cm\ s^{-1}}$ **15.** $c_{gx} = \beta(k^2 - m^2)/(k^2 + m^2)^2$, $c_{gy} = 2mk\beta/(k^2 + m^2)^2$

Chapter 8

2.
$$G(x, y) = \frac{2}{\Delta x^2 \Delta y^2} \{(\psi_{i+1,j} - 2\psi_{i,j} + \psi_{i-1,j})(\psi_{i,j+1} - 2\psi_{i,j} + \psi_{i,j-1}) - \tfrac{1}{16}[(\psi_{i+1,j+1} - \psi_{i+1,j-1}) - (\psi_{i-1,j+1} - \psi_{i-1,j-1})]^2\}$$

4. After 4 iterations with successive relaxation, residuals are all $\langle |0.02|$, $X_{m,n}$ values are as follows:

(m, n)	$X_{m,n}$	(m, n)	$X_{m,n}$
(1, 1)	1.54	(3, 2)	2.27
(2, 1)	2.79	(1, 3)	1.52
(3, 1)	1.52	(2, 3)	2.77
(1, 2)	2.29	(3, 3)	1.51
(2, 2)	4.04		

7. 0.5%, 19%, 59% **8.** $|f\,\Delta t| \le 1$
9. $u = u_0 \cos(ft)$, $v = u_0 \sin(ft)$. Ratio of amplitude of computational mode to amplitude of physical mode is 0.05 for $f\,\Delta t = 0.2$ and 0.45 for $f\,\Delta t = 1$
10. 1.4%

Chapter 9

1. 6.04×10^4 s **2.** $\psi'_3 = -\psi'_1$, $\omega'_2 = \{2i\lambda^2 k\beta\,\Delta p/[f_0(k^2 + 2\lambda^2)]\}\psi'_1$
3. $\psi'_3 = \psi'_1(\sqrt{3} - 1)/(\sqrt{3} + 1)$, $\omega'_2 = ik\beta\,\Delta p(1 - \sqrt{3})\psi'_1/(3f_0)$
4. $\psi'_3 = \psi'_1(1 - i\sqrt{3})/2$, $\omega'_2 = -k^3\,\Delta p\,U_T(1 + i\sqrt{3})\psi'_1/(\sqrt{3}f_0)$
5. $c = \bar{u} - \beta/(k^2 + m^2) - i\mu/k$ **7.** ψ'_1 lags ψ'_3 by 65.5°
8. $|B| = |A|$, $|C| = 19 \times 10^{-4}$ mb s^{-1}

Chapter 10

5. 6.25×10^5 kg s^{-2} **9.** 0.49 cm s^{-1}

Chapter 11

2. 6.697 km **3.** 87.6 m s^{-1}, 48.7 m s^{-1}, 28.0 m s^{-1}
4. $|\Phi'| = 200$ m^2 s^{-2}, $|w'| = 1.57$ mm s^{-1}
6. $w' = -(v^2/N^2)[1/(2H) + i\lambda_0]y \exp[-(1 + kv/\beta)\beta^2 y^2/(2v^2)]$ **7.** $\pm 6.14°$ **8.** 1.6 K

Chapter 12

1. 344 K **2.** 1.42 m s^{-1} **5.** (a) $2u_0 l^2 = \beta$, (b) $2u_0 l^2 > \beta$
6. $\Delta\Phi = 750$ m^2 s^{-2}, $r = 136$ km

Bibliography

Anthes, R. A. (1974). The dynamics and energetics of mature tropical cyclones. *Rev. Geophys. Space Phys.* **12**, 495–522.

Arakawa, A. (1972). "Design of the UCLA General Circulation Model." Technical Report No. 7, Department of Meteorology, Univ. of California, Los Angeles, California.

Batchelor, G. K. (1967). *An Introduction to Fluid Dynamics.* Cambridge Univ. Press, London and New York.

Blackmon, M. L., Wallace, J. M., Lau, N-C., and Mullen, S. L. (1977). An observational study of the northern hemisphere wintertime circulation. *J. Atmos. Sci.* **34**, 1040–1053.

Bourne, D. E., and Kendall, P. C. (1968). *Vector Analysis.* Allyn & Bacon, Boston, Massachusetts.

Brown, R. A. (1970). A secondary flow model for the planetary boundary layer. *J. Atmos. Sci.* **27**, 742–757.

Chang, C. P. (1970). Westward propagating cloud patterns in the Tropical Pacific as seen from time composite satellite photographs. *J. Atmos. Sci.* **27**, 133–138.

Chang, J. (ed.) (1977). *Methods in Computational Physics, Vol. 17: General Circulation Models of the Atmosphere.* Academic Press, New York.

Chapman, S., and Lindzen, R. S. (1970). *Atmospheric Tides.* Reidel, Dordrecht, Holland.

Charney, J. G. (1947). The dynamics of long waves in a baroclinic westerly current. *J. Meteorol.* **4**, 135–163.

Charney, J. G., and Eliassen, A. (1964). On the growth of the hurricane depression. *J. Atmos. Sci.* **21**, 68–75.

Climatic Impact Committee (1975). *Environmental Impact of Stratospheric Flight.* National Academy of Sciences, Washington, D.C.

Cole, J. D. (1968). *Perturbation Methods in Applied Mathematics.* Ginn (Blaisdell), Waltham, Massachusetts.

Currie, I. G. (1974). *Fundamental Mechanics of Fluids*. McGraw-Hill, New York.
Dütsch, H. U. (1971). Photochemistry of atmospheric ozone. *Advan. Geophys.* **15**, 219–322.
Fawcett, E. B. (1977). Current capabilities in prediction at the National Weather Service's National Meteorological Center. *Bull. Amer. Meteorol. Soc.* **58**, 143–149.
Gerbier, N., and Berenger, M. (1961). Experimental studies of lee waves in the French Alps. *Quart. J. Roy. Meteorol. Soc.* **87**, 13–23.
Gossard, E. E., and Hooke, W. H. (1975). *Waves in the Atmosphere*. Elsevier, Amsterdam.
Greenspan, H. P. (1968). *The Theory of Rotating Fluids*. Cambridge Univ. Press, London and New York.
Haltiner, G. J. (1971). *Numerical Weather Prediction*. Wiley, New York.
Hare, F. K. (1968). The Arctic. *Quart. J. Roy. Meteorol. Soc.* **94**, 439–459.
Hess, S. L. (1959). *Introduction to Theoretical Meteorology*. Holt, New York.
Hide, R. (1966). Review article on the dynamics of rotating fluids and related topics in Geophysical Fluid Dynamics. *Bull. Amer. Meteorol. Soc.* **47**, 873–885.
Hildebrand, F. B. (1962). *Advanced Calculus for Applications*. Prentice-Hall, Englewood Cliffs, New Jersey.
Holton, J. R. (1975). The dynamic meteorology of the stratosphere and mesosphere. *Meteorology Monograph No. 37*, American Meteorological Society, Boston, Massachusetts.
Hoskins, B. J., and Bretherton, F. P. (1972). Atmospheric frontogenesis models: Mathematical formulation and solution. *J. Atmos. Sci.* **29**, 11–37.
Iribarne, J. V., and Godson, W. L. (1973). *Atmospheric Thermodynamics*. D. Reidel, Boston, Massachusetts.
Kittel, C., Knight, W. D., and Ruderman, M. A. (1965). *Berkeley Physics Course, Vol. I: Mechanics*. McGraw-Hill, New York.
Kornfield, J., Hasler, A. F., Hanson, K. J., and Suomi, V. E. (1967). Photographic cloud climatology from ESSA III and V computer-produced mosaics. *Bull. Amer. Meteorol. Soc.* **48**, 878–882.
Leith, C. E. (1978). Objective methods of weather prediction. *Ann. Rev. Fluid Mechan.* **10**.
Lindzen, R. S. (1967). Planetary waves on beta-planes. *Monthly Weather Rev.* **95**, 441–451.
Lindzen, R. S., and Holton, J. R. (1968). A theory of the quasibiennial oscillation. *J. Atmos. Sci.* **25**, 1095–1107.
Lindzen, R. S., Batten, E. S., and Kim, J. W. (1968). Oscillations in atmospheres with tops. *Monthly Weather Rev.* **96**, 133–140.
Lorenz, E. N. (1960). Energy and numerical weather prediction. *Tellus* **12**, 364–373.
Lorenz, E. N. (1967). *The Nature and Theory of the General Circulation of the Atmosphere*. World Meteorological Organization, Geneva, Switzerland.
Lumley, J. L., and Panofsky, H. A. (1964). *The Structure of Atmospheric Turbulence*. Wiley (Interscience), New York.
Manabe, S., and Mahlman, J. D. (1976). Simulation of seasonal and interhemispheric variations in the stratospheric circulation. *J. Atmos. Sci.* **33**, 2185–2217.
Matsuno, T. (1966). Quasi-geostrophic motions in the equatorial area. *J. Meteorol. Soc. Japan* **44**, 25–42.
Matsuno, T. (1971). A dynamical model of the stratospheric sudden warming. *J. Atmos. Sci.* **28**, 1479–1494.
Newell, R. E., Kidson, J. W., Vincent, D. G., and Boer, G. J. (1974). *The General Circulation of the Trpical Atmosphere and Interactions with Extratropical Latitudes*. MIT Press, Cambridge, Massachusetts.
Oort, A. H., and Peixoto, J. P. (1974). The annual cycle of the energetics of the atmosphere on a planetary scale. *J. Geophys. Res.* **79**, 2705–2719.
Ooyama, K. (1969). Numerical simulation of the life cycle of tropical cyclones. *J. Atmos. Sci.* **26**, 3–40.

Palmén, E., and Newton, C. W. (1969). *Atmospheric Circulation Systems.* Academic Press, New York.

Phillips, N. A. (1956). The general circulation of the atmosphere: A numerical experiment. *Quart. J. Roy. Meteorol. Soc.* **82**, 123–164.

Phillips, N. A. (1963). Geostrophic motion. *Rev. Geophys.* **1**, 123–176.

Phillips, N. A. (1973). Principles of large scale numerical weather prediction. In *Dynamic Meteorology* (P. Morel, ed.), D. Reidel, Boston, Massachusetts.

Platzman, G. W. (1967). A retrospective view of Richardson's book on weather prediction. *Bull. Amer. Meteorol. Soc.* **48**, 514–551

Reed, R. J., and Rogers, D. G. (1962). The circulation of the tropical stratosphere in the years 1954–1960. *J. Atmos. Sci.* **19**, 127–135.

Reed, R. J., Wolfe, J. L., and Nishimoto, H. (1963). A spectral analysis of the energetics of the stratospheric sudden warming of early 1957. *J. Atmos. Sci.* **20**, 256–275.

Reed, R. J., Norquist, D. C., and Recker, E. E. (1977). The structure and properties of African wave disturbances as observed during Phase III of GATE. *Monthly Weather Rev.* **105**, 317–333

Richardson, L. F. (1922). *Weather Prediction by Numerical Process.* Cambridge Univ. Press, London and New York (reprt. Dover, New York).

Richtmeyer, R. D., and Morton, K. W. (1967). *Difference Methods for Initial Value Problems* (2nd ed.). Wiley (Interscience), New York.

Riehl, H. (1954). *Tropical Meterology.* McGraw–Hill, New York.

Riehl, H., and Malkus, J. S. (1958). On the heat balance of the equatorial trough zone. *Geophysica* **6**, 503–538.

Rosenthal, S. L. (1978). Numerical simulation of tropical cyclone development with latent heat release by the resolvable scale I: Model description and preliminary results. *J. Atmos. Sci.* **35**, 258–271.

Rosenthal, S. L., and Koss, W. J. (1968). Linear analysis of a tropical cyclone model with increased vertical resolution. *Monthly Weather Rev.* **96**, 858–866.

Sawyer, J. S. (1956). The vertical circulation at meteorological fronts and its relation to frontogenesis. *Proc. Roy. Soc. A* **234**, 346–362.

Schneider, E. K., and Lindzen, R. S. (1976). A discussion of the parameterization of momentum exchange by cumulus convection. *J. Geophys. Res.* **81**, 3158–3160.

Scorer, R. S. (1958). *Natural Aerodynamics.* Pergamon, Oxford.

Shuman, F. G., and Hovermale, J. (1968). An operational six-layer primitive equation model. *J. Appl. Meteorol.* **7**, 525–547.

Sinclair, P. C. (1965). On the rotation of dust devils. *Bull. Amer. Meteorol. Soc.* **46**, 388–391.

Smagorinsky, J. (1967). The role of numerical modeling. *Bull. Amer. Meteorol. Soc.* **48**, 89–93.

Stevens, D. E., and Lindzen, R. S. (1978). Tropical wave—CISK with a moisture budget and cumulus friction. *J. Atmos. Sci.* **35**, 940–961.

Tennekes, H., and Lumley, J. L. (1972). *A First Course in Turbulence.* MIT Press, Cambridge, Massachusetts.

Turner, J. S. (1973). *Buoyancy Effects in Fluids.* Cambridge Univ. Press, London and New York.

U.S. Government Printing Office (1976). *U.S. Standard Atmosphere, 1976.* NOAA-S/T76-1562. U.S. Government Printing Office, Washington, D.C.

U.S. Navy (1962). *Arctic Forecast Guide.* Navy Weather Research Facility NWRF 16-0462-058.

Wallace, J. M. (1971). Spectral studies of tropospheric wave disturbances in the tropical Western Pacific. *Rev. Geophys.* **9**, 557–612.

Wallace, J. M., and Hobbs, P. V. (1977). *Atmospheric Science: An Introductory Survey.* Academic Press, New York.

Wallace, J. M., and Kousky, V. E. (1968). Observational evidence of Kelvin waves in the tropical stratosphere. *J. Atmos. Sci.* **25**, 900–907.

Warsh, K. L., Echternacht, K. L., and Garstang, M. (1971). Structure of near-surface currents east of Barbados. *J. Phys. Oceanog.* **1**, 123–129.

Williams, K. T. (1971). "A Statistical Analysis of Satellite-Observed Trade Wind Cloud Clusters in the Western North Pacific." Atmospheric Science Paper No. 161, Dept. of Atmospheric Science, Colorado State Univ., Fort Collins, Colorado.

Index

A

Acceleration
  absolute, 30
  centripetal, 12, 30
  Coriolis, 30
  in natural coordinates, 57–58
  relative, 31
  vertical, 40
Acoustic waves, 152–155
Adiabatic cooling, 52
Adiabatic lapse rate
  dry, 49
  pseudo-, 332
Adiabatic motion, 48–49
Advection, 28
  by geostrophic wind, 129
  thermal, 70, 128
  thickness, 128, 134
  vorticity, 129–130
Advective time scale, 36
African waves, 350–352
Ageostrophic wind, 290
Angular momentum, 15–16
  absolute, 262
  conservation of, 262–264
  global balance of, 261
Angular velocity, 11–12
Annulus experiment, 275–280
Anomalous gradient wind, 63
Anticyclonic flow, 59
Atmosphere
  baroclinic, 71
  barotropic, 70
  incompressible, 44
Atmospheric predicatability, 208–211
Atmospheric statics, 17–21

B

Backing wind, 71
Balance equation, 179
Baric flow, 64
Baroclinic atmosphere, 71
Baroclinic instability, 124, 216–217
Baroclinic numerical models, 183–186
Baroclinic wave(s)
  developing, 140–143

energetics of, 227–236
parcel trajectories in, 235
vertical motion in, 223–227
Barotropic atmosphere, 70
Barotropic instability, 352–355
Barotropic numerical models, 181–183
Barotropic vorticity equation, 98, 182, 328
Basic state, 147
β effect, 166
β plane, 129, 308
Bjerknes circulation theorem, 81
Boundary conditions
for primitive equation models, 199
for quasi-geostrophic model, 250
Boundary layer
planetary (Ekman), 102, 107–112
secondary circulation driven by flow in, 113–116
Boussinesq approximation, 160
Brunt Väisällä frequency, 51
Buoyancy frequency, 51
Buoyancy oscillation, 49–50
Buoyancy wave, 159

C

Cartesian coordinates, 369
Centrifugal force, 11–12
Centripetal acceleration, 12, 30
Circulation, *see also* General circulation
absolute, 80
definition, 78
relative, 80
Circulation theorem, 78–80
Circulation, secondary, 112–115, 140
CISK, *see* Conditional instability of the second kind
Clausius Clapeyron equation, 372
Climate simulation, 281
Cloud clusters, 345
Cloudiness, tropical, 344–347
Coefficient of viscosity
dynamic, 9
eddy, 107
kinematic, 10
Computational instability, 175, 186, 194–197
Computational mode, 196
Condensation heating, 339–343
Conditional instability, 330, 332–333
of the second kind, 355–357, 359
Confluence, 239
Conservation
of absolute vorticity, 98
of angular momentum, 262–264
of energy, 44–47, 250–260
of mass, 40–44
of potential vorticity, 82–88
Constant pressure surface, 20–21
geostrophic wind on, 55
Continuity equation, 40–44
eulerian derivation, 40–42
in isobaric coordinates, 55–56
lagrangian derivation, 42–43
in log-pressure coordinates, 301
scale analysis of, 43–44
in sigma coordinates, 201
Continuum, 1
Control volume, 26
Convection, cumulus, 330–343
Convective adjustment, 285
Convergence
frictional, 112
vertical motion and, 73
Coordinates
inertial, 11, 29
isentropic, 53
isobaric, 20–21
natural, 57–58
rotating, 30–31
spherical, 31–35
Coriolis acceleration, 30
Coriolis force, 13–17
Coriolis parameter, 37
Coriolis torque, 272
Critical line, 269, 305
Critical wavelength, 220
Cumulus convection, 330–343
Cumulus friction, 343
Curl of vector, 368
Curvature
radius of, 57
terms in equations of motion, 35
Curvature vorticity, 86
Cyclogenesis, 216–227
Cyclones, tropical, 357–360
Cyclonic flow, 59
Cyclostrophic flow, 60–61
Cylindrical coordinates, 369

D

Deflecting force, 14
Deformation, 238–239

Density, 6
Differential advection
 of temperature, 134
 of thickness, 134
 of vorticity, 137, 225
Direct circulation, 270
Dishpan experiments, 274
Dispersion, 150–151
Divergence, 41–42
 of geostrophic wind, 77
 horizontal, 73
 level of nondivergence, 181
 long wave, 182
 mass, 41
 term of vorticity equation, 93
Divergence equation, 177
Divergence theorem, 368
Divergent wind component, 178
Doppler shifting, 155
Dry static energy, 332
Dynamic viscosity coefficient, 9

E

Eddy exchange coefficient, 105
Eddy flux
 of heat, 257, 267
 of momentum, 264, 267
 of potential vorticity, 268–269
Eddy stress, 104
Eddy viscosity coefficient, 107
Ekman layer, 102, 107–109
 instability of, 111
 modified, 110–111
Ekman spiral, 108
Energy
 conservation of, 44–47
 equations, 230–232
 gravitational potential, 228
 internal, 228
 kinetic, 229
 potential
  available, 227–230
  total, 228
 quasi-geostrophic system, 250–260
Energy cycle
 in stratosphere, 298–299
 in troposphere, 250–260
Entrainment, 337–339
Entropy, 48
Equation of continuity, *see* Continuity equation
Equation of state, 18
 for liquid, 275
Equations of motion, 29–31
 in absolute coordinates, 30
  in Cartesian coordinates, 38
  in isobaric coordinates, 54–55
  in natural coordinates, 58
  in spherical coordinates, 31–35
Equatorial β plane, 308
Equatorial wave disturbances, 345–351
 origin of, 351–357
 stratospheric, 307–313
Equilibrium
 geostrophic, 37
 hydrostatic, 18–19
Equivalent potential temperature, 330–331, 371–372
Equivalent static stability, 340

F

Ferral cell, 273
Field variables, 1
Filtered models, 175–286
Filtering meteorological noise, 175–177
Finite differences, 186–189
First law of thermodynamics, *see* Thermodynamic energy equation
Flux
 heat, 257, 267
 momentum, 264–267
 potential vorticity, 268–269
Flow
 cyclostrophic, 60–61
 geostrophic, 37, 58
 gradient, 62–64
 inertial, 59–60
 mean meridional, 269–273
Force
 centrifugal, 11–12
 Coriolis, 13–17
 frictional, 8–10
 gravatational, 7–8
 gravity, 12–13
 pressure gradient, 5–6
Fourier series, 149–151
Friction force, 8–10
Friction layer, 107–111
 convergence in, 112

Friction velocity, 106
Frictional torque, 266
Frontogenesis, 237–242
Fronts, 120, 236–244
  polar, 123
  slope of, 242–244

G

Gas constant, 18
General circulation, 247–249
  angular momentum balance in, 261–264
  energetics of, 250–260
  laboratory modeling of, 274–280
  numerical modeling of, 281–288
  observed, 120–123
  stratospheric, 296–298
Geopotential, 18
Geopotential height, 19
Geopotential tendency equation, 130–132
Geostrophic adjustment, 142
Geostrophic advection, 129
Geostrophic approximation, 37
Geostrophic flow (wind), 37, 58
  in natural coordinates, 64
  on isentropic surface, 53
  on isobaric surface, 55
  on isothermal surface (problem), 75
  vertical shear of, 68–71
Geostrophic vorticity, 97–98
Geostrophic momentum, 76
GFDL model, 285–288
Gradient, pressure, 5–7
Gradient flow (wind), 62–64
  anomalous, 63
  comparison with geostrophic flow, 64
Gradient wind level, 109
Gravitation, Newton's law of, 7
Gravitational constant, 7
Gravity, 12–13
Gravity waves, 155–159
  atmospheric internal, 159–165
  filtering of, 175–177
  speed of, 159
Grid points, 187–188
Group velocity, 151–152, 164

H

Hadley cell (circulation), 248
Hadley regime, 275
Heat, 48
  latent, 328
  specific, 47–48
Heat balance, 248–249
  of stratosphere, 299
  of troposphere, 259–260
Height, *see* Geopotential
High, anomalous, 63
Hodograph of wind in Ekman layer, 109
Horizontal acceleration, 57
  in natural coordinates, 57–58
Horizontal advection
  by geostrophic wind, 129
  of temperature, 128
  of thickness, 128, 134
  of virticity, 129–130
Hot towers, 344, 350
Hurricanes, 357–360
Hydrodynamic instability, 214–216
Hydrostatic approximation, 18, 39
Hydrostatic temperature changes, 139
Hypsometric equation, 19

I

Incompressible fluid, 44
Indirect circulation, 270
Inertia circle, 60
Inertia gravity oscillations, 307
Inertial frame, 11
Inertial instability, 215–216
Inertial motion, 11, 59–60
Inertial oscillations, 59–60
Initialization, 205–207
Instability
  baroclinic, 216–227
  barotropic, 352–355
  computational, 175, 186, 194–197
  conditional, 332, 333
    of the second kind, 355–357, 359
  hydrodynamic, 214–216
  inertial, 215–216
  parcel method, 51, 215–216
Integral constraints in numerical models, 189
Internal energy, 228
Internal gravity waves, 159–165
Intertropical convergence zone (ITCZ), 344–345
Irrotational wind component, 178

Isentropic coordinates, 53
Isobaric coordinates, 20–21
Isotherms, 70
ITCZ, *see* Intertropical convergence zone

J

Jet streams, 120–121, 289–291
  stratospheric, 296

K

Kelvin waves, 308–309
Kelvin's circulation theorem, 80
Kinematics of frontogenesis, 237–240
Kinetic energy, 229
  balance of in general circulation, 255
  generation of in baroclinic waves, 232

L

Lamb waves, 176
Laplacian, 131
Lapse rate of temperature, 49
  moist adiabatic, 332
  in numerical model of general circulation, 285
Latent heat, 328
  in equatorial circulation systems, 345
Lateral shear, and instability, 351
Latitude effect
  in circulation theorem, 81
  in potential vorticity conservation
Lee waves, 164–165
Leeside trough, 90
LFM model, 207–208
Linearized equations, 147
Local change, 27–28
  of temperature, 128
  of vorticity, 130
Local turning of wind, 66
Logarithmic wind profile, 107
Log-pressure coordinates, 300–301
Low, anomalous, 62–63
Lower symmetric regime, 279

M

Map factors, 190
Map projections, 190
Marginal stability, 221
Mass conservation, 40–42
Material lines, 316
Mechanical energy equation, 47
Meridional circulation, 270–273
Meridional momentum flux, 264–265
Meridional motion, mean 269–273
Meridional streamfunction, 269–270
Mesosphere, 296
Meteorological noise, 175–177
Middle latitude systems, 123–125
Mixed Rossby-gravity waves, 310–311
Mixing length, 102–105
Mixing ratio, 371
Model experiments, 274–280
Moist adiabatic lapse rate, 332
Moist static energy, 332
Molecular viscosity, 9
Momentum budget, 260–266
Momentum equation, *see* Equations of motion
Monsoon circulations, 291–293
Motion
  absolute and relative, 11
  equations of, *see* Equations of motion
  geostrophic, 37, 58
Mountain pressure torque, 264–265

N

Natural coordinates, 57–58
Neutral stability, 221
Newtonian cooling, 267
Newton's law
  of gravitation, 7
  of motion (second), 29–31
NMC six-layer model, 207–208
Noise, meteorological, 175–177
Nondivergence, level of, 181
Nondivergent wind component, 178
Normal mode solutions, 216, 219
Normal modes, 168
Numerical dispersion, 197
Numerical forecasting
  barotropic model, 181–183
  filtered models, 178–186
  history of, 173–175
  primitive equation models, 198–205
  two-level baroclinic model, 183–186
Numerical simulation of general circulation, 281–288

O

ω, 72
Omega equation, 136–139, 180
  two-level model, 186
Ω-momentum flux, 264
Orographic effect
  on angular momentum, 265–266
  on planetary waves, 89–90
Orographic leeside trough, 90
Orography, 120
Oscillations, transverse, 153
Ozone, 317–321
Ozone mixing ratio, 320

P

Parcel method
  for conditional instability, 333
  for gravitational instability, 49–50
  for inertial instability, 215–216
Partial pressure of water vapor, 371
Pendulum day, 60
Perturbation method, 147
Phase speed, 148
Physical mode, 196
Planetary boundary layer, 101, 105–112
Planetary waves, 300–303
Poisson's equation, 48
Polar front, 123
Polar night jet, 296
Polar stereographic map, 190
Potential energy
  available, 227–230, 255
  total, 228
Potential temperature, 48
  equivalent, 330–331, 371–372
Potential vorticity, 87–91
  conservation of, 87–88
  quasi-geostrophic, 136, 180
  zonal mean, 267–269
Prandtl mixing length hypothesis, 102–105
Precipitation, 328
Predictability, 208–211
Pressure, 5
Pressure coordinates, 20–21
  geostrophic wind in, 55
  governing equations in, 54–56
Pressure gradient force, 5–6
Pressure tendency equation, 202
Primitive equation model
  two level, 202–204
Primitive equations, 198
  initial data for, 205–207
  log pressure coordinates, 300–301
  sigma coordinates, 200–202
Pseudo-adiabatic ascent, 331–332

Q

Quasi-biennial oscillation, 314–317
Quasi-geostrophic system, 130–143, 180

R

Radius of curvature, 57, 65
Radius of deformation, 145, 185
Rayleigh friction, 267
Reference frame
  eulerian, 26
  geocentric, 10
  lagrangian, 26
  noninertial, 10
Relaxation, 191–194
Residual, 191
Retrogression, 133
Rossby critical velocity, 303
Rossby number, 38, 61
  thermal, 277–278
Rossby regime, 275
Rossby waves, 165–168
  barotropic, 219
  internal baroclinic, 219–220
Rotating annulus experiment, 275–280
Rotating coordinate system, 11, 30–31
Rotating dishpan experiment, 274
Roughness length, 107

S

Saturated adiabatic process, 332–333
Saturation mixing ratio, 331
Scale analysis, 4–5
  of equations of motion, 36–37, 39–40
Scale analysis
  of continuity equation, 43–44
  of thermodynamic energy equation, 51–52
  for tropical circulation systems, 324–329
  of verticity equation, 95–96
Scale height, 19

Sea breeze, 82–83
Secondary circulations
  in baroclinic waves, 140
  in Ekman layer, 112–116
Shallow water waves, 155–159
Shear
  lateral, 351
  vertical, 68–71, 220–223
Shear vorticity, 86
Shearing stress, 9–10
Sigma coordinate system, 199–202
Simultaneous relaxation, 191
Slice method, 335–337
Slope of frontal surfaces, 242–244
Solenoidal term
  in circulation theorem, 80
  in vorticity equation, 94
Sound waves, 152–155
  filtering of, 175–176
  horizontally propagating, 153–155
Specific heat, 47–48
  of water vapor, 371
Specific volume, 17
Spectrum, 60
Speed of sound, 155
Speed of waves, *see* Wave speed
Spherical coordinates, 31–35, 370
Spin-down time, 114
Spiral bands, 360
Stability, *see* Instability
Standard density, 39
Standard pressure, 39, 48
Static energy, 332
Static stability
  equivalent, 49–51
  parameter, 127
Statics, 17–21
Stationary waves, 288–291
Stokes' theorem, 369
Stratopause, 296
  general circulation of, 296–298
  Kelvin waves in, 308–309
  mixed Rossby-gravity waves in, 310–311
  polar night jet, 296
  quasi-biennial oscillation, 314–317
Stratospheric ozone layer, 317–321
Streamfunction, 178
Streamlines, 64–65
Stress, shearing, 9
Sudden warmings, 303–307
Surface frictional torque, 266
Surface layer, 101, 106–107
Surface waves, 155–159
Sutcliffe development formula, 224
Synoptic cyclone, 4

T

Temperature, 2–3, 17
  equivalent potential, 330–331, 371–372
  potential, 48–49
  zonal mean, 121
Temperature advection, 134–135, 139–140
Tendency equation, 130–132, 185–186
Thermal wind, 68–71
Thermocline, 159
Thermodynamic energy equation, 44–47, 127
  in isobaric coordinates, 56
  with latent heating, 340–341
  in log-pressure coordinates, 301
  quasi-geostrophic, 128
  scale analysis of for tropics, 327
  zonal mean, 257
Thermodynamics, 47–52
Thickness, 19, 70
Thickness advection, 134–135, 139–140
Tilt
  of internal waves, 163
  of troughs and ridges, 125, 225
Tilting effect in vorticity equation, 93
Time-longitude cloud analysis, 345
Topography, flow over, 89–90
Total derivative, 27–29
Trajectory, 64–65
Tropical cyclones, 357–360
Tropical monsoons, 291–293
Tropical wave disturbances, 345–351
  barotropic instability of, 352–355
  CISK mechanism of formation, 355–357
  Kelvin waves, 308–309
  mixed Rossby-gravity waves, 310–311
Tropopause, 120, 296
Troposphere, 120, 296
Troposphere–stratosphere exchange, 318–319
Truncation errors, 187
Turbulent flow, 103–105
Twisting term in vorticity equation, 93
Two-level model, 183–186

linearized, 217–218
energy equations for, 230–232

U

Units, SI system of, 2–3
Upper symmetric regime, 279

V

Vacillation, 275–276
Vapor pressure, 371
Vector
irrotational, 178
nondivergent, 178
Veering wind, 70–71
Velocity, *see also* Wave speed
absolute, 30
angular, 11–12
group, 151–152, 164
relative, 30
Vertical acceleration, 40
Vertical coordiantes
isobaric, 19–20
generalized, 21–22
log-pressure, 300–301
sigma, 199–202
Vertical momentum flux in cumulus clouds, 338–343
Vertical motion, determination, of,
adiabatic method, 74
kinematic method, 72–73
omega equation, 137
vorticity method, 130
Vertical propagation, 300–303
Vertical shear
baroclinic instability and, 216–222
of geostrophic wind, 68
Vertical stretching, 88
Vertical tilt
of baroclinic waves, 125–335
of internal gravity waves, 163
Viscosity, 9
kinematic coefficient, 10
Viscous stress, 9
von Karman's constant, 107
von Neumann necessary condition for stability, 195
Vorticity
absolute, 83
advection of, 132–133
circulation and, 84
conservation of, 98
differential advection of, 137
geostrophic, 97–98
in natural coordinates, 85
potential, *see* Potential vorticity
relative, 83
Vorticity equation, 92–94
barotropic, 98, 182, 328
in isobaric coordinates, 94–95, 128–129
quasi-geostrophic, 129
scale analysis of 95–96
vector form of, 94–95
zonal mean, 255

W

Warm advection, 70, 128
Warm core system, 358
Water vapor, 331
Wave equation, 168
Wave-CISK, 357
Wave number, 131, 148
Wave speed, 148
for barotropic Rossby waves, 167
for internal gravity waves, 163
for surface gravity waves, 159
in two-level model, 219
Waves
acoustic, 152–155
African, 350–352
baroclinic, 140–143, 216–222
barotropic, 219
equatorial, 307–313, 345–351
gravity, 155–159
internal, 159–165
Kelvin, 308–309
Lamb, 176
mixed Rossby-gravity, 310–311
properties of, 147–149
Rossby, 165–168
shallow water, 155–159
sound, 152–155
stationary, 288–291
transverse, 160
tilt of, 163
unstable, 216–227, 352–355

Wind
cyclostrophic, 60–61
in Ekman layer, 107–109
geostrophic, 37, 58
gradient, 62–64
inertial, 59–60
irrotational, 178
nondivergent, 178
thermal, 68–71
zonal mean, 120

Z

Zonal flow Hadley regime, 275
observed zonal mean, 120
Zonally symmetric circulations, 267–273

International Geophysics Series

EDITED BY

Volume 1 Beno Gutenberg. Physics of the Earth's Interior. 1959

Volume 2 Joseph W. Chamberlain. Physics of the Aurora and Airglow. 1961

Volume 3 S. K. Runcorn (ed.). Continental Drift. 1962

Volume 4 C. E. Junge. Air Chemistry and Radioactivity. 1963

Volume 5 Robert C. Fleagle and Joost A. Businger. An Introduction to Atmospheric Physics. 1963

Volume 6 L. Dufour and R. Defay. Thermodynamics of Clouds. 1963

Volume 7 H. U. Roll. Physics of the Marine Atmosphere. 1965

Volume 8 Richard A. Craig. The Upper Atmosphere: Meteorology and Physics. 1965

Volume 9 Willis L. Webb. Structure of the Stratosphere and Mesosphere. 1966

Volume 10 Michele Caputo. The Gravity Field of the Earth from Classical and Modern Methods. 1967

Volume 11 S. Matsushita and Wallace H. Campbell (eds.). Physics of Geomagnetic Phenomena. (In two volumes.) 1967

Volume 12 K. Ya. Kondratyev. Radiation in the Atmosphere. 1969

Volume 13 E. Palmen and C. W. Newton. Atmospheric Circulation Systems: Their Structure and Physical Interpretation. 1969

Volume 14 Henry Rishbeth and Owen K. Garriott. Introduction to Ionospheric Physics. 1969

Volume 15 C. S. Ramage. Monsoon Meteorology. 1971

Volume 16 James R. Holton. An Introduction to Dynamic Meteorology. 1972

Volume 17 K. C. Yeh and C. H. Liu. Theory of Ionospheric Waves. 1972

Volume 18 M. I. Budyko. Climate and Life. 1974

Volume 19 Melvin E. Stern. Ocean Circulation Physics. 1975

Volume 20 J. A. Jacobs. The Earth's Core. 1975

Volume 21 David H. Miller. Water at the Surface of the Earth: An Introduction to Ecosystem Hydrodynamics. 1977

Volume 22 Joseph W. Chamberlain. Theory of Planetary Atmospheres: An Introduction to Their Physics and Chemistry. 1978

Volume 23 JAMES R. HOLTON. Introduction to Dynamic Meteorology, Second Edition. 1979

Volume 24 ARNETT S. DENNIS. Weather Modification by Cloud Seeding. 1980

Volume 25 ROBERT G. FLEAGLE AND JOOST A. BUSINGER. An Introduction to Atmospheric Physics. Second Edition. 1980

In preparation

Volume 26 KUO-NAN LIOU. An Introduction to Atmospheric Radiation.